Design of Workplace Dilution and Local Ventilation Systems

Temperate buildings are comfortable, safer, and more productive places to work. Effective design, function, and management of ventilation systems ensures that extremes of cold and heat, and airborne diseases are mitigated. This book addresses the basics of ventilation systems and their use in building-based workplaces and how to ensure they are managed successfully.

This book provides comprehensive technical information regarding the design and implementation of dilution and local exhaust ventilation systems to ensure safe and productive workplaces. It presents a simple and readable summary of how to design, install, and operate industrial ventilation systems, all highlighted through illustrations and tables. This title is written to international standards on the design of dilution and local exhaust ventilation systems, including ASHRAE and ACGIH standards.

Design of Workplace Dilution and Local Ventilation Systems is an ideal reference for safety engineers, designers, specialists, engineers, architects, and contractors tasked with ventilation systems. It is also a useful reference for undergraduate and graduate students to master the complex fundamentals of designing dilution and local exhaust ventilation systems.

Design of Workplace Dilution and Local Ventilation Systems

Edited by
Amir Hossein Khoshakhlagh
with Haruna Musa Moda

CRC Press
Taylor & Francis Group
Boca Raton London New York

CRC Press is an imprint of the
Taylor & Francis Group, an **informa** business

Designed cover image: Amir Hossein Khoshakhlagh

First edition published 2026
by CRC Press
2385 NW Executive Center Drive, Suite 320, Boca Raton FL 33431

and by CRC Press
4 Park Square, Milton Park, Abingdon, Oxon, OX14 4RN

CRC Press is an imprint of Taylor & Francis Group, LLC

Library of Congress Cataloging-in-Publication Data
Names: Khoshakhlagh, Amir Hossein editor | Moda, Haruna Musa editor
Title: Design of workplace dilution and local ventilation systems / edited by Amir Hossein Khoshakhlagh and Haruna Musa Moda.
Description: First edition. | Boca Raton, FL : CRC Press, 2026. | Includes bibliographical references and index.
Identifiers: LCCN 2025051876 | ISBN 9781032800455 hbk | ISBN 9781032817262 pck | ISBN 9781003501091 eck
Subjects: LCSH: Industrial buildings--Heating and ventilation | Ventilation--Design and construction
Classification: LCC TH7392.I53 D47 2026
LC record available at https://lccn.loc.gov/2025051876

ISBN: 978-1-032-80045-5 (hbk)
ISBN: 978-1-032-81726-2 (pbk)
ISBN: 978-1-003-50109-1 (ebk)

DOI: 10.1201/9781003501091

Typeset in Times
by SPi Technologies India Pvt Ltd (Straive)

Contents

About the Editor

Amir Hossein Khoshakhlagh is an Assistant Professor in the Department of Occupational Health Engineering and serves as the chair of the Health, Safety, and Environment (HSE) Bureau. Dr. Khoshakhlagh's research focuses on chemical agent exposure in workplaces and occupational toxicology. In addition to his research, he is involved in designing ventilation systems for various industries in Iran. To date, Dr. Khoshakhlagh has authored seven books. He has published more than 85 papers in prestigious journals.

Contributor Biography

Haruna Musa Moda is a Professor and Department Head of Environmental Health and Safety at the University of Doha for Science and Technology, Doha, Qatar. Dr. Moda is actively engaged in research, consultancy, and knowledge transfer projects in the UK, West Africa, and Middle East in occupational safety and health, climate change adaption, and safety response among outdoor workers, e-waste management, and workers' safety behavior. Dr. Moda is a chartered member of the Institution of Occupational Safety and Health (IOSH) and Fellow of the Higher Education Academy and belongs to several professional bodies, including Workplace Health Without Borders (WHWB), Higher Education Academy, Association of Industrial Hygienists, Nigeria, and Environmental Health Registration Council of Nigeria.

Contributors

Agnieszka Gruszecka-Kosowska
Department of Environmental Protection
Faculty of Geology, Geophysics, and Environmental Protection
AGH University of Krakow
Poland

Amir Hossein Khoshakhlagh
Department of Occupational Health
School of Health
Kashan University of Medical Sciences
Iran

Haruna Musa Moda
Department of Environmental Health and Safety
University of Doha for Science and Technology
Qatar

Saeid Yazdanirad
Department of Occupational Health
School of Health
Shahrekord University of Medical Sciences
Iran

Foreword

This book provides comprehensive technical information on the design and implementation of industrial ventilation systems for designers, specialists, engineers, architects, and contractors. It also serves as a valuable reference for undergraduate and graduate students seeking to master the complex fundamentals of industrial ventilation system design. Each chapter presents the basics of industrial ventilation system design, installation, and commissioning in a clear and accessible manner, utilizing graphic images and tables. The book incorporates the latest international standards in industrial ventilation system design, including ASHRAE, ACGIH, and others. Additionally, it covers fundamental principles of fluid mechanics, gas and aerosol dynamics, and solid mechanics. The book comprises 11 chapters. The first five chapters address ductwork networks, general industrial ventilation, measurement of air changes using tracer gases, local exhaust ventilation, and hood design for specific activities. The remaining six chapters focus on designing laboratory hoods, single-hood system design, multiple-hood system design, quantitative evaluation of hood performance, methods of controlling air contaminants, and fans. Each chapter concludes with practical guidance to help readers understand and effectively implement the relevant designs. This book is recommended for undergraduate and graduate students, professors, and specialists in occupational health, environmental health, air pollution, environmental science, mechanics, health, safety, and environment (HSE) fields.

This book aims to provide accurate and reliable information on the topics discussed. It is understood that the editors, whether collectively or individually, do not assume any liability for errors, omissions, or any consequences resulting from the use of this publication, whether intentional or unintentional.

EDITORS OF *THE DESIGN OF WORKPLACE DILUTION AND LOCAL VENTILATION SYSTEMS*

A. H. Khoshakhlagh, Iran
H. M. Moda, Qatar
S. Yazdanirad, Iran
A. Gruszecka-Kosowska, Poland

Preface

Effective ventilation and air quality management are essential for ensuring safe, healthy, and efficient indoor and industrial environments. This book presents a comprehensive exploration of the principles, design methodologies, and practical applications of ventilation systems, combining theoretical foundations with hands-on guidance. Covering topics from the fundamentals of ductwork networks and general industrial ventilation to the design of local and laboratory exhaust systems, the book addresses the needs of students, engineers, and practitioners in HVAC and occupational hygiene fields.

The opening chapters introduce the objectives and components of ductwork networks, design procedures, and the estimation of pressure losses, establishing the foundation for advanced topics. Subsequent chapters examine general and local ventilation strategies, air change measurement using tracer gases, and the design of single- and multiple-hood systems. Specialized sections address laboratory hoods, hood performance evaluation, and methods for controlling air contaminants, including filtration technologies and adsorption processes. The final chapters provide in-depth coverage of fans, performance curves, fan laws, and systematic selection procedures.

Throughout the book, practical examples, real-world design considerations, and step-by-step procedures illustrate the application of scientific principles to real-world ventilation challenges. The material is structured to support learning, professional development, and the design of safe, efficient, and reliable ventilation systems across industrial, laboratory, and commercial environments.

This book serves as a comprehensive reference for engineers, researchers, and advanced students, effectively bridging the gap between theoretical principles and practical applications in modern ventilation engineering.

1 The Ductwork Network

A. H. Khoshakhlagh, H. M. Moda, S. Yazdanirad, and A. Gruszecka-Kosowska

1.1 OBJECTIVES OF THE LESSON

Upon completing this lesson, readers should be able to:

- Describe the design procedure of the ductwork network
- Explain the design approaches of ductwork network
- State the design parameters of the ductwork network
- Design the ductwork network

1.2 THE DUCTWORK NETWORK

When the hood or direct outlet connection maintains the emission flow, this flow is transferred to the control device through the ductwork network. The term ductwork network includes all equipment between the capture device and the control device, including: (1) direct channel; (2) connections, such as elbows and tees; (3) flow control devices (for example, dampers); and (4) channel protectors. In air pollution control systems, the fan is usually placed immediately before or after the control device. As a result, most channels (conduits) are typically under negative static pressure, ranging from a few inches to approximately 20 inches of water column. These pressure conditions determine the type of channel used as well as design parameters such as wall thickness (gauge).

For example, welded ducts are preferred over spiral-wound ducts in vacuum applications.

The duct is made of metal or plastic, and the choice of material is determined by the characteristics of the contaminated gas flow, structural considerations, purchase and installation costs, aesthetics, and other factors.

Metals used include carbon steel (uncoated or galvanized), stainless steel, and aluminum.

Commonly used plastics include polyvinyl chloride (PVC) and fiberglass reinforced plastic (FRP), although polypropylene (PP) and linear polyethylene (LPE) can also be used.

However, a major problem with using PP and LPE is that both are combustible.

Polyvinyl chloride and other plastic ducts are resistant to a variety of corrosive substances, from aqua regia (a mixture of three volumes of hydrochloric acid and one volume of nitric acid) to 95% sulfuric acid, but plastic ducts cannot withstand

DOI: 10.1201/9781003501091-1

ambient temperatures above 150°F. Metal ducts can handle temperatures up to about 1000°F, but only certain alloys can withstand corrosive flows. Structurally, the ductwork network can be rigid or flexible. As the name suggests, rigid ductwork, whether metal or plastic, has a fixed shape. Conversely, flexible ductwork can be bent to fit situations where space is limited, or the formwork is so complex that rigid connections cannot meet the needs of the structure. Usually circular in cross-section, the flexible duct can be made of metal or plastic and can be insulated.

Rigid ductwork network is made in circular, flat oval, or square/rectangular cross-section. Of these, the circular duct is mostly used in air pollution control systems. Although square and rectangular ducts are useful when space is limited, round ducts offer several advantages, including greater resistance to collapse, improved airflow conditions, and requiring less metal compared to oval, square, rectangular, or flat ducts with an equivalent cross-sectional area. The following discussion pertains to the rigid circular duct, because this type is commonly used in air pollution control.

The rigid metal circular duct is further classified based on the manufacturing method. The longitudinal seam duct is made by bending the metal sheet in a circular shape on a mold and connecting the two ends together. A spiral seam duct is made of a longitudinal strip of sheet metal, the edges of which are connected through a continuous spiral seam that runs the length of the duct. This seam is either raised or leads to the surface of the duct wall.

The construction method and the shape of the section are not the only considerations in duct design. The diameter, wall thickness, type, number and location of connections, controllers and protectors, and other parameters should also be specified. As a result, most ductwork components are custom designed and built to provide the optimal control device. Some manufacturers offer prefabricated components, but these are usually common fittings (e.g., 90° elbows) that come only in standard sizes (e.g., 3–12″ diameter).

If the gas flow temperature or moisture content is excessive, it may be necessary to insulate the duct. Insulation on the one hand prevents heat loss/gain, saving energy (and cost)—and on the other hand prevents condensation. The insulation also protects the hands of workers who may touch the duct from burns. Insulating the ducts can be done in two ways. The first method is to install insulation on the outer surface of the duct and cover it with vapor insulation of plastic or metal sheet. The type and thickness of insulation used will depend on the heat transfer properties of the material. For example, 4 inches of mineral wool insulation is required to maintain a surface (body) temperature of 140°F (the OSHA limit for the workplace) or lower, provided the exhaust gas temperature does not exceed 600°F.

The second method of insulating the ducts is to use insulated double-walled ducts and connections.

The double-walled ductwork network reduces both heat loss and noise. It is usually made of an outer pressurized shell of sheet metal and an inner lining of sheet metal with a layer of fiberglass insulation in between. The insulation layer is usually 1 inch, although 2- and 3-inch thicknesses are available for more severe applications. The thermal conductivities of these thicknesses are 0.09, 0.13, and 0.27 Btu/hr-ft^2-°F, respectively.

1.3 DUCTWORK NETWORK COMPONENTS

As mentioned in the above section, a ductwork network system consists of straight ducts, connections, flow control devices, and protection. Direct duct visualization is easy; however, the category of fittings includes a wide range of components that perform one or more functions: Change the direction of gas flow in the duct, change the flow rate, connect it to other duct(s), facilitate the connection of two or more parts to each other or in case of thermal stress, providing conditions for contraction or expansion.

The most common connections are elbows. Elbows are designed to divert gas flow, typically at 30°, 45°, 60°, or 90°, although they may be designed for other angles.

The radius of the center line of the elbow determines the extent of this change of direction (Figure 1.1). The standard radius of the central line (R_{cl}) is 1.5 times the diameter of the elbow section (D_c). However, in long-radius elbows, where the change in direction is more gradual than standard elbows, R_{cl} is greater than or equal to 2 times DC.

When two or more gas flow paths must be connected, T-shaped connections are used. In straight tees, the currents meet at a 90° angle, while in angled tees (laterals, 45° tees (Wyes)) the junction is at 30°, 45°, 60°, or another angle is formed (Figure 1.1). Tees may have one or two taps (connections) and a straight or tapered cross-section at both ends. Connections are also used to connect duct branches. Here, two branches intersect each other at a right angle.

Whenever ducts of different diameters are connected, converters (commonly called expanders or shrinkers) are required. Converters are either concentric or eccentric. In concentric converters, the diameter gradually decreases from larger to smaller cross-section. However, in eccentric converters, the diameter is completely reduced on one side of the connection.

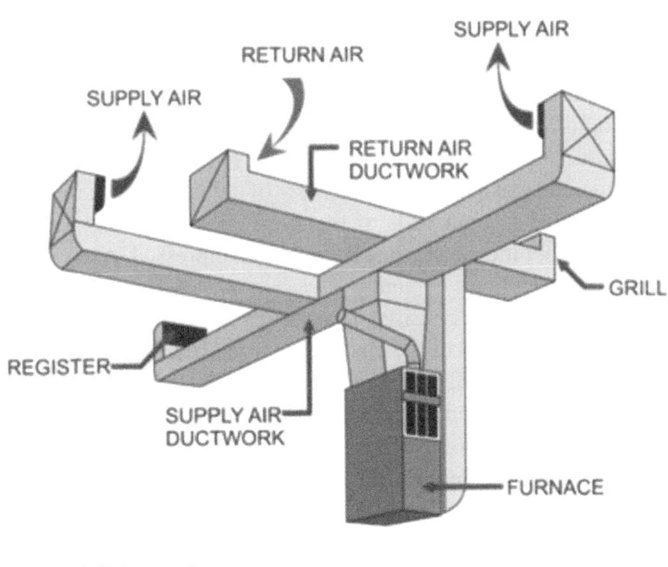

FIGURE 1.1 HVAC ductwork.

Dampers control volumetric flow through ventilation systems. They are usually characterized by flow control mechanism (single blade or multi blade), pressure rating (low/light or high/heavy), and control devices (manual or automatic). In single blade dampers, a circular plate is attached to a rod with one end outside the duct. In the most common type of single blade damper (propeller type), this rod is used to control the gas flow by rotating the plate in the damper. In the fully closed case, the damper surface is perpendicular to the gas flow path. In the completely open case, the surface is parallel to the gas flow lines. HVAC ductwork is shown in Figure 1.1.

Air regulating valve dampers control the flow by pulling the damper blade in and out of the duct.

Air regulating valves are often used to control the flow of air containing suspended solids such as those found in pneumatic conveyors. In this respect, butterfly dampers and air regulating valves are similar to ball valves and relief valves, respectively, used to regulate fluid flow in pipes. Multi-blade dampers (shutters) work the same way as single-blade dampers.

However, instead of using a single blade or plate to control gas flow, multi-blade dampers use components that open and close like a venetian blind. Shutter dampers are usually used in very large ducts where it is very difficult to move a one-piece damper blade.

Manual control dampers have a handle attached to the control rod, which is easily used to adjust the gas flow by hand. If automatic control is required, a pneumatic or electronic actuator is utilized. The actuator receives a pneumatic (pressurized air) or electrical signal from a controller and converts it into mechanical energy, which in turn is used to open/close the damper from the damper rod. In this sense, the actuated damper is similar to the automatic control valve.

For example, an automatic damper can be used to control the progressive airflow to a waste incinerator combustion chamber. This flow rate, in turn, depends on the accumulation of combustibles (i.e., %LEL) in the inlet contaminated gas flow. If this accumulation deviates from a predetermined value (set point), a signal from the measuring device is sent through the controller to the automatic damper to increase or decrease the amount of progressive airflow to maintain the desired LEL percentage.

Expansion joints are specifically installed in long metal ducts to allow the duct to expand and contract in response to thermal stresses. These connections have several designs. One type, the expansion joint, consists of a flexible piece of metal (e.g., 304 stainless steel) welded to both ends of the duct, connecting them together. As the temperature in the duct increases, the elbows are compressed. As the temperature in the duct decreases, the elbows expand.

A common expansion joint consists of two ridges between which a piece of fabric is installed. The expansion joints of the elbows are compressed as the temperature of the duct increases and vice versa. The temperature determines the type of fabric used. For example, silicone fiberglass and Aramid fiber fabric can be used for duct temperatures up to 500°F, while coated fiberglass fabric is suitable for temperatures up to 1000°F.

The last thing to be considered is the duct protection system. However, this issue is not unimportant. As presented in the Sheet Metal and Air Conditioning Contractors' National Association (SMACNA) Standard Manual for Heating, Ventilation and Air Conditioning (HVAC) Duct Construction, choosing a hanging system should not be taken lightly—not only because it involves a significant portion of the installation work, but also installing an inadequate suspension system can be disastrous. As a rule of thumb, a guard should be installed for every 8–10 feet of ductwork. The ductwork network can be hung from the ceiling or other roof structures through hangers or protected from below by beams, columns, or other supports.

The suspension system usually consists of a top link, a hanger and a bottom link. The top connection connects the pendant to the ceiling, etc., which can be a concrete valve, an eye bolt, or a fastener such as a rivet or a nail pin. A hanger is generally a strip of galvanized steel, round steel rod, or wire attached to the ceiling by a joint. The type of hanger used is determined by the diameter of the duct, which is linear in proportion to its weight in each base.

For example, wire hangers are only recommended for duct diameters up to 10 inches. For larger diameters (up to 36 inches), a belt or rod should be used. Typically, a strap hanger is wrapped around the duct from the top connection and secured by a fastener (bottom connection). A rod hanger also extends from the ceiling downwards. Unlike strap hangers, they are securely attached to the duct via a band or bands that wrap around the perimeter. A duct with a diameter of more than 3 feet should be protected with two hangers on both sides of the duct and firmly connected to two perimeter strips, one above it and the other below the duct. In addition, the protectors of larger channels should provide the possibility of axial and longitudinal expansion and contraction to be suitable for thermal stresses.

1.4 THE DESIGN PROCEDURE OF THE DUCTWORK NETWORK

Designing a ductwork network can be quite a complex process. Determining the number, location, and dimensions of duct components—straight ducts, elbows, tees, dampers, etc.—can be tedious and time-consuming. However, for the purposes of the study-oriented control system construction cost estimates, it is not necessary to perform such design procedures. Rather, a much simpler method can be used to determine the duct size.

1.5 TWO APPROACHES TO THE DESIGN OF THE CHANNELIZATION NETWORK

There are two common ways to determine duct size and price. In the first, the total weight of the duct is calculated from the number and dimensions of different components. This weight is then multiplied by a unit price ($/lb) to obtain the cost of the ductwork equipment. In order to determine the weight of the ductwork network, the diameter, length, and wall thickness of each component in the system must be known. As mentioned in the above section, obtaining this data is not an easy task.

The second method is a modification of the first method. In this method, the duct components are individually sized and priced. Direct ducts are usually priced based on length, diameter, wall thickness, and construction materials. Elbows, tees, and other fittings are priced according to all of these factors, except length. Other variables, such as the amount and type of insulation, also affect the price. Since the second method provides more detail and precision, it is used in this section.

1.6 DESIGN PARAMETERS OF DUCTWORK NETWORK

Again, the main variable to determine the size of the duct is its length, diameter, and wall thickness. Another parameter is the amount of insulation required.

Length: The length of the ductwork network required by the air pollution control system depends on the distance of the source from the control device and the number of directional changes required. Without knowing the location of the source, it is impossible to determine the length accurately. Its range can be from 20–2000 feet or more. It is better to consider the cost of the direct channel based on ft/$. This length should be part of the emission source specifications where the ductwork network is installed.

Diameter: Circular duct is preferred over rectangular, oval, or other duct shapes. For a circular duct, the cross-sectional area, A_d, is obtained from the following equation:

$$A_d = \frac{\pi D_d^2}{4} \tag{1.1}$$

where, A_d = duct cross-sectional area $\left(\text{ft}^2\right)$; D_d = duct diameter $\left(\text{ft}\right)$;

The cross-section of the duct, volume flowrate coefficient $\left(Q\right)$, and transfer speed in the duct $\left(u_t\right)$ are:

$$A_d = \frac{Q}{u_t} \tag{1.2}$$

Combine Equations 1.1 and 1.2 and calculate D_d:

$$D_d = 1.128\left[\frac{Q}{u_t}\right]^{1/2} \tag{1.3}$$

The key variable in Equation 1.3 is the transmission speed in the duct.

The value of this variable must be chosen carefully. If the amount selected is too low, the duct will be too large and, most importantly, the velocity will not be high enough to transport the suspended particles in the contaminated gas flow to the control device. If u_t is too high, the static pressure drop (which is proportional to the square of u_t) will be excessive, as well as the power consumption of the corresponding fan.

Cost is also considered in determining the optimal diameter of the duct. The cost of the equipment increases with the increase in the diameter of the duct. However, the changes in fan power cost have an inverse relationship with the diameter. However, for estimation purposes in studies, the optimal diameter of the duct need not be determined. Calculating the diameter of the duct is only using the transmission speed values.

The transfer speed usually varies from 2000–6000 ft/min, depending on the composition of the contaminated gas.

Lower velocities in the duct will be appropriate for a contaminated gas containing gaseous contaminants or very fine dusts, while higher velocities are required for conveying a stream with a large amount of metal or other heavy or wet material. The speeds given in Table 1.1 may be used as a general guide.

In Table 1.2, these values are presented with suggested velocities in the duct for conveying various materials.

TABLE 1.1
Minimum Transmission Speed for Material Conduction

Transferred Materials	Minimum Transmission Speed $(u_t, \text{ft / min})$
Gases: very fine light dusts	2000
Fine light powders and dusts	3000
Medium industrial dusts	3500
Coarse dusts	4000–4500
Wet or heavy load of dust	≥4500

TABLE 1.2
Minimum Duct Speeds for Selected Materials

Material	Minimum Transmission Speed (ft/min)
Aluminum dust (coarse)	4000
Brass turnings	4000
Cast iron boring dust	4000
Clay dust	3500
Coal dust (pulverized)	4000
Cocoa dust	3000
Flour dust	5000[a]–3000
Wheat dust	3000–2500
Lead dust	4000
Limestone dust	3500
Magnesium dust (coarse)	3000
Metal chips	5000–4000
Plastic dust (smoothing by a special sanding stone machine)	3000
Leather dust	(Coarse) 4000–(Fine) 2500
Silica dust	4500–3500
Soap dust	3000
Soapstone dust	3000
Spray paint	2000
Starch dust	3000
Stone dust	3500
Tobacco dust	3500

[a] Transfer speed that varies depending on the casting operation.

Wall thickness: The wall thickness of a duct depends on various factors including internal pressure, diameter, structural material, and other structural parameters. However, the specified diameter duct can be used for a wide range of wall thicknesses. For example, 24-inch diameter "all-welded longitudinal seam duct" 304 stainless steel is manufactured in thicknesses ranging from 14 to 22 gauge (0.0313 to 0.0781 inches). The same range of gauges is used with duct diameters in the range of 3–36 inches.

Note that as the wall thickness increases, the number of gauges decreases. This unit (gauge), which is commonly used in the metal fabrication industry, is more convenient than the thickness expressed in inches, because the latter is usually a value less than 0.25. In addition, the gauge value varies according to the metal used, such as carbon steel (galvanized or non-galvanized), stainless steel, or aluminum.

The gauges of these metals are presented in Tables 1.3a and 1.3b for a wide range of nominal thicknesses.

The unit of gauge is not used for plastic duct, as the wall thickness for these ducts is normally expressed in inches. In any case, it is usually not necessary to know the wall thickness in estimating the cost of the duct because this parameter is already included in the cost equations.

Insulation: The insulation can be installed on the outer surface of the ductwork network, or the duct word network itself can be made with internal insulation. In the first case, the amount of insulation required depends on several heat transfer variables, including the temperature, velocity, composition, and other characteristics of

TABLE 1.3a
Wall Thickness of Aluminum and Steel Duct (American Standard)

	Nominal Thickness (inch)			
	Carbon Steel			
Gauge Number	Galvanized[a]	Non-galvanized[b]	Stainless Steel (316 or 304)	Aluminum
28	0.0187	0.0149	0.0156	-
26	0.0217	0.0179	0.0188	0.02
24	0.0276	0.0239	0.0250	0.025
22	0.0336	0.0299	0.0313	0.032
20	0.0396	0.0359	0.0375	0.04
18	0.0516	0.0478	0.0500	0.05
16	0.0635	0.0598	0.0625	0.064
14	0.0785	0.0747	0.0781	0.08
12	0.1084	0.1046	0.1094	0.1
10	0.1382	0.1345	0.1406	0.125
8	0.1681	-	0.172	0.16
7	-	-	0.188	0.19

[a] To obtain equivalent strength and hardness, the nominal thickness of aluminum is assumed to be approximately 150% of the nominal thickness of galvanized carbon steel of the same gauge.

[b] Colorable galvanized carbon steel and galvanized.

TABLE 1.3b
Proposed Gauges of Metal Duct

Rectangular Duct		Round Duct		
The Largest Dimension	Galvanized Steel	Diameter	Galvanized Steel	Aluminum
to 30 inches	24	to 8 inches	24	22
31–60	22	9–24	22	20
61–90	20	25–48	20	18
91 to above	18	49–72	18	16

Note: Galvanization is based on the American standard.

the contaminated gas, ambient temperature, duct diameter, wall thickness and thermal conductivity, and the temperature of the desired surface (skin).

Determining these variables involves performing a series of complex calculations.

The second approach is to choose a pre-insulated ductwork network. As mentioned earlier, it can be equipped with any type and thickness of insulation. However, 1, 2, or 3 inches is common.

1.7 PRESSURE LOSSES IN THE CANALIZATION NETWORK

Energy losses in a ventilation system due to friction are typically expressed as a fraction of the velocity pressure, V_P.

Technically, pressure loss calculations apply only to those areas in the ventilation system where there is no change in velocity pressure (for example, where the duct diameter is constant). These areas include straight ducts, hoods, and other fittings such as couplings and simple elbows.

But, with tees, 45-degree tees, and other flow-split fittings, the velocity and velocity pressure are not constant between the inlet and outlet of the fittings. The corresponding friction loss (F_b) is a function of upstream (inlet) and branch *VPs*, as given in the following equation:

$$F_b = VP_u\left(k_b - 1\right) + VP_b \tag{1.4}$$

Where:

F_b: The corresponding friction loss
VP_u: Upstream pressure (pa)
k_b: compressibility factor (unitless)
VP_b: Base pressure

However, flow split fittings (flow split fittings are required in conjunction with more complex control systems that collect polluted gases from multiple emission points) are generally not used with simple air pollution control ventilation systems, except where tee connections may be required, for example, for the purpose of adding progressive air conditioning.

Friction loss for a ductwork network is a complex function of several variables: duct diameter and length, transport velocity, and gas viscosity and density. In particular, the Darcy-Weissbach and Klebrock equations are used for this equation—the latter being used to calculate the Reynolds number.

Typically, the friction loss is obtained from a calculation chart or, more recently, by computer programs. Also, to simplify the calculation, empirical equations derived for specific types of commercially available ductwork networks are presented. For example, to estimate the friction loss per 100 feet (F_d/100 ft) under standard conditions for a round, spiral, galvanized duct with 10 connections per 100 feet, the following equation is used:

$$\frac{F_d}{100\,\text{ft}} = 0.136\left[\frac{1}{D_d}\right]^{1.18}\left[\frac{u_t}{1000}\right]^{1.8} \tag{1.5}$$

Where, D_d = duct diameter (ft), and: $0.25 \leq D \leq 5$.

It is clear that this equation calculates the total friction loss, not the coefficient of loss (k). However, it is possible to calculate k for a given diameter (D_d) and discharge (Q) by dividing Equation 1.5 by VP and multiplying by 100.

To estimate the friction loss for other duct materials, the value obtained from Equation 1.5 is multiplied by a roughness correction factor (which accounts for the energy losses in the fluid). Approximate values for this factor are provided (Table 1.4):

Based on the experimental data, the loss coefficients of the connections are presented. Mostly, 90° elbows, which are the most used in air pollution control systems, are preferred. "k_{90}" values for elbows vary according to the diameter and radius of curvature, which are expressed as a multiple of the diameter of the elbow. The usual ranges of these values are as follows (Table 1.5):

As these values show, the larger the radius of curvature, the lower the friction loss. This seems logical because, the greater the radius of curvature, the gas flow gradually changes direction. For an angle less than 90°, multiply the value of k_{90} above by an adjustment factor (θ/90), so:

$$K_\theta = \left[\frac{\theta}{90}\right]k_{90} \tag{1.6}$$

Where, K_θ = loss factor for $\theta < 90$.

TABLE 1.4
Roughness Correction Factor of Different Duct Materials

Material	Roughness Correction Factor
Non-spiral screw galvanized	0.9
Fiberglass (polished finish)	0.8
Acrylonitrile butadiene styrene plastic (ABS) and polyvinyl chloride (PVC)	0.8
Concrete	1.4
Soft corrugated and coated channel	2.3

TABLE 1.5
Typical Ranges of Friction Loss Coefficient (k_{90})

Radius of Curvature	Friction Loss Coefficient (k_{90})
0.50	0.80
1	0.35
1.25	0–30.55
1.50	0–27.39
2	0–24.27
2.50	0–22.24

Example: A control device in a cosmetic manufacturing plant is connected to a source with a 250-ft round spiral duct. The intended duct consists of three 90° elbows and two 45° elbows, each of which has a radius of curvature of 1.50. The volumetric flow rate (Q) of contaminated gas (which contains face powder) is 1500 ft³/min under standard conditions. Calculate the friction loss for the ductwork network.

Solution: Since the material transported in the ductwork network (face powder) is light, a suitable transfer speed (u_t) in this case is 2000 ft/min (see Table 1.1). After that, by substituting this value and volume flow in Equation 1.3, the duct diameter will be obtained (D_d):

$$\left[\frac{15000}{2000}\right]^{1/2} = 3.09\text{ft}\, 1.128 = D_d$$

Then, replace the diameter and speed in Equation 1.5 to calculate the friction loss (static pressure), F_d of the straight duct:

$$F_d = 0.136\left[\frac{1}{3\cdot 09}\right]^{1/18} = \left[\frac{2000}{1000}\right]^{1/8}\left[\frac{250}{100}\right] = 0.313 \text{ in. w.c.}$$

A factor of 100/250 in this expression adjusts the friction loss from 100 feet (based on Equation 1.5) to 250 feet (the length of the duct system in this example). Other friction losses are created in five elbows (three 90°, two 45°), each with a radius of curvature of 1.50.

These losses (F_c) are calculated through Equation 1.7:

$$F_c = K_\theta VP \tag{1.7}$$

Where, $VP = (2000/4016)^2 = 0.248$ inches of water.

For 90° elbows, $K_\theta = K_{90} = 0.33$ and:

$$F_c = 3 \times 0.33(0.248) = 0.246 \text{in. w.c.}$$

For 45° elbows, $K_\theta = (45/90)\,K_{90} = 0.165$ and:

$$F_c = 2 \times 0.165(0.248) = 0.0818 \text{ in. w.c.}$$

Therefore, the total friction loss will be:

$$F = 0.313 + 0.246 + 0.0818 = 0.641 \text{ in. w.c.}$$

From this example, two conclusions can be drawn: (1) the static pressure loss of the straight duct is not large, even at this length (250 feet) and (2) the loss in the elbows, which total 0.328 inches of water, is greater than direct ducts. Although it may be tempting to ignore connection loss for expediency, this can understate the static pressure loss of the ventilation system (Table 1.6).

1.8 ESTIMATION OF DUCT PRESSURE LOSS IN HVAC SYSTEMS BASED ON DESIGN BY MECHANICAL ENGINEERS

For ducting HVAC systems, standards include such things as speed in the duct, maximum length of flexible duct, type of supply and return devices, materials of duct structure, etc.

1.8.1 Method

- One-line duct location diagram: There is no need to determine the duct size, but the velocity in the duct and the design flow loss (DFL) (DFL - pressure loss/100 feet of duct) are necessary
- Selection of the possible critical path (duct path with the highest loss)
- Determining the total duct length (TDL), including the input and return paths along the critical path
- Determining the type and number of duct connections along the critical path
- To determine the total equivalent length of the channel (TEL), use the connection loss diagrams for all connections; the corresponding tables are for a velocity of 1350 fpm and a loss of 0.1 inch of water/100 feet of duct
- To convert the equivalent length for other system design conditions except for 1350 fpm velocity and 0.1 inch of water/100 feet of duct loss, use the fitting loss conversion factor table (Table 1.17)
- The estimation of duct system losses is done as follows:

$$\frac{(\text{TDL} + \text{TEL})}{100} \times \text{DFL} \tag{1.8}$$

- Adding losses of other components

TABLE 1.6
The Maximum Duct Speeds for Different Locations

	Control Factor				
	Noise Production	Duct Friction			
		Main Channels		Branch Channels	
Application	Main Channels	Input (Supply)	Return	Input	Return
Residential, personal	600	1000	800	600	600
Apartments, hotel rooms, hospital rooms	1000	1200	1300	1200	1000
Private administrative offices, libraries	1200	2000	1500	1600	1200
Theaters, amphitheaters	800	1300	1100	1000	800
Government administrative offices, upscale restaurants, luxury shops, banks	1500	2000	1500	1600	1200
Normal shops, self-service restaurants	1800	2000	1500	1600	1200
Industrial	2500	3000	1800	2200	1500

1.8.2 An Example of the Design of the Ductwork Network (Figure 1.2)

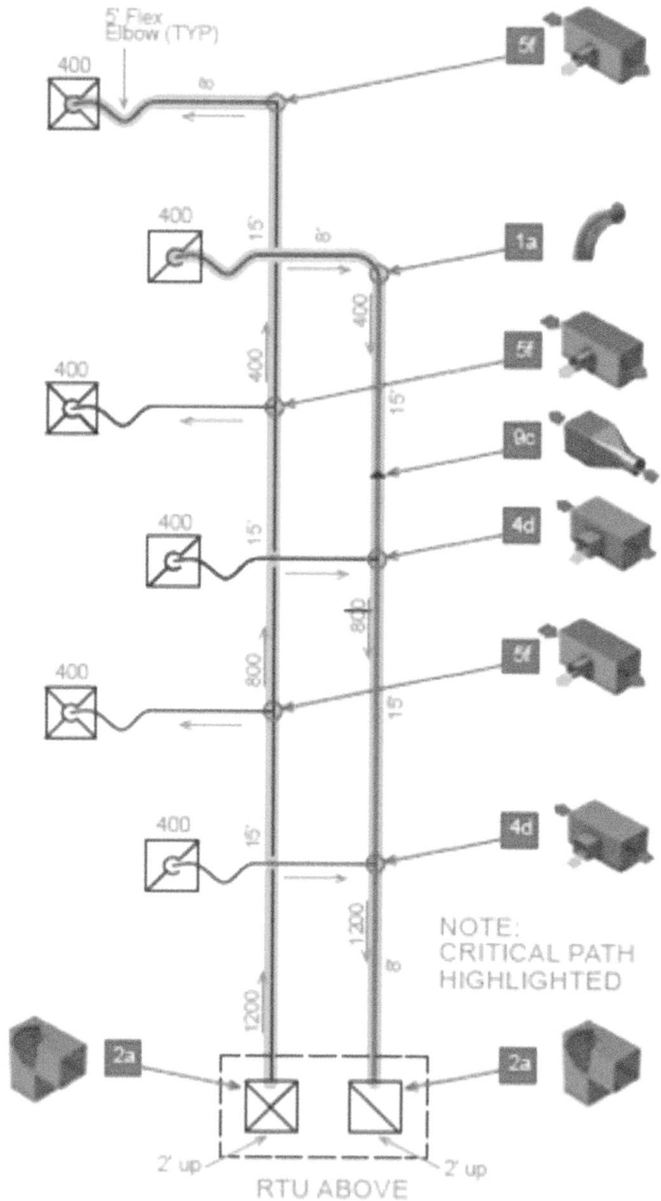

FIGURE 1.2 Layout of the sample system.

1.8.3 Example of Duct Loss Estimation

A three-ton of rooftop package unit (RTU) can be considered an independent, highly energy-efficient device within HVAC systems. Its structure combines a condensing unit (a chiller without an operator and expansion valve) and an air conditioning system with a DX cooling coil. The unit includes all necessary components required to provide air conditioning, with a flow rate of:

1200 cfm
Velocity in duct - 1000 fpm @100′/0.1”
Calculation of total duct length (incoming TDL and return TDL) along the critical path
Input TDL = 2′ + 15′ + 15′ + 15′ + 8′ = 55′
Returned TDL = 2′ + 8′+ 15′ + 15′ + 8′ = 48′
Total TDL = 55′ + 48′ = 103′
Connection losses
Connections of the entrance path
Type 2a (H/W = 1): 45′
Type 5f (where $Q_b / Q_t = 0.3$): 19′
Type 5f (where $Q_b / Q_t = 0.5$): 10′
Type 5f (where $Q_b / Q_t = 1$): 0′
Stainless steel flexible hose: 120′
Input TEL = 45′ + 19′+ 10′ + 0′ + 120′ = 194′
Connections of the return path
Flexible elbow: 120′
Type 1a ($R / D = 1$): 25′
Type 9c (where $A_1 / A_2 = 0.3$ and transition angle = 90): 36′
Type 4d (where $Q_b / Q_t = 0.5$): 60′
Type 4d (where $Q_b / Q_t = 0/3$): 43′
Type 2a: 45′
Return TEL = 120′ + 25′ + 36′ + 60′ + 43′ + 45′ = 329′
Total TEL = 194′ + 329′ = 523′

Since the design speed in the duct is 1000 fpm instead of 1350 fpm, correct the design criteria by using the connection loss conversion coefficient table. For 0.1@1000 fpm, read 0.55. Therefore, 0.55 × 523′ = 288′ (TEL correction for system design criteria).

$$\text{TDL} + \text{TEL} = 288' + 103' = 391'$$

Therefore,

$$\frac{(\text{TDL} + \text{TEL})}{100} \times \text{DFL} = 391/100 \times 0.1 = 0.391' \text{or } 0.4''\text{wg}$$

Add supply diffuser, return diffuser, and filter to the estimated losses:

$$0.1 + 0.1 + 0.1 + 0.4 = 0.7 \text{ inches of water}$$

TABLE 1.7
Table of Round Elbows

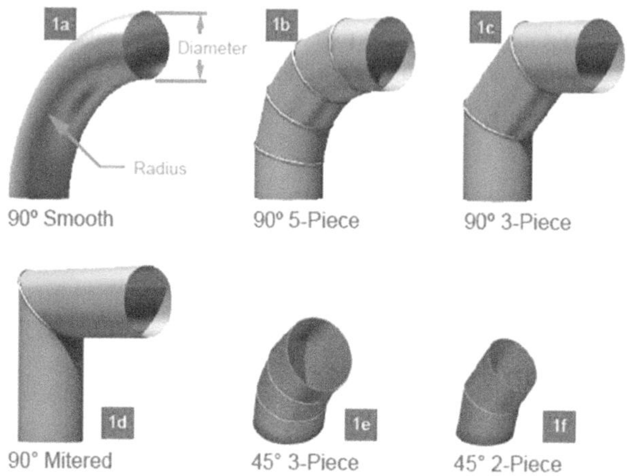

Radius/ Diameter	90°				45°	
	1a	**1b**	**1c**	**1d**	**1e**	**1f**
0.75	37′	52′	61′	136′	15′	20′
0.1	25′	37′	48′		15′	20′
+1.5	17′	27′	39′		15′	20′

Therefore, the estimated external system pressure loss for the RTU is 0.7 inches of water.

Connection loss tables – equivalent length of straight duct (Table 1.7)

1.8.4 Mitered Elbows

2c- Tee (Equivalent to 2 Elbows)

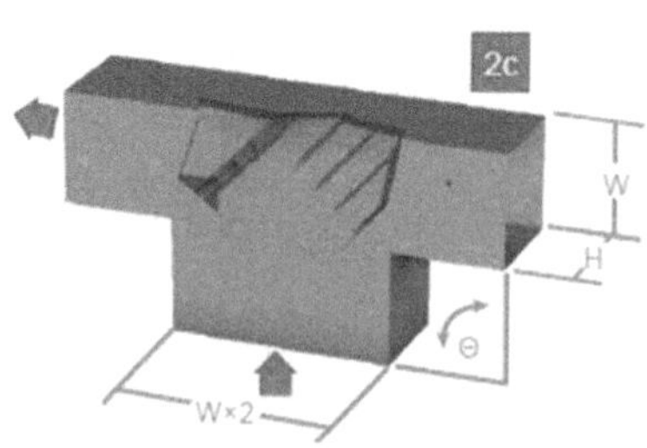

The flow can be divergent or convergent, with or without vanes; use Tables 1.8–1.14 for similar elbows.

TABLE 1.8
Table of Mitered Elbows (45 Degrees) Rectangular: A) 2a: With Blade; B) 2b: Without Blades

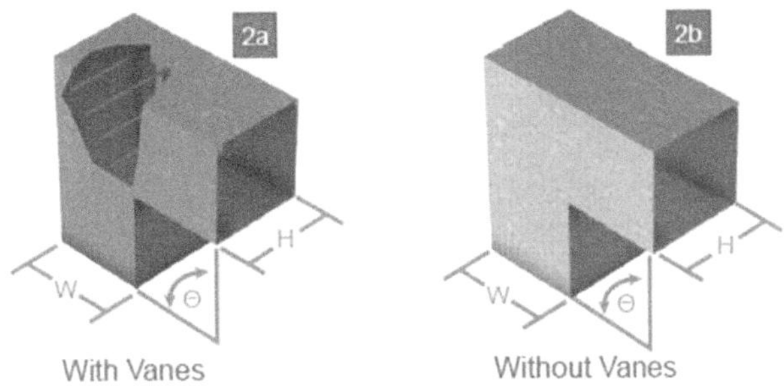

H/W = 0.25	H/W = 1	H/W = 4
A		
53′	45′	72′
B		
148′	136′	105′

Angles other than 90	
Angle	**Coefficient**
30°	0.45
45°	0.6
60°	0.78

1.8.5 Radial Elbows Rectangular Shape

3c- Tee (equivalent to 2 elbows)

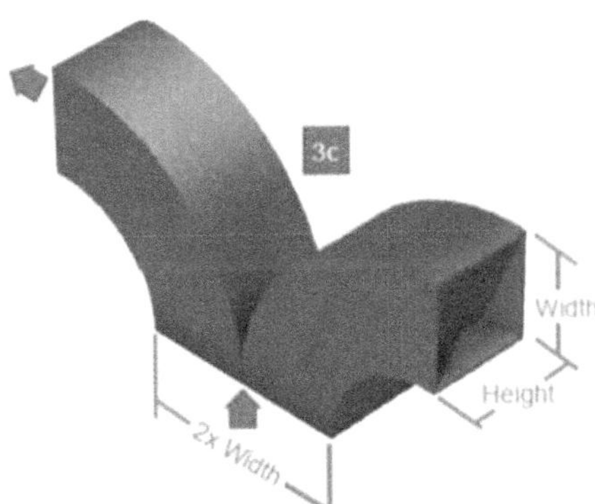

The flow can be divergent or convergent, with or without vanes; use Table 1.9 for similar elbows.

TABLE 1.9
Radial Elbows of Rectangular Shape - A) 3a: With Blade; B) 3b: Without Blades

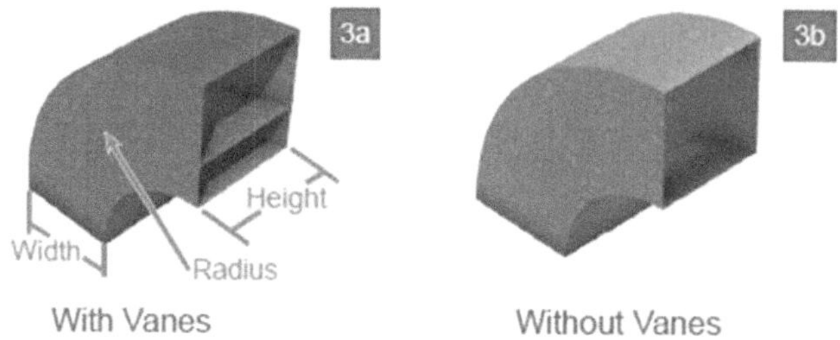

Radius/Diameter	H/W = 0.25	H/W = 1	H/W = 4
A			
0.25	9′	5′	6′
0.5	3′	2′	1′
B			
0.5	170′	136′	125′
1	31′	24′	22′
2	23′	17′	16′

1.8.6 Converging Tee Rectangular Trunk (Table 1.10)

TABLE 1.10
Converging Tee Rectangular Trunk: A) 4a: Rectangular Branching; B) 4b: Rectangular Branching With 45 Degree Entrance; C) 4c: Round Branching; D) 4d: Rectangular Branching; 4e: Rectangular Branching With 45° Inlet; 4f: Round Branching

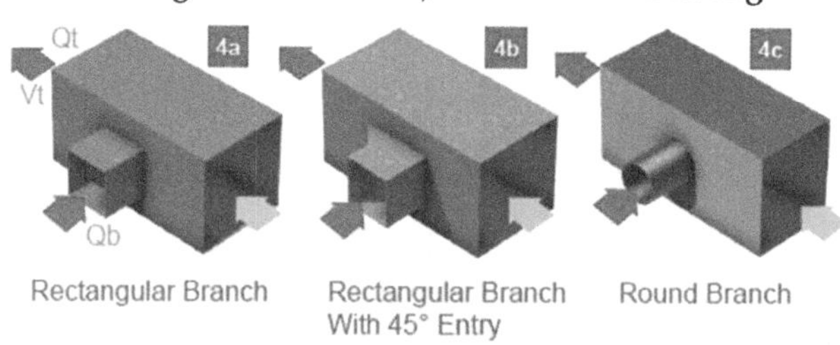

Use the following values when evaluating the branching path. To clarify the issue, airflow is marked with arrows.

(*Continued*)

TABLE 1.10 (Continued)

Velocity in Body (V_t)	Volumetric Flow Rate (Branching)/Volumetric Flow Rate (Body) Q_b / Q_t								
	0.1	0.2	0.3	0.4	0.5	0.6	0.7	0.8	0.9
A									
fpm 1200>	0′	3′	10′	37′	117′	125′	244′	333′	475′
fpm 1200<	1′	8′	26′	76′	133′	189′	303′	382′	447′
B									
fpm 1200>	0′	8′	16′	32′	62′	117′	170′	219′	284′
fpm 1200<	2′	10′	20′	39′	86′	130′	208′	228′	330′
C									
fpm 1200>	0′	7′	15′	26′	89′	148′	219′	352′	554′
fpm 1200<	2′	12′	26′	68′	144′	234′	312′	420′	560′

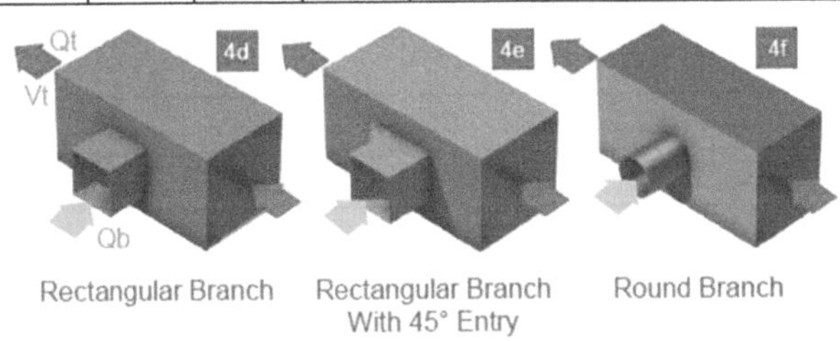

D									
EL	18′	31′	43′	52′	60′	65′	67′	68′	67′

Note: EL: equivalent length: equivalent length of the channel in feet.

1.8.7 Divergent Tee Rectangular Trunk (Tables 1.11 and 1.12)

TABLE 1.11
Divergent Tee Rectangular Trunk: A) 5a: Rectangular Branching; B) 5b: Rectangular Branching with 45° Entrance; C) 5c: Round Branching; D) 5d: Rectangular Branching; 5e: Rectangular Branching with 45° Entrance - 5f: Round Branching

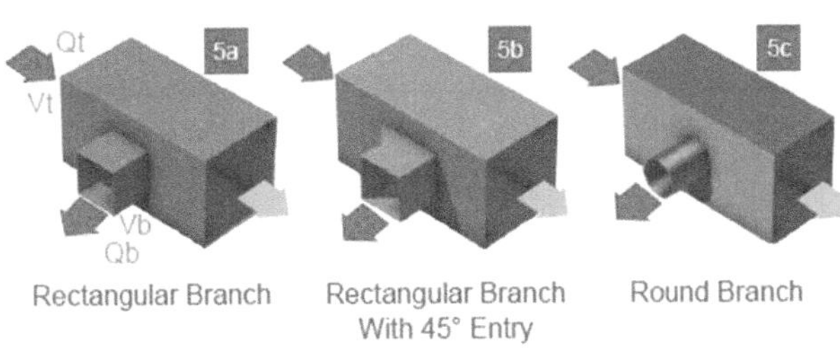

Use the following values when evaluating the branching path. For clarity, the airflow is marked with arrows.

(Continued)

TABLE 1.11 (Continued)

V_b / V_t	Q_b / Q_t				
	0.1	**0.2**	**0.3**	**0.4**	**0.5**
A					
0.2	117′				
0.4	118′	115′			
0.6	126′	117′	119′		
0.8	132′	137′	133′	127′	
1	157′	159′	148′	155′	144′
B					
0.2	103′				
0.4	92′	90′			
0.6	87′	82′	80′		
0.8	89′	83′	78′	75′	
1	89′	111′	97′	90′	84′
C					
0.2	114′				
0.4	115′	122′			
0.6	130′	125′	123′		
0.8	134′	149′	127′	128′	
1	148′	157′	136′	140′	143′

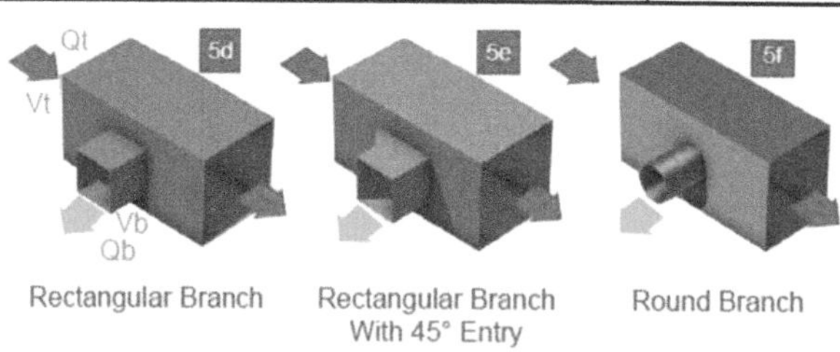

D

V_b/V_t	Equivalent Length (Body Path)							
	0.1	0.2	0.3	0.4	0.5	0.6	0.8	1
EL	32′	25′	19′	15′	10′	7′	2′	0′

TABLE 1.12
90° Round Tee: A) 6a: Convergent; B) 6b: Divergent; C) 6c: Convergent; T) 6d: Divergent

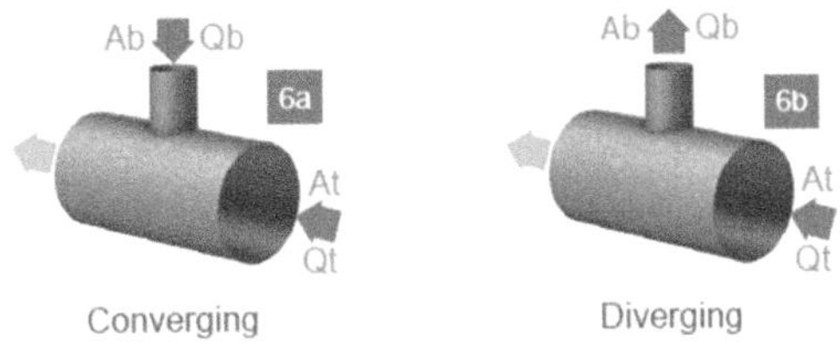

Use the following values when evaluating the branching path. For clarity, the airflow is marked with arrows.

A							
Q_b / Q_t	**Branching Surface/Body Surface $\left(A_b / A_t\right)$**						
	0.1	**0.2**	**0.3**	**0.4**	**0.6**	**0.8**	**1**
0.2	432′	82′	19′	17′	15′	13′	12′
0.4		489′	239′	107′	61′	45′	36′
0.6			534′	182′	105′	78′	65′
0.8				307′	170′	125′	98′
1				454′	239′	159′	125′

B									
A_b / A_t	**Volumetric Flow Rate of Branching Flow $\left(Q_b\right)$/Volumetric Flow Rate of Body Flow $\left(Q_t\right)$**								
	0.1	**0.2**	**0.3**	**0.4**	**0.5**	**0.6**	**0.7**	**0.8**	**0.9**
0.8	89′	70′	56′	45′	39′	35′	36′	40′	45′
0.6	84′	64′	50′	42′	40′	41′	49′	61′	77′
0.4	75′	53′	45′	49′	61′	78′	108′	148′	193′
0.2	64′	64′	114′	205′					

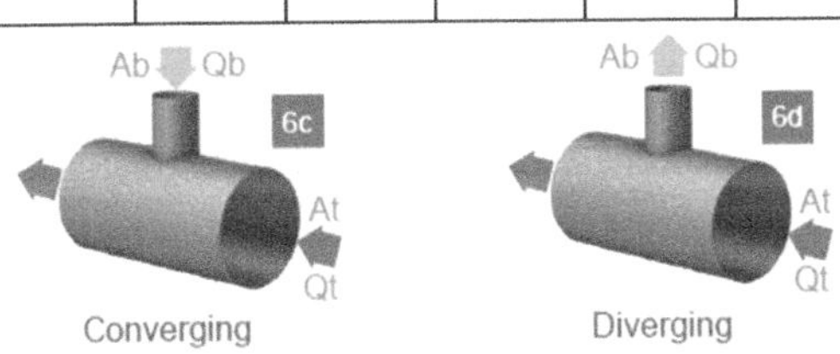

Use the following values when evaluating the branching path. For clarity, the airflow is marked with arrows.

C								
Q_b / Q_t	**Volumetric Flow Rate (Branching)/Volumetric Flow Rate (Body) Q_b/Q_t**							
	0.1	**0.2**	**0.3**	**0.4**	**0.5**	**0.6**	**0.8**	**0.9**
EL	18′	31′	43′	52′	60′	65′	68′	67′

(Continued)

TABLE 1.12 (Continued)

D								
V_b / V_t	**Volumetric Flow Rate (Branching)/Volumetric Flow Rate (Body) Q_b/Q_t**							
	0.1	**0.2**	**0.3**	**0.4**	**0.5**	**0.6**	**0.8**	**1**
EL	32′	25′	19′	15′	10′	7′	2′	0′

Note: EL: equivalent duct length in feet.

1.8.8 Divergent Tee Round Body (Table 1.13)

TABLE 1.13

Divergent Tee Round Body: A) 7a: With A 45° Round Branching Elbow; B) 7b: with Conical Round Branching; C) 7c: Elbow with 90° Round Branching; T) 7d: 45° Rolled Elbow with Round Branching; 7e: Conical Round Branching; 7f: Elbow with 90° Round Branching

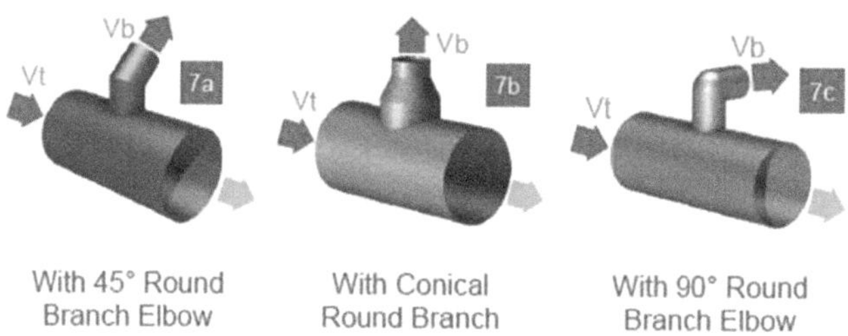

Use the following values when evaluating the branching path. For clarity, the airflow is marked with arrows.

V_b / V_t	**0.2**	**0.4**	**0.6**	**0.8**	**1**
A					
EL	108′	102′	98′	92′	90′
B					
EL	97′	84′	70′	59′	48′
C					
EL	117′	123′	134′	151′	177′

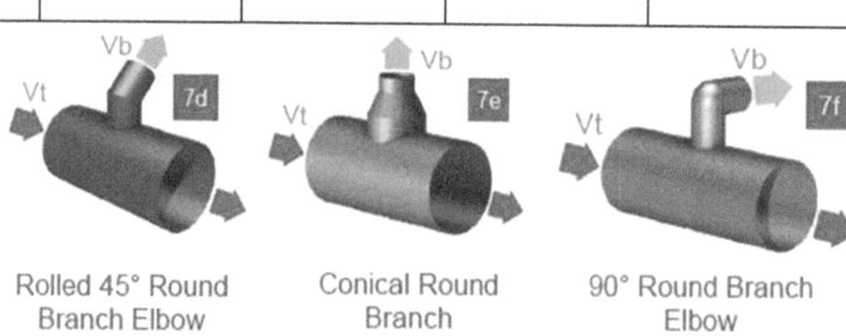

Use the following values when evaluating the branching path. For clarity, the airflow is marked with arrows.

(*Continued*)

TABLE 1.13 (Continued)

D								
V_b / V_t	**0.1**	**0.2**	**0.3**	**0.4**	**0.5**	**0.6**	**0.8**	**1**
EL	32′	25′	19′	15′	10′	7′	2′	0′

1.8.9 Rectangular or Round Tee (Pair of Pants) (Table 1.14)

TABLE 1.14
Rectangular or Round Tees: A) 8a: Divergent; B) 8b: Convergent

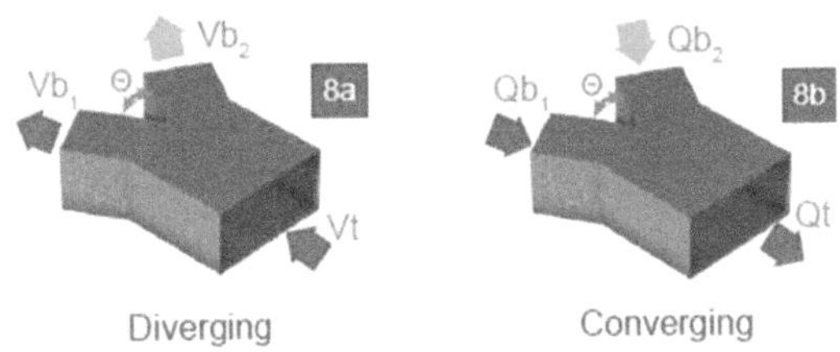

The values presented in the following section are used to evaluate branching.

A							
	V_{b1} / V_t **or** V_{b2} / V_t						
Angle	0.1	0.2	0.3	0.4	0.5	0.6	0.8
15°	92′	70′	58′	43′	32′	23′	12′
30°	95′	78′	64′	50′	39′	32′	22′
45°	99′	84′	72′	61′	51′	43′	33′
60°	102′	93′	90′	75′	67′	60′	49′

B							
	Volume Flow Rate Q_{b1}/Q_t **or** Q_{b2}/Q_t						
Angle	**0.1**	**0.2**	**0.3**	**0.4**	**0.5**	**0.6**	**0.8**
15°	0′	1′	2′	4′	11′	47′	97′
30°	2′	4′	8′	16′	32′	78′	125′
45°	4′	8′	10′	23′	64′	105′	182′

1.8.10 Transitional Expansion Flow (Intermediate) (Table 1.15)

TABLE 1.15
Transitional Expansion Flow: A) 9a: Conical to Round; B) 9b: Rectangular to Rectangular; 9c: Round to Rectangular; 9d: Rectangular to Round

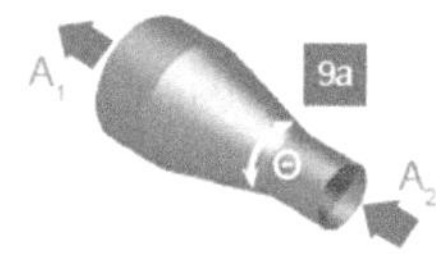

A

A_1 / A_2	Transmission Angle							
	16°	20°	30°	45°	60°	90°	120°	180° (Fast)
2	16′	22′	36′	37′	37′	36′	35′	34′
4	26′	34′	52′	69′	77′	73′	72′	70′
6	31′	37′	55′	75′	87′	84′	83′	82′
10	33′	43′	67′	86′	91′	94′	95′	94′
+16	35′	43′	68′	95′	100′	100′	100′	100′

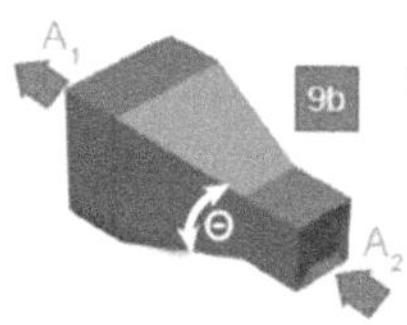

Rectangular to Rectangular

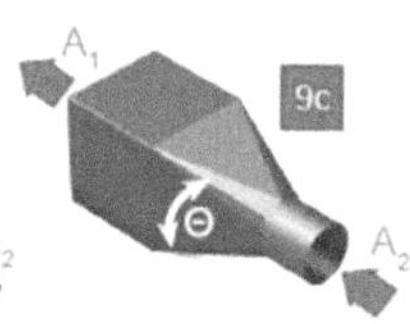

Round to Rectangular

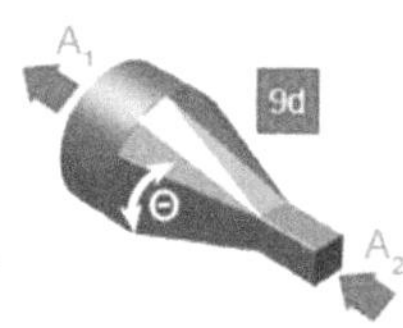

Rectangular to Round

B

A_1/ A_2	Transmission Angle							
	16°	20°	30°	45°	60°	90°	120°	180° (Suddenly)
2	20′	25′	28′	33′	35′	36′	37′	34′
4	41′	49′	57′	64′	69′	72′	72′	72′
6	48′	53′	66′	77′	82′	86′	86′	85′
10	48′	56′	67′	80′	91′	99′	97′	98′

1.8.11 Transitional Contraction Flow (Tables 1.16 and 1.17)

TABLE 1.16
Transitional Contraction Flow: A) 10a: Expansion Flow, Rectangular with Vertical Sides; B) 10b: Contraction Flow, Rectangular; 10c: Contraction Flow, Round

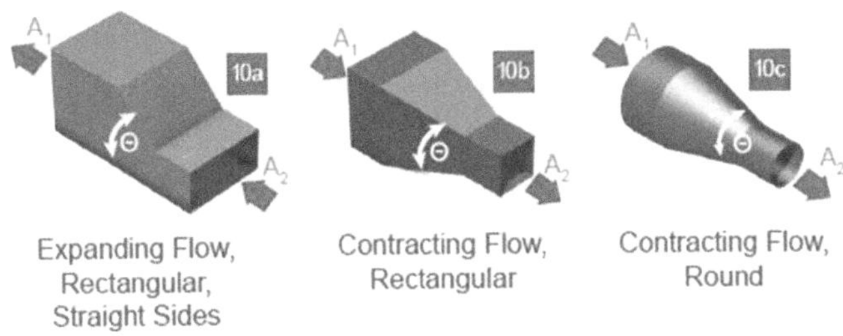

A							
A_1 / A_2	Transmission Angle						
	14°	20°	30°	45°	60°	90°	180° (Suddenly)
2	10′	14′	23′	39′	42′	43′	40′
4	18′	28′	48′	68′	77′	80′	75′
6	22′	34′	55′	74′	86′	94′	91′
B							
A_1 / A_2	Transmission Angle						
	10°	15°–40°	50°–60°	90°	120°	150°	180° (Suddenly)
2	6′	6′	7′	14′	20′	27′	30′
4	6′	5′	8′	19′	31′	40′	47′
6	6′	5′	8′	20′	32′	41′	48′
10	6′	6′	9′	22′	33′	42′	49′

TABLE 1.17
Coefficients of Connection Loss Conversion

	The Amount of Design Loss Due to Friction ("/'100 Ducts)	
Design Speed in Duct (Fpm)	**0.08**	**0.1**
900	0.56	0.44
1000	0.69	0.55
1100	0.83	0.66
1200	0.99	0.79

(Continued)

TABLE 1.17 (Continued)

Design Speed in Duct (Fpm)	The Amount of Design Loss Due to Friction ("/'100 Ducts) 0.08	0.1
1300	1.16	0.93
1400	1.34	1.08
1500	1.54	1.23
1600	1.76	1.40
1700	1.98	1.59
1800	2.22	1.78
1900	2.48	1.98
2000	2.74	2.19

TABLE 1.18
Typical Design Speeds for HVAC Components

Part	Speed (fpm)
Inlet blowers (7000 cfm and above)	400
Outlet blowers (5000 cfm and above)	500
Page filters	
Viscous collision	200–800
Regular folded media, dry type	
Low efficiency	350
Medium efficiency	500
High efficiency	500
HEPA	250
Renewable filter media	
Viscose collision with moving curtain	500
Media with movable curtain	200
Electronic air cleaners	
Type of ionizing plate	300–500
Non-ionizing charged media	250
Charged media ionizing	150–350
Steam and hot water coils	500–1000 Minimum 200–maximum 1500
Electrical coils	
Open wire	Based on the manufacturer's data
Finned tubular	
Dehumidifier coils	400–600
Air washers	Based on the manufacturer's data
Spray type washers	
Cellular type washers	1200–1800
Spray type air washers, high speed	

Search for the appropriate coefficient for the conditions related to the system design and multiply it by the equivalent length of the total connections (Table 1.18 and Figures 1.3 and 1.4).

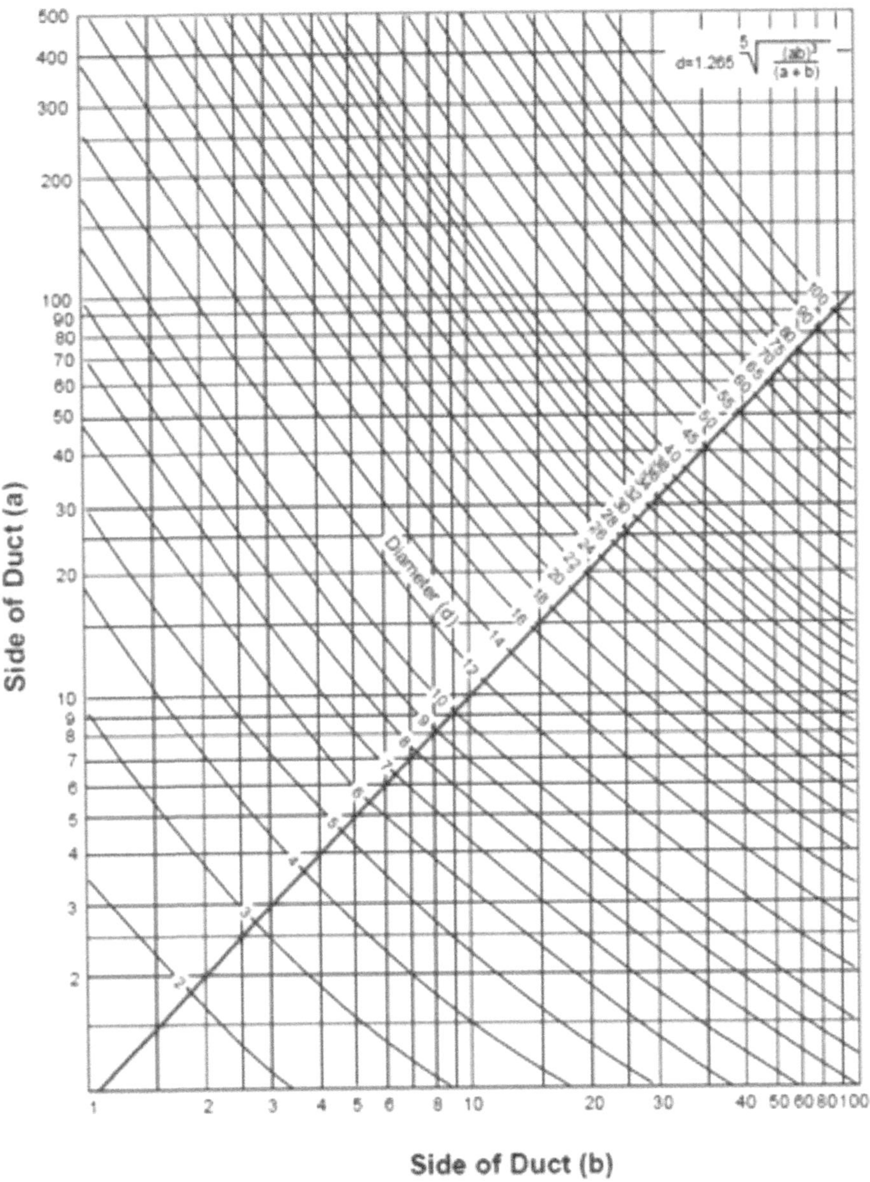

FIGURE 1.3 The rectangular equivalent diameter of round ducts.

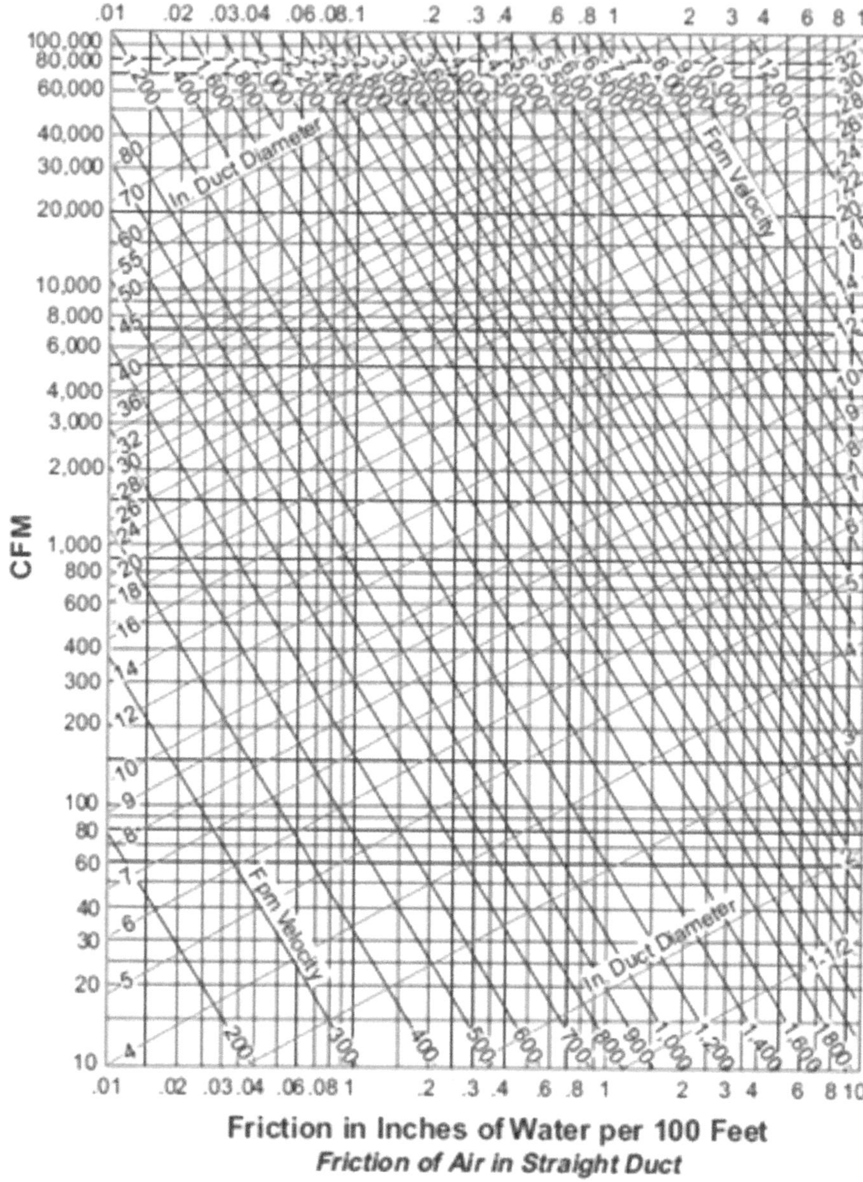

FIGURE 1.4 Duct resistance.

TABLE 1.19
Classes of Sealing and Leakage in the Duct

Duct Class	**1″, 1.2″, and 2″ Waters**		**3″ Water**	**4″, 6″, and 10″ Waters**
Sealing Class	Without	C	B	A
Application of sealing	N/A[a]	Only transverse seams	Only transverse seams and gaps	Seams, cracks, and all wall penetrations
	Leakage Class (CFM/100 sq ft @ 1″ w.g.)			
Rectangular metal	48	24	12	6
Oval and round metal	30	12	6	3
Rectangular fiber glass	N/A	6	N/A	N/A
Round fiber glass	N/A	3	N/A	N/A

a Not applicable.

Note: Structural classes upstream of the box in any variable air volume (VAV) system with 1.2-inch duct and 1-inch water are Class C.

1.9 CLASSES OF LEAKAGE AND PRESSURE IN THE DUCT

Sheet Metal and Air Conditioning Contractors' National Association (SMACNA) has provided duct leakage classes based on the amount of leakage in cfm allowed per 100 square feet of duct cross-sectional area under pressure.

Once the pressure test and leakage class are defined, the allowable leakage is calculated as follows:

$$L\max = Cl \times P^{0.65} \tag{1.9}$$

Where, $L\max$ = maximum allowable leakage in cfm per 100 square feet of duct cross-sectional area; Cl = duct leakage class in cfm per 100 square feet per 1 inch of water; P = test pressure (channel pressure class gauge in inches of water). Duct sealing classes also correspond to static pressure classes, which indicate which duct components must be sealed (Table 1.19 and Figures 1.5a and 1.5b).

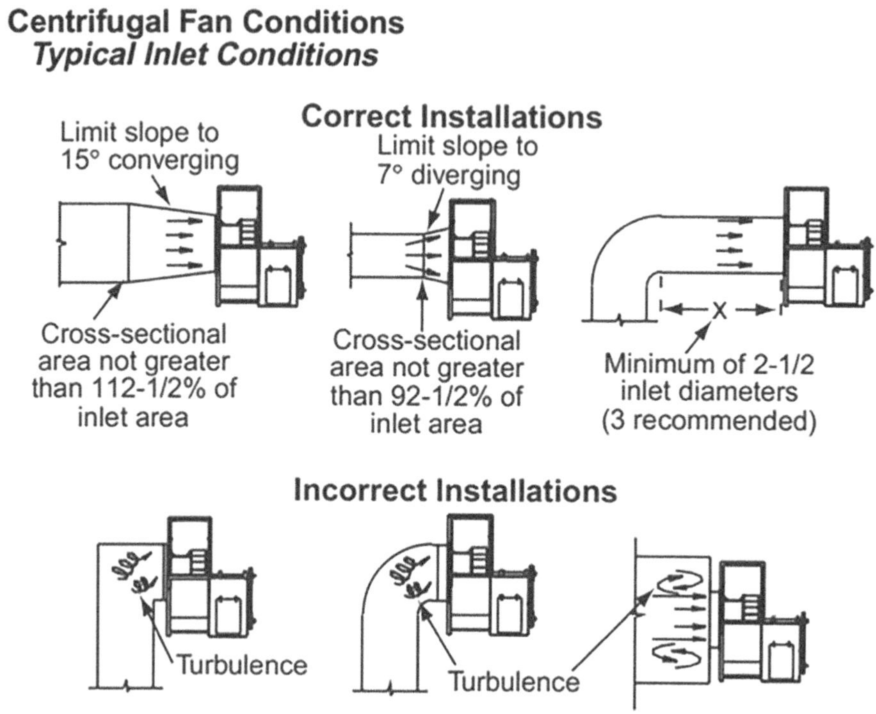

FIGURE 1.5a Typical input conditions: examples of correct installation and examples of incorrect installation.

1.10 STRUCTURAL MATERIALS

Duct components, such as hoods and other structures, are typically made of black iron or welded galvanized steel plate (including flanged gaskets), unless the presence of corrosive gases, vapors, mists, or other conditions makes the use of these materials impossible. In such cases, stainless steel, PVC, special coatings, or other materials compatible with the gas flow should be used. Electric arc welding is not recommended for black iron lighter than No. 18 gauge. Galvanized construction is not recommended for temperatures above 400°F [204°C]. Table 1.20 provides guidelines for selecting plastic materials for corrosive conditions. For non-corrosive activities, there are four classifications for exhaust systems:

Typical Outlet Conditions

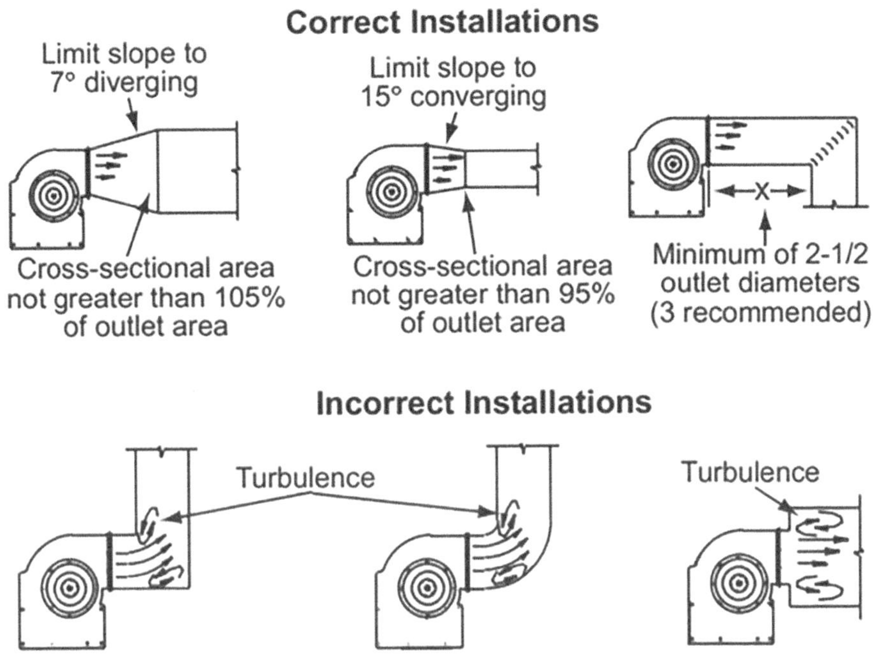

FIGURE 1.5b Typical outlet conditions: examples of correct installation and examples of incorrect installation.

Class 1 (light activity): for non-abrasive activities (e.g., air replacement, general ventilation, control of gaseous emissions without oil mist and condensable vapors); **Class 2 (moderate activity)**: for activities with moderate abrasive particles in low concentrations (e.g., grinding and polishing, woodworking); **Class 3 (heavy activity)**: for activities where abrasive particles are present in low, very high concentrations (for example, abrasive cleaning operations, dryers and furnaces, boilers, working with foundry sand); **Class 4 (extra heavy duty)**: for activities with very abrasive particles in high concentrations (for example, conveying materials with a high concentration of particles in all the examples listed in class 3; usually used in large industrial plants such as steel factories, foundries, mines, and smelters).

TABLE 1.20
Common Physical and Chemical Properties of Structural Plastics and Other Materials

Type of Chemical Material	Maximum Operating Temperature °F [°C]	Flammability	Resistance Against: Gasoline	Mineral Oil	Strong Alkali	Weak Alkali	Strong Acid	Weak Acid	Solvents
Urea formaldehyde	170 [77]	Self-extinguishing	Good	Good	Unacceptable	Relatively good	Weak	Weak	Good
Melamine formaldehyde	210–300 [99–149]	Self-extinguishing	Good	Good	Weak	Good	Weak	Good	Good
Phenolic	250–450 [121–232]	Self-extinguishing	Relatively good	-	Weak	Relatively good	Weak	Relatively good	Relatively good
Alkyd	-	Self-extinguishing	Good	-	Unacceptable	Weak	-	Good	Good
Silicon	550 [288]	-	Good	Good	-	-	Good	Good	Unacceptable
Epoxy	50–200 [10–93]	Self-extinguishing	Good	-	Good	Good	Good	Good	Good
Phenolic cast resin	-	Self-extinguishing	-	-	Unacceptable	Relatively good	Good	Good	Good to relative good
Allyl & polyester	300–450 [149–232]	Self-extinguishing	-	-	Weak	Relatively good	Weak	Relatively good	Relatively good
Acrylic	140–200 [60–93]	0.5–2 In/min [0.5–9 mm/sec]	-	-	-	Good	-	Good	Good to unacceptable
Polyethylene	140–200 [60–93]	Slow burning	-	-	-	-	-	-	-
Tetrafluoroethylene	500 [77]	Non-flammable	Good	-	Good	Good	Good	Good	Good
Chlortrifluoroethylene	-	-	-		-	-	-	-	-
Polyvinyl formal & butyral	-	Slow burning	Good	Good	Good	Good	Unacceptable	Unacceptable	Unacceptable

Vinyl chloride	130–175 [54–79]	Slow burning	-	-	Good	Good	Good	Good	Unacceptable
Polymer & copolymer	-	-	-	-	-	-	-	-	-
Vinylidene chloride	160–200 [71–73]	Self-extinguishing	Good	Good	Good	Good	Good	Good	Relatively good
Styrene	150–165 [66–74]	-	Unacceptable	Relatively good	Good	Good	-	-	Relatively good
Polystyrene reinforced with fiberglass	-	-	Unacceptable	Relatively good	Good	Good	-	-	Weak
Cellulose acetate	-	0.5–2 In/min [0.5–9 mm/sec]	Good	Good	Unacceptable	Unacceptable	Unacceptable	Relatively good	Weak
Nylon	250 [121]	Self-extinguishing	Good	Good	Good	Good	-	Good	Good
Glass	450 [232]	Non-flammable	Good	Good	Good	Good	Good	Good	Good

2 General Industrial Ventilation

A. H. Khoshakhlagh, H. M. Moda, S. Yazdanirad, and A. Gruszecka-Kosowska

2.1 OBJECTIVES OF THE LESSON

Upon completing this lesson, readers should be able to:

- Explain the general principles of ventilation
- Explain the types of ventilation systems
- Describe the effective conditions in general ventilation design
- Explain central exhaust systems
- Design dilution ventilation system

2.2 GENERAL INDUSTRIAL VENTILATION

2.2.1 Total Principles of Ventilation

The importance of clean air in the workplace is clear. In modern industry, many compounds and chemicals are used along with the complexity of operations and processes, most of which are toxic. The use of such materials leads to the release of suspended particles, gases, vapors and/or mists in the air of the workplace. Also, heat stress can result in an unsafe or unpleasant work environment. In this context, effective ventilation with proper design is the ideal solution to achieve the health and comfort of the workforce. Ventilation can also be effective in control of odors, humidity, and other adverse environmental conditions.

The potential for health hazards for an airborne substance is determined by the Threshold Limit Value (TLV). TLV refers to the airborne concentration of a substance, which represents the conditions that almost all workers can be exposed to every workday without causing adverse health effects. Time-weighted average (TWA) is introduced as the time-weighted average concentration in a typical 8-hour workday and a 40-hour work week without side effects for workers' health. TLV-TWA is commonly used to determine safe exposure levels. TLVs for chemicals are provided by the American Conference of Governmental Industrial Hygienists (ACGIH).

Ventilation systems used in industries are of two general types: a supply system that provides air in a workplace, and an exhaust system that removes pollutants produced by an operation.

DOI: 10.1201/9781003501091-2

In a basic ventilation program, both supply and exhaust systems should be considered. If the total amount of air discharged from a working space is more than the amount of air supplied from the outside to the inside space, the air pressure in the working environment is lower than that outside. Awareness of this issue is important when dilution ventilation is used to control contaminants in a specific area of the plant. Usually, these conditions occur because local exhaust systems are installed in the work environment, but no attention is paid to the air replacement systems. As a result, air enters the workplace under uncontrolled conditions through cracks, walls, windows and doors, which typically results in: (1) worker discomfort during winter months; (2) reduced exhaust system performance, and (3) higher heating and cooling costs.

2.2.2 Supply Systems

Supply systems are used for two purposes: (1) to create a comfortable environment in the workplaces (HVAC system) and (2) to replace the air removed from the workplaces (replacement system). Supply and exhaust systems are often used in conjunction with dilution control systems.

A supply system with proper design consists of an air entrance section, filters, heating and/or cooling equipment, ventilators, ducts, and networks/heat conditioning devices to distribute air within the workplaces. Heating or cooling equipment, filters, and ventilation are often integrated into a complete unit called an air house or air supply unit. If part of the air supplied by the system is recirculated, a return system is used to return the air to the air house.

2.2.3 Exhaust Systems

Exhaust ventilation systems are classified into two general groups: (1) general exhaust system and (2) local exhaust system. A general exhaust system can be used to control heat and/or remove pollutants produced in space. When this system is used for heat control, air may be moderated and purified. When it is applied for pollutant control (dilution system), outside air must be sufficiently mixed with the pollutant to reduce the average concentration to a safe level. Contaminated air is normally discharged into the atmosphere. A supply system in conjunction with a general exhaust system is usually used to replace the exhausted air.

Dilution ventilation systems are typically applied for pollutant control when the use of a local exhaust system is not possible. In this ventilation system, the large amounts of conditioned replacement air necessary to replace the exhaust air can lead to high operating costs.

Local exhaust ventilation systems operate based on the principle of trapping a pollutant at or near the source. In fact, this control method is preferred because it is more effective. The lower exhaust flow rate for this ventilation system compared to the high flow rate required by the general exhaust system leads to lower heating costs. In the conditions of air pollution control, efficient air cleaning devices are needed in industrial ventilation systems, and the lower flow rate of the local exhaust system has lower costs for air cleaning devices.

Local exhaust systems consist of four main components: hood(s), duct system (including exhaust vent and/or recirculation duct), air cleaner, and ventilator. The hood collects the produced pollution. Then, a duct system transports the polluted air to the air cleaning device—or to the ventilator if available. In an air purifier, pollution is removed from the air stream. The ventilator must overcome all losses caused by friction, hood entrance, and system connections, as well as produce the desired flow rate. The exhaust duct typically discharges air to the atmosphere in such a way that it is not re-entered by replacement and/or HVAC systems. In some situations, clean air is returned to the workplace.

2.2.4 General Exhaust Ventilation (Dilution)

General exhaust ventilation (GEV), which is also called dilution ventilation, has differences from local exhaust ventilation (LEV). LEV is based on the principle of pollutant collection and removal at the emission source and prevents pollution emission into the indoor environment. In practice, complete pollutant capture and removal is rarely achieved with LEV, resulting in minor emissions to workplace air. General ventilation minimizes pollutant concentrations based on the dilution principle. As workplace air is exhausted by mechanical systems, air is replaced into the building from outside, either through air replacement systems and/or through building openings. When clean air enters the workplace, air mixes with polluted air and dilutes it, which results in lower concentrations. In the design or evaluation of the GEV system, it is assumed that all materials emitted from the process enter the air of the workplace. Therefore, a sufficient amount of clean air must enter to dilute the concentration to an acceptable level.

The main component in a general ventilation system is an exhaust fan or a blower, usually a propeller unit installed on a ceiling or wall. A GEV system does not include an extensive hood and exhaust duct, although the replacement air supply system may be quite complex. Replacement air is often provided by natural ventilation. However, rather than uncontrolled infiltration of replacement air through cracks and openings, fresh air should be supplied through properly designed ducts and vents.

In general exhaust ventilation, to decrease the contaminant concentration to a safe level, sufficient air must be provided in any place with probability of pollution exposure. This design should be performed with the awareness of the following items:

- Physical properties of pollutants
- The amount of pollutant production
- Guidelines of health hazards (for example, threshold limit value)
- Relative positions of pollution production points, work areas, and exhaust and air supply points
- Available ventilation (natural and mechanical)

One common use for general exhaust ventilation is the prevention of flammable gases or vapors' accumulation in concentrations with hazardous conditions. For these conditions, the "safe" concentration is usually determined by 10% of the lower flammable limit (LFL) concentration for the vapor. Because this concentration (10% LFL) is almost always higher than any health hazards (for example, threshold limit values, permissible exposure limits, and maximum acceptable concentrations), this approach is limited to areas where worker exposure is short and intermittent. Some of these workplaces include drying ovens and storage tanks. This method can also be applied in warehouses and production lines to prevent vapor accumulation in case of leakage or spillage.

2.2.5 Application Limitations

General exhaust ventilation should be considered as an acceptable control method only after a thorough understanding of the GEV limitations. Often, one or more factors prevent the use of general ventilation. The first factor is the physical nature of the substance (gas, vapor, or suspended particles (dust, mist, or fume). General ventilation is usually not suitable for controlling particulate pollutants for several reasons. Because air velocities in general exhaust ventilation are low, large particles are not effectively transported from their production point to the exhaust point. As the second limitation, the proper design of a GEV system requires awareness of the pollutant production rate. It is difficult to determine this parameter for production processes of particulate matters. Many particle emission processes occur diffusely. So, the production rate, even if known, will not be constant unless the GEV system is designed to control maximum emissions. As the final reason, particulate pollutants are usually produced and emitted in a small area, so the use of local exhaust ventilation is much more appropriate for health and economic reasons.

2.3 PRINCIPLES OF DILUTION VENTILATION

The principles of dilution ventilation system design are as follows:

- From the available data, choose the amount of air required for proper dilution of polluted air; the values presented in Tables 2.1a and 2.1b provide the complete distribution and dilution of air and solvent vapors—these values should be multiplied by the selected K value
- If possible, place exhaust vents near the pollution sources to achieve spot ventilation
- Determine the location of the air supply and exhaust outlets in such a way that the air passes from the contaminated area; the operator should be placed between the air supply source and the pollutant source
- Replace the exhausted air using the air replacement system; this replacement air must be heated in cold weather; dilution ventilation systems usually control large amounts of air by low-pressure ventilators—if the system works satisfactorily, replacement air should be provided
- Prevent re-entry of exhaust air by placing the exhaust vents above the roofline and away from any window, outdoor air entrance, and/or other vents near the exhaust discharge

TABLE 2.1a
Air Dilute Volumes for Vapors

Liquid (TLV in ppm)	Air (STP) ft^3 Required for Dilution to Reach TLV Evaporation per pint
Acetone (500)	11025
N-amyl acetate (100)	27200
Benzene (0.5)	Not recommended
N-butanol (butyl alcohol) (50)	88000
N-butyl acetate (150)	20400
Butyl cellulose (2-butoxyethanol) (25)	Not recommended
Carbon disulfide (10)	Not recommended
Carbon tetrachloride (5)	Not recommended
Cellulose (2-ethoxyethanol) (5)	Not recommended
Cellosolve acetate (2-ethoxyethyl acetate) (5)	Not recommended
Chloroform (10)	Not recommended
2-1 dichloroethane (ethylene dichloride) (10)	Not recommended
2-1 dichloroethylene (200)	26900
Dioxane (25)	Not recommended
Ethyl acetate (400)	10300
Ethyl alcohol (1000)	6900
Ethyl ether (400)	9630
Gasoline (300)	Requires special considerations
Isoamyl alcohol (100)	37200
Isopropyl alcohol (400)	13200
Isopropyl ether (250)	11400
Methyl acetate (200)	25000
Methyl alcohol (200)	49100
Methyl n-butyl ketone (5)	Not recommended
Methylcellulose (2-methoxyethanol) (5)	Not recommended
Methyl cellulose acetate (2-methoxyethyl acetate) (5)	Not recommended
Methyl chloroform (350)	11390
Methyl ethyl ketone (200)	22500
Methyl isobutyl ketone (50)	64600
Methyl propyl ketone (200)	19900
Heavy gasoline (coal tar) (Naphtha (coal tar))	Requires special considerations
VM&P Heavy Gasoline (300)	Requires special considerations
Nitrobenzene (1)	Not recommended
N-propyl acetate (200)	17500
Stoddard solvent (100)	30000–35000
2,2,1,1 tetrachloroethane (1)	Not recommended
Tetrachloroethylene (25)	159400
Toluene (50) (through the skin)	75700
trichloroethylene (50)	90000
Xylene (100)	33000

The Following Values Are Represented using the TLV Values Shown in Parentheses (ppm). The Values in the Table Should Be Multiplied by the Evaporation Rate (Pts/Min) to Produce the Effective Ventilation Flow Rate (Q') (Equation 2.7) (Table 2.1a).

TABLE 2.1b
Effective Volumetric Flow Rate (*Q'*) for Vapors per Pint of Vaporized Liquid

Material	TLV (ppm)	*Q'* (acfm)	*Q'* (am^3/s)
Acetone	250	22000	1310
N-butyl acetate	150	20400	1240
2-1 Dichloroethylene	200	26900	1600
Ethyl acetate	400	10300	610
Ethyl alcohol	1000	6910	410
Ethyl ether	400	9700	580
Gasoline	300	10310	610
Isoamyl alcohol	100	37100	2210
Isopropyl alcohol	400	13000	770
Isopropyl ether	250	11500	689
Methyl acetate	200	25300	1500
Methyl alcohol	200	49850	2970
Methyl chloroform	350	11500	680
Methyl ethyl ketone	200	22500	1340
N-propyl acetate	200	17500	1040
Tetrachloroethylene (perchloroethylene)	25	158000	9400
Xylene	100	33500	1990

2.3.1 Dilution Ventilation to Control Health Hazards

The use of dilution ventilation for health has four limiting factors:

1. The amount of pollutant produced should not be too high or the airflow rate required for dilution should not be unreasonable
2. Workers should be far from the pollutant source, and pollutant emission concentrations should be very low so that workers are not exposed to concentrations higher than TLV
3. The toxicity of the pollutant should be low
4. The emission of pollutants from the source should be completely uniform

Dilution ventilation is often used to control organic liquid vapors with a TLV of 100 ppm or higher.

2.3.2 General Dilution Ventilation Equation

The amount of ventilation required to maintain a constant concentration at a uniform production rate is obtained by Equation 2.1. In this equation, it is assumed that no contamination is in the air source.

$$\text{Accumulation rate} = \text{production rate} - \text{removal rate}$$

Or

$$V_r dC_g = Gdt - Q'C_g dt \tag{2.1}$$

V_r = room volume, ft^3 [m^3]
G = pollutant production rate, lbm/min [g/s]
Q' = effective volume airflow rate, acfm [am^3/s]
C_g = gas or vapor concentration at time t, ppm
t = time

In steady state:

$$dC_g = 0$$

$$Gdt = Q'C_g dt$$

$$\int_{t_1}^{t_2} Gdt = \int_{t_1}^{t_2} Q'C_g dt$$

At a constant concentration, C_g and uniform production rate, G:

$$G(t_2 - t_1) = Q'C_g(t_2 - t_1) \tag{2.2}$$

$$Q' = (G/C_g)(10^6)$$

Due to incomplete mixing, the value of K is added to the ventilation rate, thus:

$$Q' = \frac{Q}{K} \tag{2.3}$$

In which,
Q = actual ventilation flow rate, cfm
Q' = effective ventilation flow rate, cfm
K = coefficient of incomplete mixing

$$Q_a = KQ_i = (G/C_g)(10^6) \tag{2.4}$$

In which,
Q_i = air exhaust flow rate under ideal conditions
Q_a = air exhaust flow rate under actual conditions
K = multipurpose safety factor (dimensionless)

2.3.3 Inlet and Outlet Points

Since the physical mixing coefficient K_m depends on the places of exhaust outlets and supply inlets (Figure 2.1), those should be determined in the design process. Note that the rational selection of the mixing coefficient will be associated with the cleaner workplace and lower cost. In the equations described in the steady state conditions, it is assumed that the concentration is uniform at the workplace. It means that pollution is evenly distributed. Of course, there is always a concentration gradient. Concentrations will be higher near emission sources and slightly lower near air replacement inlets. This fact has great importance at the design stage for two reasons. First, the position of the workers should be considered in relation to the source. The workers close to the point of production are exposed to higher concentrations of pollutants. In choosing the safety coefficient S_f, this issue should be considered. Second, the concentration difference can be used to effectively remove the pollutant using special concepts of local exhaust ventilation. If the exhaust point is as close as possible to the emission point, the removal rate of pollutants from the workplace QC will increase. Thus, it provides a lower overall concentration throughout the workplace. This assumption relies on reasonable and logical hypotheses that the highest concentrations are near the exposure source.

Figure 2.2 shows that a source emits pollutants with a production speed G. The concentration near the source C_s is higher than the background concentration of Cr in the room. There is an intermediate region of concentration C_i. Although only in this approach, these three items are assumed together. In real life also, there is no clear demarcation between them. Based on the general principle, there is a continuous concentration gradient with distance away from the source. For design purposes, these three regions can be considered as: $C_s > C_i > C_r$.

As a common approach to determining the place of the exhaust outlet, the vent can be placed in the ceiling, considering its place relative to the three zones (source, intermediate, or room). Therefore, usually with air drawn from the room area, pollutant removal is performed at a QC_r rate. However, note that by placing the vent in the middle region where the concentration is higher than the background concentration, more effective pollutant removal is achieved. Hence, the pollutants are removed at a QC_i rate greater than QC_r. As a result, transition of the exhaust point to the source area of pollutant production provides more effective removal.

"Point" ventilation, referred as hybrid, has properties of both general and local ventilation systems. The effectiveness of a GEV system can be increased by placing the exhaust point as close as possible to the emission point.

Several matters should be considered with use of this K coefficient:

1. Efficiency of mixing and distribution of replacement air entered into the room or space under ventilation (Figure 2.2)
2. Solvent toxicity: Although TLV and toxicity are not synonymous, the following guidelines are suggested for selection of the appropriate K coefficient value:
   a. Substances with low toxicity: TLV > 500 ppm
   b. Substances with moderate toxicity: TLV ≤ 500–20 ppm
   c. Substances with high toxicity: TLV < 20 ppm

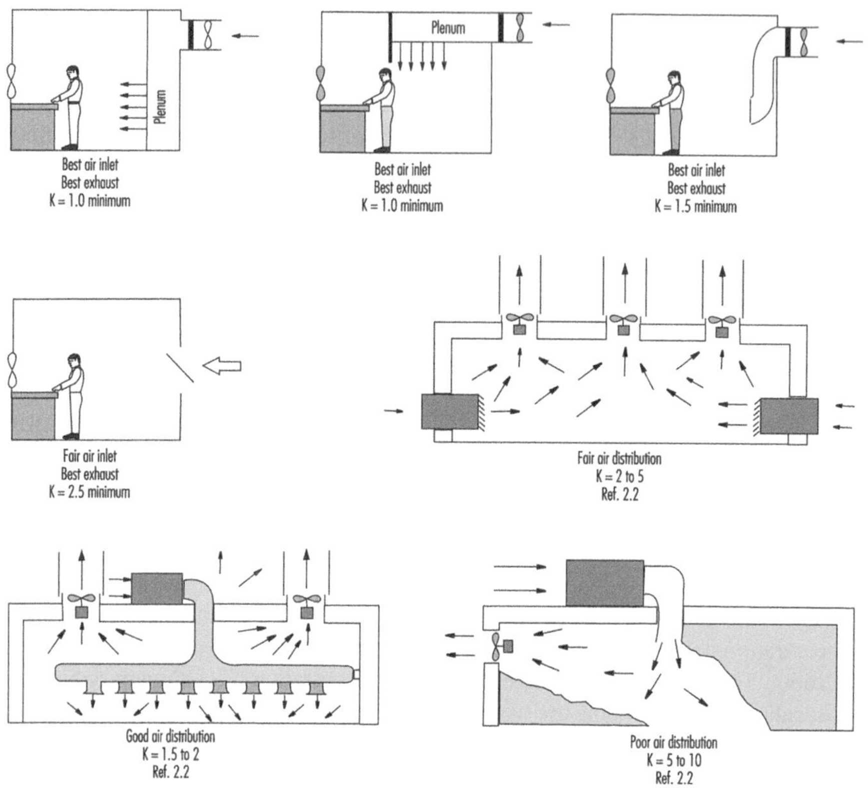

The values of K, a safety factor cited here take only air inlet and air extraction vents into account, and are meant as rough guidelines. The proper value of K to be used for the equation should be determined keeping the following things in mind: a) the number and placement of the workers, b) the source(s) of pollution and c) the toxicity of the contaminant(s).
Ref 2.2: Air Force AFOSH Standard 161.2.

Source: ACGIH 1992.

FIGURE 2.1 (a) Suggested K coefficients for inlet and outlet places; (b) the obstacle effect; (c) the obstacle effect in weak conditions; (d) mixing with uniform exhaust and supply ($1 \approx K$); (e) mixing with non-uniform exhaust and uniform supply ($2 < k < 5$); (f) mixing with non-uniform exhaust and non-uniform supply in two corners ($2 < k < 5$); (g) mixing with exhaust and supply only near the floor ($4 < k < 8$); (h) mixing with exhaust and supply only near the ceiling ($8 < k < 10$); (i) mixing with exhaust and supply only at the end of the room ($10 < k$).

3. Evaluation of occupational health specialists in other important matters based on individual experience. Among these concerns, the following points can be mentioned:

    a. Process duration, operating cycle, and workers' usual places in relation to the pollution sources
    b. Place and number of pollution production points in the workplace
    c. Seasonal changes in the amount of natural ventilation
    d. Decreased operational efficiency of mechanical air moving devices
    e. Other conditions affecting the concentration of hazardous substances in the respiratory zone

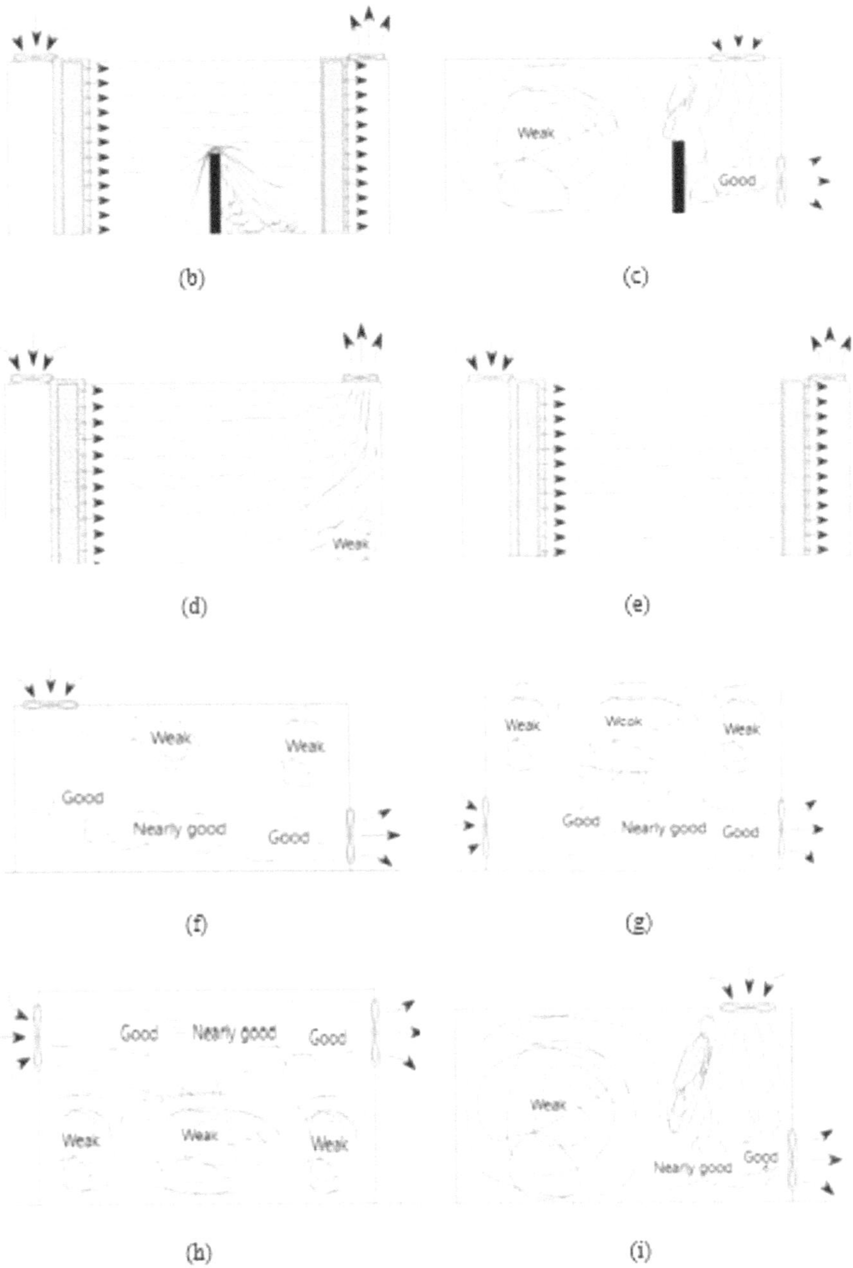

FIGURE 2.1 (Continiued)

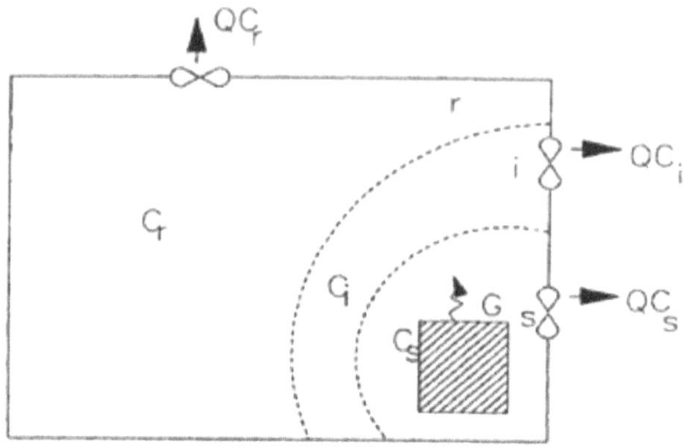

FIGURE 2.2 Three regions (source, *s*; intermediate, *i*; and room, *r* are shown. If the vent discharges the air from the room to the outside, the mass flow rate of pollutant removal is expressed as *QCr*. If the vent is placed in the intermediate area, the pollutant is removed at a higher rate; hence, $Ci > Cr$. By placing the vent in the near to the source, more effective removal is achieved. So, the removal rate of *QCs* is higher than *QCr*.

The selected *K* coefficient value varies from 1 to 10, which depends on the above considerations.

The air supply inlet places are determined in the same way as the exhaust outlet places. Since the temperate replacement air is cleaner than the general air in the room, this air should be transported directly to the area occupied by the workers before passing through the contaminated area. In fact, it has double effects. It not only provides clean air for the workers but also prevents the entrance of polluted air into this area. In fact, it will help create and maintain an airflow pattern, as shown in Figure 2.3. This air supply requires conditioning and a delivery system (for example, ventilators, ducts, and diffusers).

2.4 GENERAL EXHAUST VENTILATION EQUATIONS

To facilitate calculations using GEV performance prediction equations, several assumptions should be considered:

- There should be a complete mixing in the room
- The production rate should be constant
- Diluting air contains low amounts of pollutants
- The pollutant enters the workplace only through the process
- Contamination should be removed from the workplace only through general exhaust ventilation

FIGURE 2.3 Clean and moderate air supply in the crowded places and directly polluted air exhaust from the process surface prevents the penetration of polluted air into the crowded area (Figure 2.3). The supplied air (inlet) must be properly ventilated, and a delivery system (for example, ventilators, ducts, and diffusers) must be provided for it.

The first assumption, complete mixing, applies both spatially and temporally. Complete mixing means that when the pollutant is emitted, it is completely and immediately mixed throughout the room. This means that the replacement air is completely and immediately mixed. Incomplete mixing makes problems for the designer in calculation of the system capacity that effectiveness of the GEV system can be optimized using the concentration gradient. To calculate the concentration C in the air as a function of time t, the production rate G, the ventilation rate Q, and the room volume V_r (Figure 2.4a) should be determined or estimated. Any change in concentration ΔC over a time period Δt will be due to the addition of substances to the room air during production process and/or the removal of substances by exhaust ventilation. The amount of introduced substances during a time period Δt is equal to $G\,\Delta t$, while the amount of removed substances is equal to $Q\,C_g\,\Delta t$. Therefore, the net change in the total amount of pollutant in the atmosphere will be ΔM:

$$\Delta M = G\,\Delta t - Q\,C_g\,\Delta t \tag{2.5}$$

Also, this change can be expressed as a change in air concentration ΔC_g via dividing both sides of the equation by the volume V_r:

$$\Delta C_g = \frac{\Delta M}{V_r} = \frac{G\,\Delta t}{V_r} - \frac{QC_g\,\Delta t}{V_r} \tag{2.6}$$

Figure 2.4b shows dilution ventilation in a process.

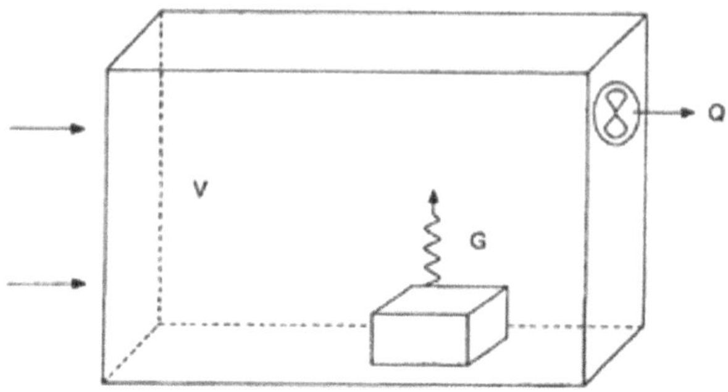

FIGURE 2.4a The workplace is shown in the form of a simple box with a pollutant-producing unit process with a mass flow rate G. The room volume is V_r and the air is removed by the general exhaust system with a volumetric flow rate Q. No air replacement system is shown. Hence, it is assumed that air replacement is performed by natural ventilation.

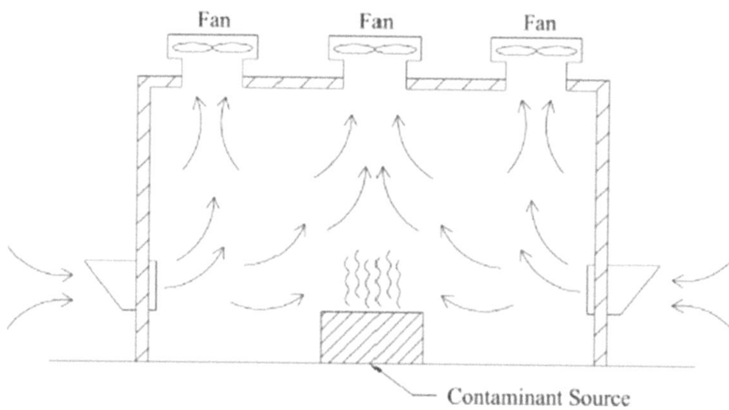

FIGURE 2.4b General or dilution ventilation.

2.4.1 Calculation of Dilution Ventilation for Steady State Concentration

The concentration of gas or vapor in a steady state can be expressed by the material balance equation:

$$Q' = \frac{G}{C_g}$$

Therefore, the clean airflow rate required to maintain the atmospheric concentration of the hazardous substance at an acceptable level can be easily calculated if the production rate is determined. Usually, acceptable concentration (C_g) (ppm: parts per million) is considered for TLV. For liquid solvents, the production rate will be:

$$G = \frac{\text{Constant} \times SG \times ER}{MW} \tag{2.7a}$$

In which,

G = production rate, acfm [am³/s]

Constant = vapor volume resulting from evaporation of 1 pint of liquid at STP conditions, 403 ft³ -lbm/pt-lbmol [24 m³ -g/l-gmol]

SG = specific gravity of volatile liquid

ER = liquid evaporation rate, pts/min [l/s]

MW = molecular weight of liquid, lbm/lbmol [g/gmol]

Therefore, $Q' = (G / C_g)(10^6)$ can be expressed as:

$$Q' = \frac{403 \times 10^6 \times SG \times ER}{MW \times C_g} \tag{2.7b}$$

In the SI system:

$$Q' = \frac{24 \times 10^6 \times SG \times ER}{MW \times C_g}$$

2.4.1.1 Accumulation of Pollutant Concentration

The concentration of a pollutant can be calculated after each change in time (Figure 2.5). By readjusting the material balance difference, it will be:

$$\frac{dC_g}{G - Q'C_g} = \frac{dt}{V_r}$$

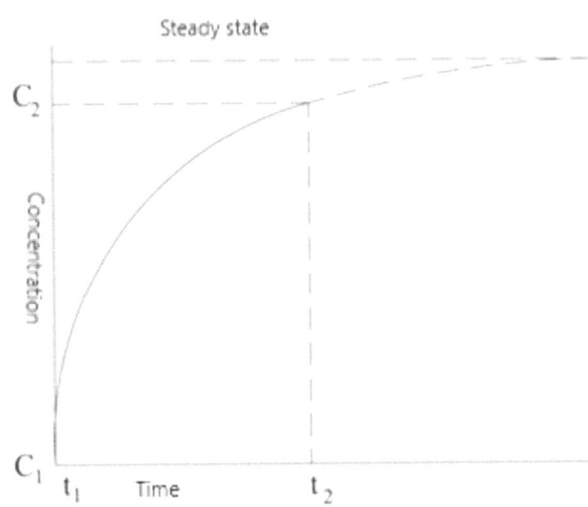

FIGURE 2.5 Accumulation of contaminant concentration.

which can be expressed as the following equation:

$$\ln\left[\frac{G - Q'C_{g2}}{G - Q'C_{g1}}\right] = -\frac{Q'(t_2 - t_1)}{V_r} \tag{2.8}$$

where subscript 1 refers to initial conditions, and subscript 2 refers to final conditions. To calculate the time required to reach a certain concentration, $t_2 - t_1$ or Δt should be rewritten as the following equation:

$$\Delta t = -\frac{V_r}{Q'}\left[\ln\left[\frac{G - Q'C_{g2}}{G - Q'C_{g1}}\right]\right] \tag{2.9}$$

If $C_1 = 0$, then Equation 2.9 becomes:

$$\Delta t = -\frac{V_r}{Q'}\left[\ln\left[\frac{G - Q'C_{g2}}{G}\right]\right] \tag{2.10}$$

Note: Concentration is C_{g2} (ppm) (for example, if $C_{g2} = 200\,\text{ppm}$, enter C_{g2} as $200 \div 10^6$).

For determining the concentration level (C_{g2}) after a certain time interval, $t_2 - t_1$ or Δt, and if $C_{g1} = 0$, then the equation is as follows (Figure 2.6).

$$C_{g2} = \frac{G\left[1 - e\left[-\frac{Q'\Delta t}{V_r}\right]\right]}{Q'} \tag{2.11}$$

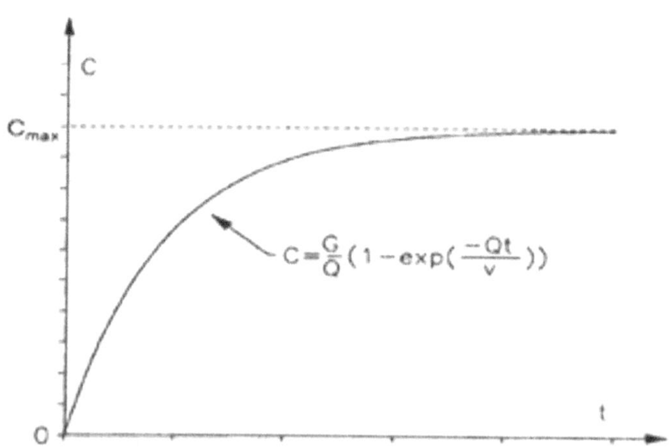

FIGURE 2.6 The general shape of the curve related to Equation 2.11 is shown. The concentration C asymptotically reaches the steady state concentration C_{max}. This diagram assumes that the concentration at time $t = 0$ is zero.

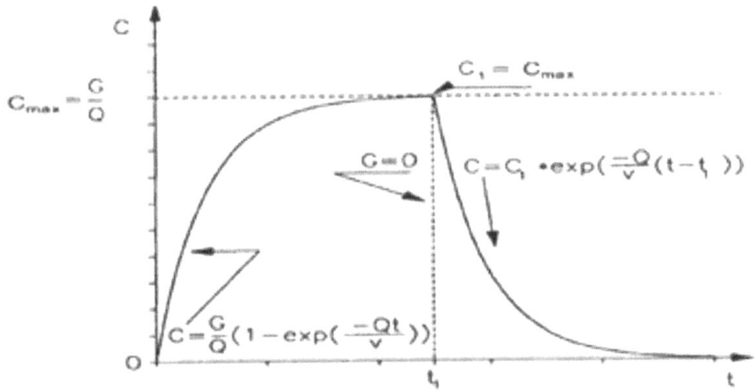

FIGURE 2.7 Concentration decay following cessation of production at time t_1 (G_0).

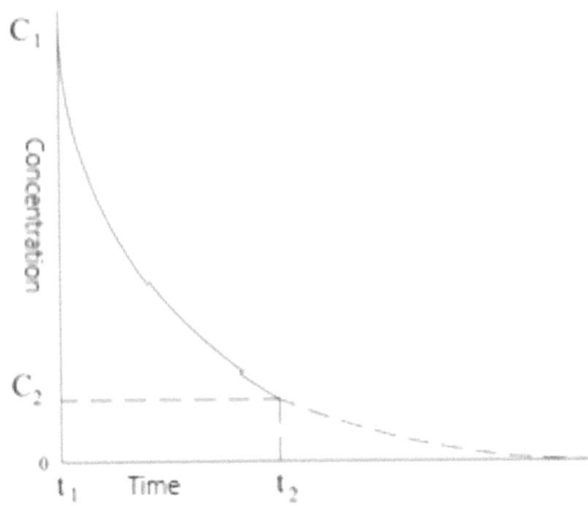

FIGURE 2.8 Cleaning rate.

Note: To convert C_{g2} to ppm, multiply the answer by 10^6.

As shown in Figure 2.7, the concentration decreases exponentially after stopping production, and it approaches zero asymptotically.

Cleaning rate (Figure 2.8): Where the air is polluted but further pollution or production has stopped, the reduction rate of concentration over a time period is calculated as follows:

$$V_r dC_g = -Q'C_g dt$$

$$\int_{c1}^{c2} \frac{dC_g}{C_g} = -\frac{Q'}{V_r} \int_{t1}^{t2} dt$$

$$\ln\left(\frac{C_{g2}}{C_{g1}}\right) = -\frac{Q'}{V_r}(t_2 - t_1)$$

OR

$$C_{g_2} = C_{g_1} e^{\left[-\frac{Q'(t_2 - t_1)}{V_r}\right]} \tag{2.12}$$

2.4.2 Dilution Ventilation for the Control of Health Hazards from Material Mixtures

In many cases, the vaporized liquid, for which dilution ventilation rates are designed, contains a mixture of solvents. The common procedure used in this situation is as follows.

When there are two or more dangerous substances, their combined effect should be considered. If there is no information about the interaction of substances, the effects of different substances should be considered additively (cumulatively). If the sum of the following fractions is more than one, the threshold limit of the mixture should be considered higher than the permissible limit.

$$\frac{C_{g1}}{TLV_1} + \frac{C_{g2}}{TLV_2} + \ldots + \frac{C_{gn}}{TLV_n} \tag{2.13}$$

C indicates the observed atmospheric concentrations, and "*TLV*" is the threshold limit. If there is no information about the interaction of substances, dilution ventilation should be calculated on the basis that the effects of different substances are additive. The amount of air required to dilute each component of the mixture to the required safe concentration is calculated, and the sum of the air amounts is used as the dilution ventilation required for the mixture.

If there is a convincing reason for non-interaction of the substances, the effects of the substances are independently considered. It means that each of these substances influences different parts of the body. In this case, the concentration of the mixed pollutant is considered higher than the permissible limit when any of the following ratios is greater than 1:

$$\frac{C_{g1}}{TLV_1} \text{ or } \frac{C_{g2}}{TLV_2}$$

2.4.3 Dilution Ventilation to Prevent Fire and Explosion

Another function of dilution ventilation is the decrease of the vapors' concentration to values lower than the explosion limit in a chamber. This concept never applies in cases where workers are exposed to vapor. In such cases, dilution rates are always applied to control health hazards. Comparison of threshold limit values (TLVs) and lower explosive limits (LELs) can explain it. The TLV of xylene is 100 ppm, while the LEL of xylene is 1% or 10,000 ppm. To prevent fire and explosion caused by xylene vapor, xylene LEL should be lower than 25% or 2500 ppm. Exposure to this concentration of xylene may lead to severe illness or death. However, dilution ventilation is used to prevent fire and explosion and decrease the vapor concentration to lower values than the LEL in baking and drying ovens, confined spaces, ventilation duct network, etc. Equation 2.7a can be modified to achieve values for air dilution below the LEL. By replacement of LEL with TLV, it will be:

$$Q = \frac{403 \times SG \times 100 \times ER \times S_f}{MW_{\text{liquid}} \times \text{LEL} \times B} \tag{2.14}$$

Note:

1. Since LEL is expressed as % (parts per 100) instead of ppm (parts per million) for TLV, the coefficient of 1,000,000 is 100
2. S_f is the safety coefficient depending on the percentage of LEL required for safe conditions; in most ovens and drying chambers, vapor concentrations must always be below 25% LEL throughout the oven, in continuous furnaces with proper ventilation, the coefficient S_f, 4 (LEL 25%) is used, in batch ovens with good air distribution due to the presence of large amounts of desiccant, a S_f coefficient of 10 or 12 is needed to always maintain safe concentrations, and in continuous or batch furnaces with poor ventilation or no recirculation, larger S_f coefficients may be required
3. B, as a constant, indicates that the lower explosion limit of a solvent vapor or air mixture decreases at high temperatures: $B = 1$ for temperatures up to 250°F and $B = 0.7$ for temperatures higher than 250°F

2.4.4 Dilution Ventilation to Prevent Fire and Explosion of Mixed Substances

If vapors of several combustible substances are present in the environment, the required dilution air should be calculated separately for each substance, with the highest value taken as the necessary air volume for diluting the entire mixture.

2.4.4.1 Cyclic Activities

The production of pollutants is usually cyclical, especially in non-continuous activities. If the cycles are short compared to the total accumulation time, the concentrations may be only slightly higher and lower than the limits if the same amount of pollutants is released with a constant speed (Figure 2.9). However, if the cycles are too long compared to the accumulation period, the peaks and valleys deviate significantly from the exponential curve. In examining the exposure value based on the 8-hour time-weighted average (TWA), these deviations are not so worrisome. For short-term averages, cycles can be very important, especially if the cycles are long relative to the measurement period. For example, if the peak-to-peak time in Figure 2.10 is one hour, then a 15-min TWA spent at the peak can be much higher than a 15-min TWA spent between two troughs.

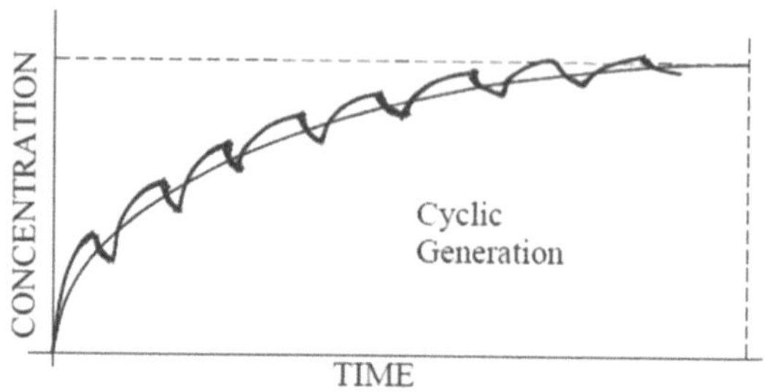

FIGURE 2.9 Cyclical production, short cycles.

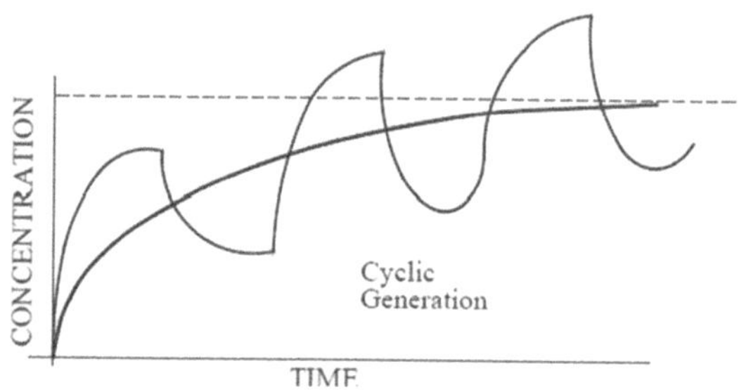

FIGURE 2.10 Cyclical production, long cycles.

2.4.4.2 Confined Space Ventilation

A confined space is a space where: (1) concentrations of hazardous air pollutants can rapidly accumulate due to the relatively small size and limited air exchange; and (2) a person can enter the space. Using this definition, the chlorine cylinders in the service room with dimensions of 10 × 10 feet are considered a closed space. Likewise, a 7-foot-deep pit can also be considered an enclosed space. Space can become even more dangerous if it is difficult to evacuate.

Monitoring, cleaning, and ventilation are key elements to protect workers from air pollution in confined spaces. However, ventilation alone is not enough to keep safety in closed spaces. OSHA standards for confined spaces (29 CFR 1910.146 and 1926 subpart AA) serve as a standard operating procedure for such situations.

Equation 2.15 is used to calculate the time required to achieve the air change rate ($t_{7AC:}$ seven air changes):

$$t_{7\mathrm{AC}} = 7\left(\frac{V_r}{Q}\right) \tag{2.15}$$

t_{7ac}: Time for 7 air changes (minutes)
V_r: Room volume (ft_3)
Q: Volumetric flow rate (CFM)

The time calculated using Equation 2.15 is a useful criterion and a reasonable estimate of the minimum time for safe entry in most conditions without air contamination.

If the concentrations reach a significant non-zero value in the exhaust air and do not drop for a significant time period, steady state conditions have been achieved. Time-dependent expressions become zero and are expressed as follows:

$$C_{exh} = \left[(G/Q')(10^6) + C_{\text{sup}}\right] \tag{2.16}$$

where, C_{exh} = outlet gas concentration (ppm); G = production rate (acfm) [am³/s]; Q' = effective volumetric flow rate (acfm) [am³/s]; C_{sup} = vapor concentration in inlet air (ppm).

If C_{sup} is confirmed to be zero at the monitoring time, the production rate (G) should be greater than zero. It means that the concentration of this pollutant is still increasing in the confined space. If concentration is unacceptable, increase of airflow to a higher level should decrease the concentration linearly:

$$Q_{new} = Q_{original}\left(\frac{C_{exh}}{C_t}\right) \tag{2.17}$$

where, C_{exh} = main concentration of exhaust gas (ppm); Q = effective volumetric flow rate (acfm) [am³/s]; C_t = target outlet gas concentration (ppm).

2.4.4.3 Decreased Production Rates

If a system doesn't operate optimally, exposure can be reduced to an acceptable level by: (1) Draining the liquid completely; (2) Minimizing the leakage potential from the workplace; (3) Installing local exhaust hoods in the space to collect pollutants released by the work; (4) Determining the air inlet and outlet places and discharge points. If the initial concentration is very high or the room has a relatively small confined space and/or its ventilation is poor, it is recommended to first utilize negative pressure ventilation, and then apply positive pressure ventilation; (5) Transmitting flexible duct to place for a part of the initial cleaning period if, based on experience, it is determined that the concentrations are higher than the average value in some parts; (6) Locating clean air inlet points where there is no pollution; (7) Cleaning the space with minimum airflow and minimum time duration determined by experience; (8) Continuing air ventilation with minimum airflow; (9) Increasing the airflow rate (Q) if the concentration is still high after a long period of time. So, highest concentration in space is decreased to value lower than the target concentration (C_T).

2.4.4.4 Clean Inlet Air Supply

Air blown or sucked into a confined space must be free of hazardous pollutants. In general:

1. Locate fresh air inlets for blowers away from potential sources of contamination, including air exhausted from the confined space if air is blown into space (Figure 2.15)

2. Keep exposure sources away from the enclosure access vent if air is drawn into it; if that is not possible, air must be blown into the enclosure and ensured that the vent inlet draws clean air

When blowing air into space, it is necessary to place the ventilator in an area with uncontaminated air (clean area) or connect it to the entrance of the ducting network and place its open end in a clean area (Figure 2.15).

When air is discharged from space, the inlet vent of the space becomes a path of clean air. If the ventilator is placed near the inlet vent (Figure 2.11) to transfer the contaminated air to an unoccupied space or large enough space with adequate ventilation, the outlet ducting networks should be used so that excessive concentrations are not created.

In cases where a gasoline blower is used (Figure 2.12), engine exhaust is a serious hazard. The motor outlet should not be near the blower air inlet for positive pressure systems or near the confined space inlet for negative pressure systems.

The concentration range in a ventilated space is strongly influenced by the mixing coefficient (m_i). It is difficult to predict the mixing coefficient for a volume without extensive sampling data over time. Mixing coefficients are usually applied to precise places in space rather than the entire space.

The range of concentrations in a space can be very high. This is especially true in the usual cases where the pollution is not uniformly distributed in the space. The lower emission from the source can be associated with higher concentration gradients in space. Likewise, if the worker is very close to the source, the mixing coefficient increases.

Mixing coefficients (m_i) usually vary from low (<2) to high (>5) in the same space. Lower ranges are sometimes achieved by placing exhaust and supply points near sources. Blowing air into space at high speed can provide better mixing due to high kinetic energies.

FIGURE 2.11 Avoid placing the ventilator near the opening.

FIGURE 2.12 Gasoline ventilator.

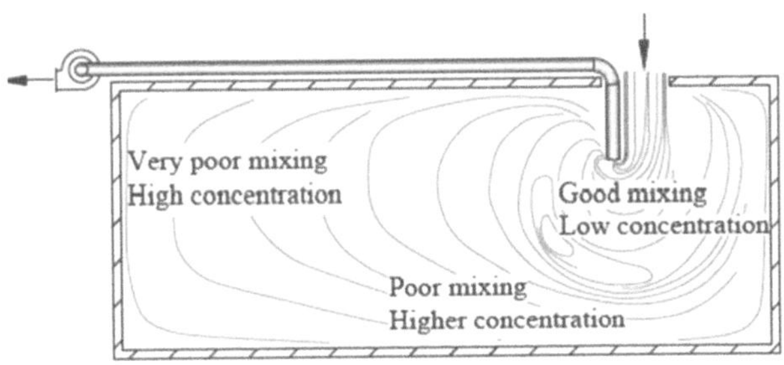

FIGURE 2.13 Contaminated output and poor mixing.

In most cases, airflow should be blown into the confined space, not sucked out of it. In general, air transfer is recommended because the air transmitted to the confined space will probably not mix well with the contaminated air in the space, which will increase the mixing coefficient (m_i) and required flow rate (Q) and/or cleaning time. Also, blowing air into the confined space (container, tank, or chamber) is very important if the source of pollution is placed near the opening of this space. As shown in Figure 2.13, the air drawn from the room has low kinetic energy to mix with the surrounding air. Note that even the system pressure can cause poor mixing in parts of the space (Figure 2.14). In the conditions described in the following section, the exhaust design presented in Figure 2.15 is preferred:

1. During initial cleaning of highly polluted spaces, if a worker must be outside of the confined space near the vent where the air is exhausted
2. During the initial cleaning of highly polluted spaces, if the confined space is small and has poor ventilation that outlet air of this place will be highly polluted
3. If a tent or other shelter is installed above the exhaust vent of the confined place; the air concentration in the shelter will be approximately the same as the air discharged from the confined space if positive pressure ventilation is applied

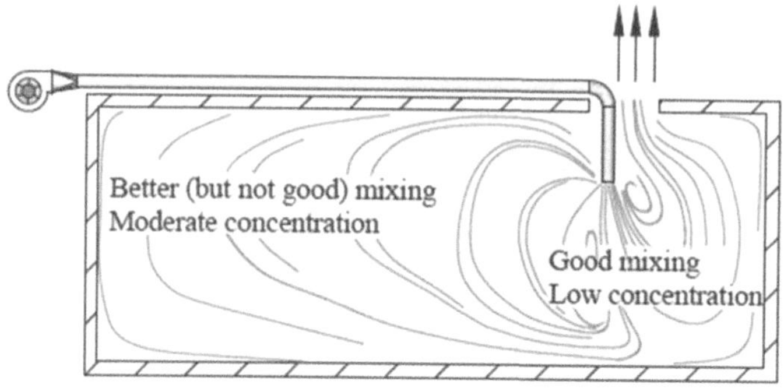

FIGURE 2.14 Contaminated inlet and better mixing.

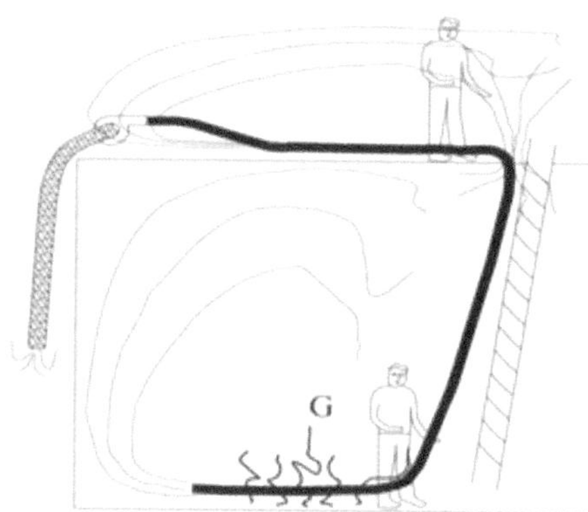

FIGURE 2.15 Air blown into a confined space.

The air should be discharged from space until the concentration level has decreased. Outlet air must be discharged into an unoccupied or large room or an adequately ventilated room to prevent excessive concentration accumulation. If it is not possible, the exhaust system should be used to clean the area.

2.5 GENERAL VENTILATION GUIDELINES

- Install the inlet and outlet vents in the proper place to use the prevailing winds
- Install intake ventilators in the proper place for maximum sweeping effect in the workplace
- Select proper size of intake ducts to keep the inlet drops below 1.8 inches of water
- Avoid placing blowers against each other. If it is not possible, separate them at least six times the diameter of the ventilator
- Use explosion-proof motors in the airflow path if the air passing through the motors contains hazardous chemicals or particles
- Use ventilators with anti-sparking structure for hazardous atmosphere applications

2.5.1 Ventilation Process Guidelines

- Fumes and heat are collected as close as possible to their point of production
- All channels should be as short and direct as possible
- Velocity in the channel should be maintained as much as possible at the operational level to transit collected fumes or particles
- When rotation is required in the channel system, use elbows with a long radius to keep the low resistance (preferably two times the channel diameter or more)
- After calculating the channel resistance, choose the ventilator with a capacity higher than the determined static pressure
- Use intake motors and ventilators in accordance with the general ventilation guidelines mentioned in the above section
- Install the exhaust fan in a place to eliminate any recirculation to other parts of the work environment

When a hood is used, the flow rate must be sufficient to collect all fumes or pollutant particles produced by the process.

2.6 CENTRAL EXHAUST SYSTEMS

Facilities with many fume hoods usually have a central exhaust system. These systems consist of a common exhaust equalizing chamber, a main ventilator, a standby ventilator, and branch connections to multiple exhaust ends. Exhaust devices are grouped based on type, proximity to other devices, and resistance to fire or contaminated areas.

Compared to a single system with equivalent size, a central exhaust system requires less initial investment and lower operating costs. The air is diluted before it is discharged into the atmosphere. The central system has a standby ventilator for safety considerations and provides more flexibility for future development.

Central systems are more difficult to balance and require periodic readjustment to ensure proper airflow. Air balancing in central systems is more difficult since different types of exhaust devices are installed on common ducts.

2.7 GENERAL VENTILATION SYSTEM DESIGN CONSIDERATIONS

1. Consider full cross-ventilation of space – cross-ventilation is based on the natural movement of air from high pressure to low pressure, for example, naturally warm air movement (natural ventilation)
2. Air exchange should be performed with a minimum flow rate of 600 cfm per 1 square meter of free area; air exchange networks usually have a free area of approximately 60%
3. Place the units at high levels and the inlet networks above the head level to prevent improper ventilation for the residents; the natural upward convection streams and secondary air movements generated by the intake networks (caused by the horizontal rapid airflow) will provide sufficient air mixing around the breathing zone to maintain a sense of freshness

4. Intake and exhaust ventilators must be correctly placed to use prevailing winds; place fans and intake vents on the work surface for maximum reciprocating motion—if filters are used on gravity inlets, size inlet vent to keep inlet drops below 1.8 inches of water
5. In large industrial buildings, temperature differences lead to upward airflow; if the building contains hot processes, dust and fumes will quickly move upwards to be discharged through vents or roof vents; always perform exhaust as much as possible to prevent warm air accumulation under the ceiling or from downwards near the breathing area—in such cases, vents at or near the edge with air exchange by side windows above head level are ideal
6. When checking the air exchange, the place of suitable air points and the size of the outlet units are important—note that only with ventilation, the temperature inside the room cannot be reduced to the outside shade temperature; it will always be a few degrees higher than this temperature; however, it is important to draw replacement air from the north-facing "cold" side of the building (north of the equator)
7. The air movement from the inlets is directional; therefore, maintain high speeds considering the appropriate distance; while air movement into the extract openings is not directional, the air speed near to the extract openings is usually low; the rapid movement of the airflow from the inlet has a greater effect on the air movement in the room or workplace, thus causing turbulence over a wide area—this principle sometimes allows the input and output vents to be located near each other on a limited surface
8. The distance between inlet and outlet points in the building must be at least 6 feet; Figure 2.16 shows the separations and free spaces required for inlet and outlet vent openings: Vents should be placed at a minimum of 1 meter (40 inches) from each building corner to minimize wind effects, inlet and outlet vents should be placed by a minimum of 450 mm (18 inches) and 200 mm (8 inches) above ground level to prevent blockage by snow and contamination such as herbicides, and inlet and outlet vents should be separated vertically or horizontally by a minimum of 1800 mm (6 ft) to minimize the potential for contaminant passage

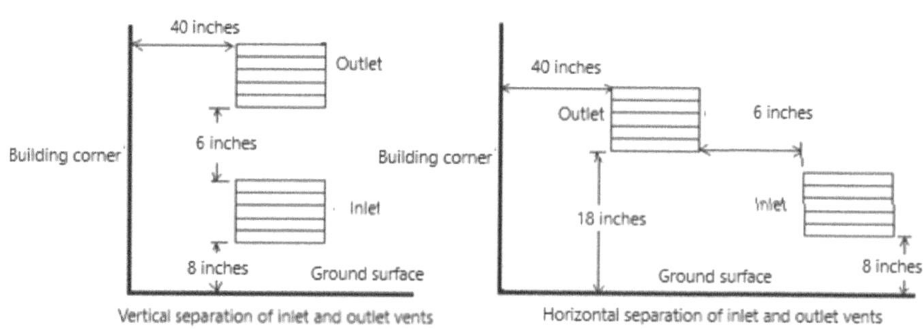

FIGURE 2.16 Place of inlet and outlet vents.

9. These vents should be in a place protected from direct wind and away from potential sources of contamination, such as vehicle parking lots, loading bays, and fuel or chemical storage—in addition, intake vents can generate noise, and those should be placed away from noise sources such as mechanical equipment, ventilators, air conditioning condensing units, and cooling towers
10. Avoid placing the ventilators in front of each other and if necessary, those should be separated from each other by a distance of at least six times of the ventilator diameter
11. Exhaust ducts should be installed with backdraft dampers; all inlet and outlet openings should be equipped with ventilators, and resistant screens should be installed to protect them from the air, insects, and small animals
12. Proper distribution and air circulation is very important—despite an adequate supply of ventilation air, there may be insufficient supply for some rooms due to poor distribution; continuous positive air movement is very suitable, as it balances the temperature, carbon dioxide, and humidity levels inside the building; it is recommended that the minimum speed is 40 fpm—below this level, the airflow is not predictable, and complete mixing is not achieved
13. The level of air movement near the worker is important—in stationary work situations with light activity, the speed should not exceed 200 fpm for continuous activities; with high work surfaces and intermittent exposures, speeds of 400 to 800 fpm may be used; when the air velocity is high, it is important to avoid the adverse effects of hot air convection and disruption of local exhaust ventilation systems—in Table 2.2, some acceptable flow rates of air movement are listed
14. Use Class B insulated motors
15. If the air moving over the engines contains chemicals or dangerous particles, use explosion-proof engines in or out of the airflow path; for hazardous atmosphere applications, use ventilators with anti-spark structure

TABLE 2.2
Acceptable Flow Rates of Air Movement in the Vicinity of the Worker

Exposure	Air Speed (fpm)
Continuous	
Air-conditioned space	50–75
Fixed workstation, general ventilation or spot cooling	
Sitting	75–125
Standing	100–200
Rest station, spot cooling, and intermittent	
Light activity and low heat loads	1000–2000
Activity and moderate heat loads	2000–3000
High activity and thermal loads cooling	3000–4000

16. If the room is very large—for example, more than 80 feet—it may be necessary that the exhaust is centrally done, and the replaced air is entered
17. Design the system for zero temperature stratification, which is the vertical temperature difference between the head and the foot; if the air is not properly mixed by the ventilation system, the temperature near the ceiling can be several degrees warmer than that at floor level
18. Absence of air draughts: Large vertical surfaces can produce a significant flow of natural convection air, which leads to discomfort; generally, the horizontal temperature difference should not exceed 10°C, and the vertical temperature difference should not exceed 5°C—to avoid air draughts, ASHRAE recommends that air velocities should be less than 40 fpm in ventilation systems with 30 to 60% of turbulence intensity; use ceiling diffusers with 45° or 90° angles—10° angles are preferable, because at 10° angles, air draught is created in the head; circulating air should not differ more than 2°C from the ambient temperature
19. To provide effective general ventilation for thermal comfort (whether natural or mechanical), air must be transported to work areas with significant air velocity (maximum 10 feet above the floor; sufficient exhaust is necessary to remove the heat released in space—local comfort systems may require a supplemental supply air to remove heat
20. Modifications in a room or workplace, especially the internal installation of partitions, may affect air circulation and reduce the ventilation effectiveness; where a part of the workplace is isolated—for example, to create internal offices—it is important to ensure that the ventilation system extends to all levels; if necessary, it is better that appropriate sizes of air grilles along with separating walls or doors are used—in some cases, supplemental hoods may be required to improve air movement and prevent localized stagnation

3 Measurement of Air Change Using Tracer Gases

A. H. Khoshakhlagh, H. M. Moda, S. Yazdanirad, and A. Gruszecka-Kosowska

3.1 OBJECTIVES OF THE LESSON

Upon completing this lesson, readers should be able to:

- Explain methods for measuring air change with tracer gas
- Perform air change modeling in indoor spaces
- State the requirements of tracer gases in measuring air change
- Explain the effects of wind and temperature on natural airflow
- State the rate of air change under different conditions

3.2 MEASUREMENT OF AIR CHANGE USING TRACER GASES

Creation of comfortable and healthy indoor air conditions is only possible with a continuous supply of fresh air. However, the minimum air change to achieve this goal depends on different perspectives. It is important to adapt the measures taken in the field of thermal insulation and energy saving to the desired air quality, as well as to provide ventilation facilities in buildings and safety aspects of building structures. Moreover, structural deficiencies and defects reduce the air quality in buildings. Meteorological conditions, including pressure differentials arising from thermal effects and airflow, together with the behavioral patterns of occupants, exert significant influence on air exchange processes. For various reasons, it is necessary to examine the flow of fresh air indoors and outdoors under specific conditions.

Indoor air quality has long been the subject of scientific investigation. Key aspects of this research include:

- In countries with cold or temperate climates, residents spend more than 50% of their time in indoor environments; in some population groups (e.g., infants and the elderly), this proportion is even more than 90%
- Indoor air contains a wide range of different organic and inorganic components; therefore, the concentration of carbon dioxide cannot always be

DOI: 10.1201/9781003501091-3

considered as an lone indicator of air quality—furthermore, organic compounds released from buildings, furniture, household appliances, as well as from the daily activities of residents such as cooking and smoking, contribute to air pollution caused by volatile organic compounds

The energy crisis of the 1970s caused an increase in energy costs and created an urgent need to reduce heat energy consumption. Prevention of heat loss in indoor spaces is a very effective way to save energy and related costs. To protect the climate and reduce CO_2 emissions in the world, the economical and cost-effective use of energy resources is vital.

Building thermal insulation is effective against heat loss, but it increases indoor pollutant concentrations and reduces air exchange. To assess indoor air quality under natural ventilation conditions, the air change rate (ACR) can be measured through standard tracer gas measurements.

For this purpose, a small amount of tracer gas is released into the indoor space, and its concentration is then recorded as a function of time. Then, using appropriate evaluation algorithms, the ACR can be calculated by the data obtained.

The first part presents the physical principles underlying air exchange processes between indoor and outdoor environments, outlines the methodologies employed to determine the air exchange rate (ACR), and describes the tracer gases commonly used for this purpose. Furthermore, it compares the application of carbon dioxide as a tracer gas with results obtained using hexafluorobenzene or sulphur-hexafluoride. The influence of weather conditions on the data obtained from ACR under natural ventilation is also explained. This section will subsequently review the ACR under passive ventilation in selected residential facades in Berlin, Germany and, finally, examines the relationship between air exchange and the concentration of selected volatile compounds (VOCs) under worst-case conditions.

3.3 MODELING AIR CHANGE IN INDOOR SPACES

Based on VDI 4300, Part 7 (2001), the air change rate is defined as the ratio of the air source $Q(t)$ to the considered area (i.e., room or space), i.e., room volume (V_R), and is generally expressed as air change per hour (h^{-1}) or ACH. The following equation is based on this definition:

$$\lambda(t) = Q(t) / V_R \tag{3.1}$$

$\lambda(t)$ = ventilation rate or the air change rate [h^{-1}]
$Q(t)$ = source of air input into the room [m^3/h]
V_R = room volume [m^3] and t = time [h]

In this model, which is designed to calculate the time duration of tracer gas concentration, the following assumptions are considered:

- The tracer gas is chemically stable and inert, with no reactions occurring in the room that alter its concentration
- No surface adsorption processes occur on walls, ceilings, or furnishings that could reduce the tracer gas concentration
- The indoor air is assumed to be fully mixed, resulting in a uniform tracer gas concentration throughout the room at any given time
- Air exchange with the ambient environment occurs only through openings in contact with the outdoors, while exchange with adjacent indoor spaces is considered negligible; accordingly, the room in which the tracer gas is released is treated as a single-environment system
- Air exchange processes are assumed to remain constant during the measurement period; thus, both the air supply rate $Q(t)$ and the air change rate $\lambda(t)$ can be regarded as constant, denoted by Q and λ, respectively

3.3.1 Model Equations

The mass balance equation is the basis for description of the relationship between the mass or the concentration of a gaseous substance in space as a function of time. This equation states that the mass in a volume is constant, and concentration of the tracer gas can change only when either more tracer gas is added to the primary amount, or the tracer gas is eliminated by removal processes. Considering the above assumptions, the supply and removal processes in the following section are significant for tracer gas concentrations in room air (the physical dimensions of these variables are presented in parentheses):

- Transfer of tracer gas from room air to outside $Q^* C_i$ [mass per time unit]
- Transfer of tracer gas from outside air to room air $Q^* C_a$ [mass per time unit]
- Constant emission (E) of the tracer gas into the space from a defined tracer gas source [mass per time unit]

Therefore, the mass balance equation can be formulated as the following differential equation:

$$V_R^* \, dC_i(t) / dt = -C_i^* Q + C_a^* Q + E$$

or

$$V_R^* dC_i(t) / dt = -(C_i - C_a)^* Q + E \tag{3.2}$$

After dividing both sides of the equation by the volume V_R, a simple differential equation is obtained to express the change in the concentration of the tracer gas in the room per unit of time:

$$dCi(t) / dt = -(C_i - C_a)^* Q / V_R + E / V_R \tag{3.3}$$

C_a: tracer gas concentration in the outside air (mass/volume)
C_i: tracer gas concentration in indoor air (mass/volume)
Q: airflow exchange between the room and the outside space (volume/time unit)
E: the amount of tracer gas released per unit of time (mass/unit of time)
V_R: room size
t: time

As mentioned in the above section, Q / V_R is defined as the air change rate. When this value is replaced in Equation 3.3 with λ (Equation 3.4), Equation 3.4 will be obtained:

$$dCi(t)/dt = -(C_i - C_a)^* \lambda + E / V_R \tag{3.4}$$

Equation 3.4 shows that in the case of constant homogeneous mixing, the changes in tracer gas concentration with the concentration difference in the internal and external spaces ($C_i - C_a$) at time t, the air change rate λ, and the amount of tracer gas released per unit of time (measure the emission (E) are proportional.

With integrating these items, the initial conditions at time 0 $\left(C(t=0) = C_0\right)$ are obtained:

$$C_i(t) = C_a + E/\left(\lambda^* V_R\right) + \left[C_0 - C_a - E/\left(\lambda^* V_R\right)\right]\exp(-\lambda t_i) \tag{3.5}$$

By replacing and expanding, the following equation is obtained:

$$C_i(t) = (C_0 - C_a)\exp(-\lambda t_i) + C_a + E/\left(\lambda^* V_R\right)\left[1 - \exp(-\lambda t_i)\right] \tag{3.6}$$

Integration of the differential equation results in an exponential function with the air change rate λ. Equation 3.7 is the basis for mathematical analysis with the aim of measurement of the tracer gas, which is recorded as a time-concentration curve. This function is characterized by the following properties: if no tracer gas enters the room (i.e., $E = 0$) and there is a non-zero tracer gas concentration C_0 at time $t = 0$ that is greater than the ambient tracer gas concentration C_a, the expression $(C_0 - C_a)\exp(-\lambda_t) + C_a$ is formed. It indicates the removal of the tracer gas from the room. The curve starts with an initial concentration of C_0 and decreases exponentially to reach the ambient tracer gas concentration C_a or any other constant background concentration. If a tracer gas is used that is not present in the outside air (i.e., Ca = 0), its concentration decreases over time and reaches a value of zero. Equation 3.6 can be simplified as follows:

$$C_i(t) = C_0 \exp(-\lambda t_i) \tag{3.7}$$

If a source in the room emits the tracer gas with a constant rate E, and if the initial concentration of the tracer gas in the room at the beginning of the C_0 measurement is equal to the external concentration C_a $(C_0 = C_a)$, the tracer gas will accumulate in the room. With start from the C_a concentration, the mathematical function moves towards C_{eq} with an asymptotically increase of t, which reaches the final value of $C_a + E / (\lambda^* V_R)$ at infinity.

$$C(t) = C_a + \left[E / (\lambda^* V_R)\right]\left[1 - \exp(-\lambda t)\right] \quad (3.8)$$

If we ignore the concentration of the outside air, it will be as follows:

$$C(t) = \left[E / (\lambda^* V_R)\right]\left[1 - \exp(-\lambda t)\right] \quad (3.9)$$

For $t >> 1 / \lambda$, Equation 3.9 can be simplified as follows:

$$C_{eq} = E / (\lambda^* V_R) \quad (3.10)$$

C_{eq}: Equilibrium concentration

Equation 3.10 indicates that at equilibrium conditions (i.e., when emission equals removal), the tracer gas concentration C_{eq} is proportional to the emission rate E. But it has an inverse relationship with the air change rate λ and the room volume V_R. This relationship is the basis for determination of the air change rate by measuring the tracer gas in equilibrium conditions.

3.3.2 Determination of the Air Change Rate (ACR) Using Tracer Gases

In addition to the physical properties described in the previous section (hypothetical models), tracer gases must meet certain requirements to be properly used in operations. These requirements include aspects of safety, low environmental load, high accessibility, and optimal practical applicability with the lowest possible cost. Also, these gases should be well traceable with measurement methods in a wide concentration range with high selectivity.

3.3.3 Requirements for Common and Ideal Tracer Gases

In the past, some gaseous materials, such as helium, hydrogen, oxygen, carbon monoxide, methane, acetone, and the radioactive noble gases argon-41 and krypton-85, have been tested to determine the ACR.

Until the early 1990s, krypton-85 was used to measure weather changes; however, its use was subsequently discontinued due to safety concerns related to radiation protection.

Today, the following noble gases are mainly used:

- Dinitrogen monoxide (N_2O)
- Sulfur hexafluoride (SF_6)
- Halogenated hydrocarbons, such as hexafluorobenzene (C_6F_6) and perfluorocarbons (PFC)

3.3.3.1 Dinitrogen Monoxide

Previously, dinitrogen monoxide (N_2O) was widely used to measure air change in buildings in Europe. In America, it was rarely used due to the low threshold limit value (TLV) of 50 ppm. The equivalent of TLV in Germany or the maximum concentration of the workplaces (MAK: Maximale Arbeitsplatzkonzentration) is equivalent to 100 ppm. For safety reasons, it should not be used in residential buildings. Another disadvantage of N_2O is its high adsorption capacity on surfaces at concentrations lower than 1000 ppm and high solubility in water, which can mean that the air change rate in air-conditioned rooms is estimated to be higher than the actual amount.

3.3.3.2 Sulfur Hexafluoride (SF_6)

Since Gregory's observation in 1962, sulfur hexafluoride (SF_6) has been reliably measurable at the nanogram scale using electron capture detection (ECD). Since the early '70s, it has been used to measure air infiltration in buildings. Among the various tracer gases available, sulfur hexafluoride (SF_6) most closely meets the criteria for an ideal tracer gas. Consequently, SF_6 remains the most widely used tracer gas globally.

SF_6 is very stable and decomposes only at temperatures above 550°C. The background concentration of SF_6 in ambient air is approximately1 ppt (6 ng/m^3). Although its density is about five times that of air, this difference does not cause any systematic effect on the results of air change measurement with typical concentrations. SF_6 can also be used in residential buildings and can be measured in a wide range of concentration with high accuracy. The concentrations that are usually used to measure indoor air changes are not more than one hundredth of the MAK value, which is 1000 ppm or 6100 mg/m^3. It should be noted that this limit value is not toxic, which is defined as the upper controllable analytical limit for gases with low toxicity. Due to its high stability, SF_6 degrades very slowly in the air and is a greenhouse gas similar to perfluorinated hydrocarbons. Therefore, to avoid environmental hazards, it should be used with high precision and with the lowest possible concentration, like other tracer gases.

3.3.3.3 Hexafluorobenzene (C_6F_6) and Hydrocarbon-Perfluorocarbon (PFT: Perfluorocarbon-Hydrocarbon Tracers) Tracers

Hexafluorobenzene and hydrocarbon-perfluorocarbon (PFT) tracers such as perfluorodimethylcyclobutane, perfluoromethylcyclobutane, perfluorodimethylcyclohexane, and perfluoromethylcyclohexane are also considered suitable tracer gases. Those are preferably used with the fixed injection method to determine ACR in the passive sampling method. Unlike SF_6, these compounds exhibit strong adsorption onto activated carbon or Tenax, making them particularly suitable for field studies and surveys aimed at determining indoor air change rates.

However, a limitation of these compounds is their tendency to adsorb onto room surfaces, with their emission rates being highly dependent on temperature. Therefore, a long ventilation period as well as careful temperature control in storage containers is necessary.

3.4 ACR DETERMINATION METHODS

Basically, there are three suitable methods for determining ACR using tracer gases. According to the VDI 4300 standard, part 7, it includes concentration decay method, constant injection method, and constant concentration method.

3.4.1 Concentration Drop Method

Tracer gas is injected into the chamber for a short time, either from a gas bottle with a pressure reducer or manually from gas tanks. After mixing with room air, the concentration of the tracer gas is measured at regular intervals. Since the tracer gas concentration decay curve C follows an exponential trend when it is completely mixed with room air (see Equation 3.8), each sample removes a certain amount of fresh air in addition to the initial air removal. This essentially means that tracer gas has been added by 37% ($100/e^1$) after a complete air change cycle. After the air change cycle, the time ($\lambda / 1$), is known as the nominal time constant τ. After $(\lambda/3)3\tau, (\lambda/4)4\tau$, and $(\lambda/4/6)4.6\tau$, the concentration of the tracer gas in the studied room volume was 5%, 2%, and 1% of its initial value, respectively (Figure 3.1a):

Using the concentration drop function, it will be:

$$C(t) = C_o^* e^{-\lambda * t} \tag{3.11}$$

Using non-linear regression analysis, ACR λ (C_0: tracer gas concentration at time $t = 0$) can be estimated. However, in most cases, logarithmic values of concentration are applied to obtain a linear relationship between the logarithm of the tracer gas concentration $C_i(t)$ and time t (Figure 3.1b):

$$\ln C_i(t) = \ln C_0 - \lambda^* (t_i) \tag{3.12}$$

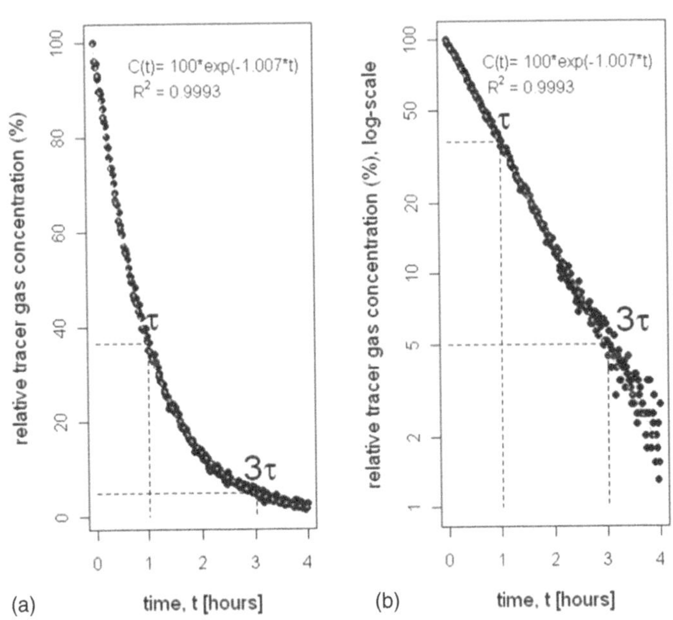

FIGURE 3.1 The relationship between tracer gas concentration and time: (a) before logarithmic transformation; (b) after logarithmic transformation.

Then, ACR λ is calculated through linear regression analysis based on Equation 3.12. Both evaluation options, linear and non-linear regression analysis, are particularly suitable for examination of the time-concentration curve recorded over a longer period. If only a few measurements are taken, for example, when sampling is done with a syringe or other device, the ACR can be determined using the following equation:

$$\left[\ln C_i\left(t = t_i\right) - \ln C_i\left(t = t_{i+1}\right)\right] / \left(t_{i+1} - t_i\right) = \lambda \tag{3.13}$$

$C_i\left(t_i\right)$ is the tracer gas concentration at time t_i.
$t_{i+1} - t_i$ is the time interval between two measurements.

An illustration of this procedure is shown in Figure 3.2a. SF_6 was released in an office space (volume of 48 square meters, equipped with two windows with double-glazed frames). The initial concentration of the tracer gas was set at 5 mg/m³. The concentration was measured for 7 hours and at 15-minute intervals using an infrared acoustic light detector with a selectable filter. During the measurement, the office was empty. The doors and windows were kept closed. The tracer gas and room air were not continuously mixed by a ventilator or similar device. For the analysis, the SF_6 concentration values were chosen in such a way that each concentration-time couple $\left(c\left(t_{i+1}\right) \cdot c\left(t_i\right)\right)$ covered exactly one time interval of one hour, which means

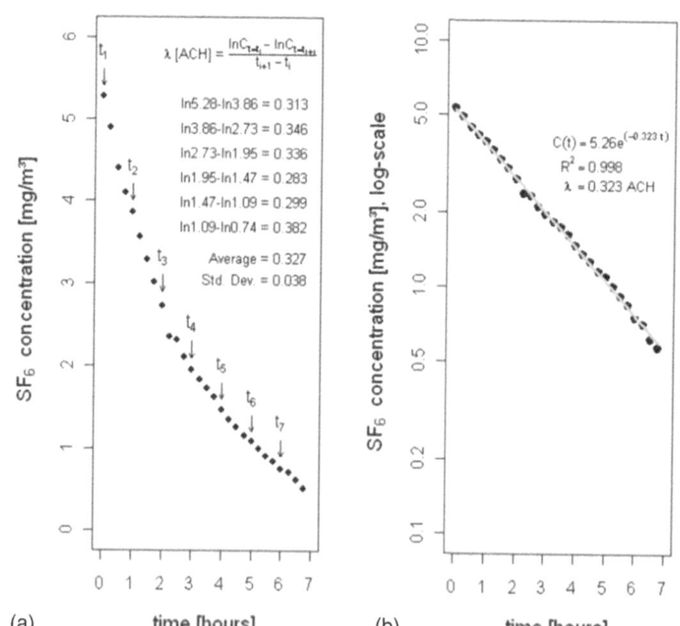

FIGURE 3.2 Determining ACR using concentration drop with two-point method, time interval $t_i + 1 - t_i$ between two measurements: one hour; (a) linear scale; (b) logarithmic scale and linear regression.

that at the time of ACR estimation, the denominator was equal to 1. After selection of the six measurement points shown in Figure 3.4a, the average ACR h^{-1} was 0.327. The standard deviation was 0.1038 h with a coefficient of variation of ≈ 12%. When the same measurement points were analyzed by linear regression analysis, ACR h^{-1} was obtained by 0.323 (Figure 3.2b). Both values are almost equal.

Determination of ACR using the concentration-time couple is also known as the two-point method. For this method, the number of at least five concentration-time couples is recommended, which should be equally distributed over the entire measurement period. To reduce the statistical error to ≤ 10%, the time interval between the first and last measurement point should be proportional to the nominal time constant. This precondition is hardly applicable for ACR between 0.1 and 0.2 h^{-1} or even less, because the minimum required reduction time will be between 5 and 10 h or more. Most likely, in the case of very low ACR, a statistical error of more than 10% should be considered. The minimum recommended time intervals between the first and last measurement and the corresponding measurement intervals based on the ASTM E 741-00 and VDI standards are presented in Table 3.1.

The concentration drop method is useful, especially for determination of ACR up to 10 h^{-1} (10ACH) in indoor air with a volume of less than 500 m^3. If data loggers are used to record the decrease of concentration, the measurement intervals can be reduced to seconds, and as a result, ACR > 10 h^{-1} can also be determined.

3.4.2 Fixed Injection Method

A diffusion tube containing C_6F_6 or PFT is often used as the tracer gas source to determine the ACR by this method. A certain amount of tracer gas is continuously emitted for a certain time. Therefore, the concentration of the tracer gas increases with time and reaches a stable value (equilibrium concentration) that depends on

TABLE 3.1
Samples for the Minimum Time Intervals between the First and the Last Sampling Based on the Recommendations of ASTM E 741-00 and the Recommended Sampling Intervals for Air Change Measurement with the Concentration Drop Method

Air Change Rate	Minimum Tracer Gas Measurement Time (Hours)	Air Change Rate	Sampling Interval (Minutes)
0.05	20	-	-
0.125	8	-	-
0.25	4	< 0.5	30 to 40
1	1	0.5 to 1	20 to 30
2	0.5	1 to 2	10
4	0.25	2 to 5	5
10	0.1	> 10	> 2

the room volume V_R, air change rate λ, and the diffusion speed. In this method, one or more air samples are gathered, and the tracer gas concentration is determined for each sample. ACR can be calculated by solving Equation 3.10 for λ. Figure 3.3a shows an example of a fixed injection method.

In this method, it is very important that sampling is only started when the concentration of the tracer gas is close to equilibrium condition. If sampling begins too early, the ACR will be overestimated. The time to reach the approximate equilibrium condition depends on λ. Based on the ASTM E 741-00 standard, the tracer gas concentration must be at least 95% of the equilibrium concentration before start of the measurements.

The time to reach approximate equilibrium can be estimated using Equations 3.9 and 3.10. However, under worst conditions in rooms with very little air change, it will be difficult to accurately measure ACR with this method, because the accumulation time to reach concentrations close to equilibrium condition is too long (Table 3.2).

The time intervals to reach the tracer gas concentration to 95%, 98%, or 99% of the target equilibrium value $\left(C_{eq}\right)$ are $\lambda / 3$, $\lambda / 4$, and $\lambda / 4 / 6$, respectively (Table 3.2).

As shown in Table 3.2, the required time periods to reach approximate equilibrium can be several days, in which case the ACR can be 0.1 h^{-1} or even less. For this reason, when emission tubes are used as the tracer gas source, the fixed injection method is only suitable for short-term measurements of ACR above 0.3 h^{-1}. In the usual conditions of "worst case" (the last active ventilation of the window has been performed 10–12 h before sampling) and ACR < 0.3 h^{-1}, tracer gas concentrations are still far from equilibrium conditions. As an advantage of the fixed injection method, it can be used for indoor health studies to examine the daily relationship between pollution and air changes in rooms. In this case, tracer gas measurement can be done with passive sampling, and the exposure times are from a few days to a few weeks.

TABLE 3.2
The Required Time Intervals [h] to Reach 95%, 98%, or 99% of Equilibrium

	Time Period (Hours) to Reach a Nearly Balanced Composition		
Air Change Rate[a]	**95% C_{eq}**	**98% C_{eq}**	**99% C_{eq}**
0	-	-	-
0.01	300	400	460
0.05	60	80	92
0.1	30	40	46
0.2	15	20	23
0.3	10	13	15
0.4	7.5	10	11.5
0.5	6	8	9.2
1	3	4	4.6
1.5	2	2.7	3
3	1	1.3	1.5

[a] Concentration (C_{eq}) for a wide range of air change rates.

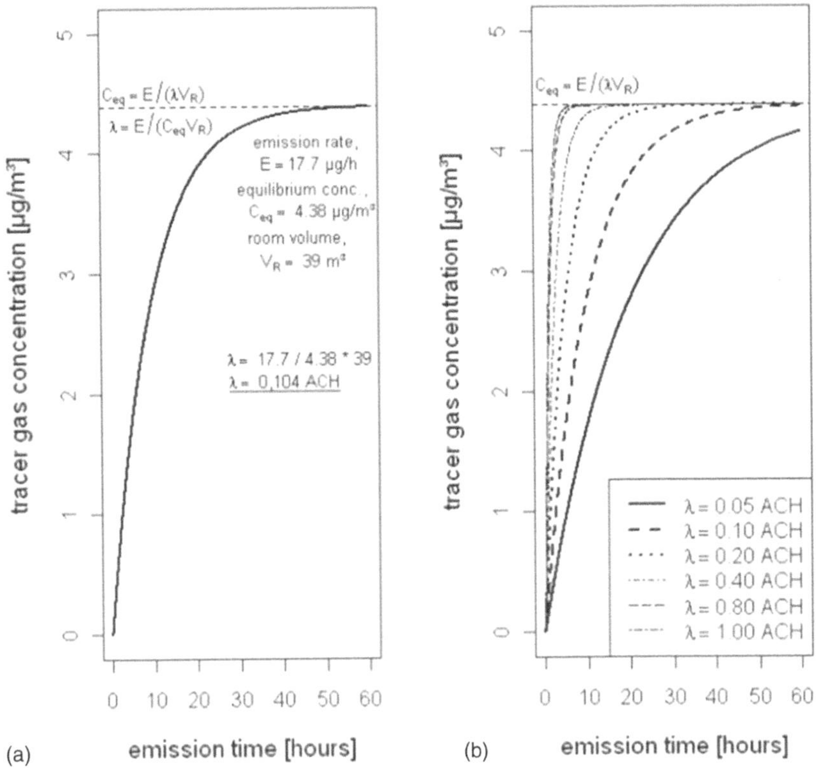

FIGURE 3.3 Determination of ACR: (a) fixed injection method; (b) concentrations after fixed injection for different ACR.

3.4.3 Fixed Concentration Method

During continuous and thorough mixing with the indoor air, the tracer gas is released into the room until it reaches a predetermined concentration. During the entire measurement time, the tracer gas concentration is kept constant by the automatic dose and control system. In the condition of constant tracer gas concentration, air supply is proportional to the amount of tracer gas supply. Then the amount of air supply can be calculated from the ratio of supply of tracer gas to the concentration of tracer gas. If the volume of the room is determined, the ACR can be calculated from this ratio. As an advantage of the fixed concentration method, it can even detect short-term changes in air supply. Compared to the previously described methods, the technical equipment required for this method is very expensive and this method is rarely used for indoor air quality assessment.

3.4.3.1 Use of Carbon Dioxide as a Tracer Gas

Carbon dioxide (CO_2) is one of the gaseous organic compounds that can always be detected in indoor air. Since humans exhale a significant amount of metabolic carbon dioxide, it can increase to thousands of ppm (ml/m³ of room air) in a short time. CO_2 concentration is often used to evaluate the air quality of occupied rooms. In this context, Pettenkofer's reference concentration is the standard. Previously, in 1858, the German chemist and health expert pointed out that the CO_2 concentration of 1000 ppm (0.1% by volume) is the highest tolerable limit in indoor environments. Today, CO_2 measurements are often used to determine indoor ACR because they can be easily performed and the required devices are reasonably priced and easy to use. In addition, CO_2 has some of the characteristics listed for a good tracer gas.

Many studies have been published by experiments related to the feasibility of human-exhaled CO_2 as a tracer gas in the context of air change. In addition to the study conducted by Pettenkofer, other studies are also mentioned. The results of the studies show that the ACR cannot be reliably determined by the point, peak, and average concentration values of CO_2 inside the buildings, because these values are strongly influenced by the number of occupants in the room and the residence times of the people—and hence, the supply of CO_2 is constantly changing. Based on the amount of natural ventilation, the ACR is sometimes overestimated by up to two times the actual value because modern air-tightness buildings and the usual length of residents' occupancy prevent approaching to the equilibrium concentration in most cases. Therefore, it is better to derive the ACR from the decline or accumulation curve, as mentioned earlier.

3.4.3.2 Statistical Evaluation of CO_2 Drop Curves with Examples

After the release of CO_2 in a room, either as exhalation or through a gas chamber, if CO_2 is not supplied, its concentration drops exponentially. For example, the CO_2 drop recorded in a bedroom of an old building is shown in Figure 3.4. This bedroom is equipped with a double-glazed window. The volume of the room is 30 m². The chamber is filled by exhaled CO_2, then the measurement is started, and the chamber is emptied. During the entire measurement process, the room was not empty, and the door and window were closed. CO_2 concentration was measured by a probe sensitive to CO_2 with infrared absorption. A Testo 400 device was used to record data and as a control unit.

Unlike other tracer gases, there is always a certain amount of CO_2 in the outdoor air, which means that the background concentration cannot be ignored. In general, the concentration of CO_2 in outdoor air is between 350 and 450 ppm or even higher, depending on the season. Therefore, the CO_2 decline curve will not be zero. Instead, the CO_2 concentration drops to values close to outside air. This should be taken into account when analyzing CO_2 loss curves. Equation 3.11 cannot be used for this case. ACR should be determined using the term derived from Equation 3.14.

$$C(t) = (C_0 - C_a)^* e^{-\lambda * t} + C_a \tag{3.14}$$

C_0 = initial concentration, C_a = concentration in ambient air (background concentration)

When the background concentration decreases before linearization, a linear curve can be checked. Therefore, background concentration should be determined along with another measurement.

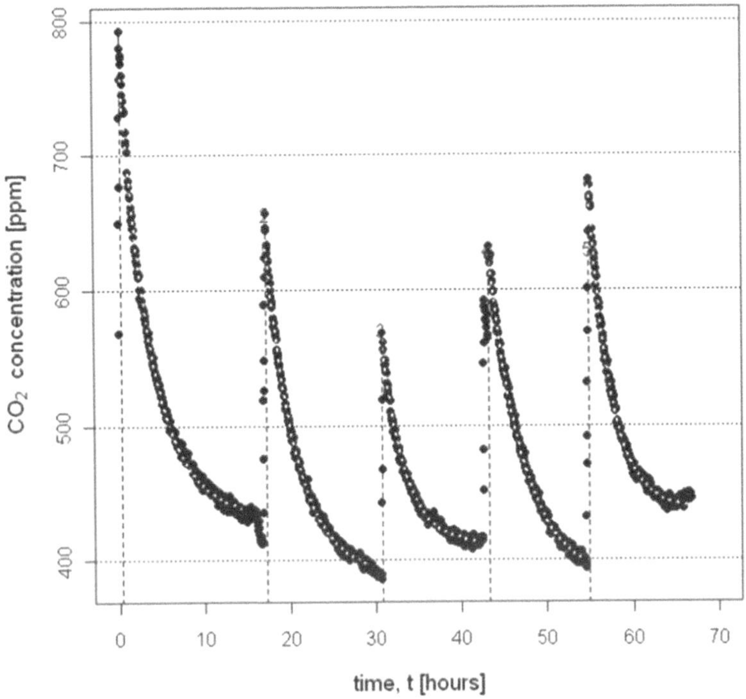

FIGURE 3.4 CO_2 concentration drop curves (black symbols) in a bedroom equipped with double-glazed windows and located in a building, estimated by non-linear regression (white curves).

$$\ln\left(C(t)-C_a\right)=\ln\left(C_0-C_a\right)-\lambda^* t \tag{3.15}$$

With this relationship, ACR can be determined by linear regression analysis. However, in practice, it will be difficult in most cases, because it is not possible to record the CO_2 concentration until the decline curve reaches the background level. Modern buildings are generally designed to achieve high ventilation with a very slow decrease in concentration, and therefore linearization of the concentration decrease curve using the logarithm of the concentration difference is not possible. In addition, background concentrations are often uncertain or unavailable for direct measurement. In these cases, the air change rate λ can be only determined by nonlinear regression with repeated calculation methods based on the model determined in the equation. Iterative calculation methods start with user-specified initial values (raw values) for the function parameters. Then, using the least squares method, those are iteratively improved until the model function is optimally fitted to the measured curve.

The results obtained from the CO_2 drop curves shown in Figure 3.4 have been evaluated using the nonlinear regression method and are presented in Table 3.3. The calculated ACR ranges from 0.25 h^{-1} to 0.4 h (on average 0.32 h^{-1}). The fit of the regression model is good, because more than 99% of the change in CO_2

TABLE 3.3
The Air Change Rate in a Bedroom Located in an Old Building Equipped with a Wooden Frame Double-Glazed Window, Reduction Method and CO_2 as a Tracer Gas

Curve	Time (h)	Air Change λ (95% Confidence Interval)	Background Concentration (ppm) (95% Confidence Interval)	Curve Fit (R^2)
1	14.5	0.290 (0.245–0.254)	428 (427–429)	0.997
2	12.8	0.301 (0.295–0.306)	391 (390–392)	0.997
3	11.3	0.394 (0.383–0.406)	412 (411–413)	0.992
4	11.0	0.249 (0.243–0.254)	384 (382–385)	0.998
5	10.8	0.400 (0.393–0.407)	438 (437–439)	0.998
95% CI Average	-	0.319 (0.226–0.411)	-	-

concentration can be explained by the corresponding regression functions. This is also proven by the high estimation accuracy for different ACRs. The 95% confidence intervals are very close and deviate from only 2–3% below and above each value. These results could be due either to the low variance of the CO_2 values from the curves (white curves in Figure 3.4) or to the long measurement periods (several hours) that produced many nodes (concentration-time data pairs) at 3-minute measurement intervals. However, the difference between daily measurements is much greater. The 95% confidence interval for the average ACR ranges from 0.226 to 0.411 h^{-1}. In the following section, additional explanations for this relatively large day-to-day variation are provided. ACR, which is determined under specific joint ventilation conditions (e.g., closed windows and doors), varies considerably. In Figures 3.5a and 3.4b, six examples of CO_2 drop curves are depicted that can be normally recorded inside the room. For obvious reasons, the decline curves with approximately the same initial concentration of 1800–1500 ppm were chosen and were recorded for a measurement period of more than 10 hours (Figure 3.5a). Another selection criterion was the presence of stable weather conditions throughout the measurement period. Both figures show that the CO_2 decline curves can decrease exponentially in a period of 50 hours and longer under stable weather conditions (Figure 3.5a).

This indicates that the tracer gas and room air can be homogeneously mixed for such long periods without the need for ventilation. The estimated ACR from the drop curves is presented in Table 3.4, and the curves' analysis is done by non-linear regression analysis. In addition to the air change rate λ, the background concentration of C_a

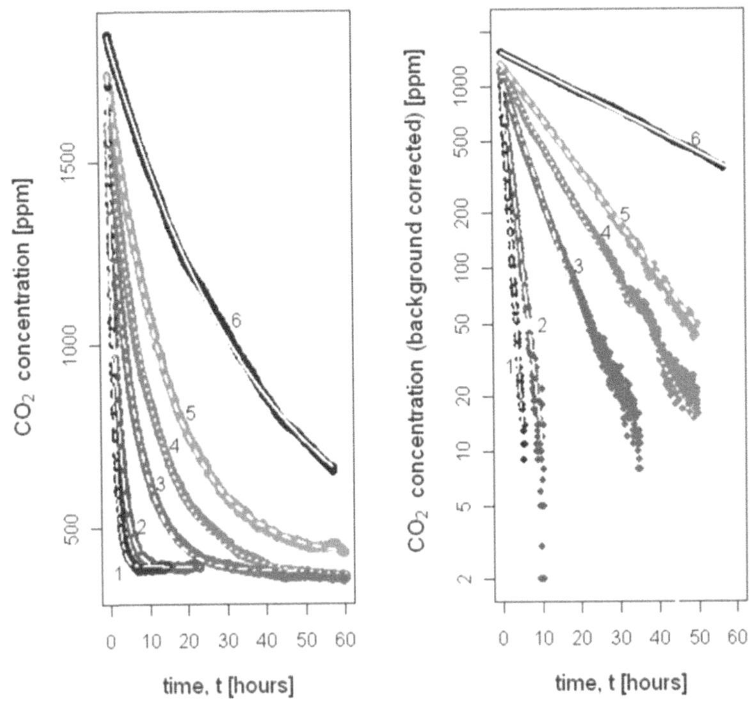

FIGURE 3.5 CO_2 drop curves (black) in six rooms with different ventilation. White curves show the results of regression analysis. (a) non-linear regression; (b) logarithmic values of linear regression.

was also determined by the regression model and presented in Table 3.4. After subtracting the relevant background concentration from the measured CO_2 concentration values, the logarithmic diagram of the concentration difference shows an almost completely linear time-concentration relationship (Figure 3.5b).

When CO_2 is supplied at a constant rate, ACR can be determined by the increasing concentration values (Figure 3.6). Evaluation of the concentration curve is the same as the fixed injection method. The air change rate λ is calculated by the nonlinear regression model approach $\left[y = a_0 + a_1^* \left[1 - \exp\left(-a_2 * t\right)\right]\right]$, where $a_0 = C_a$, $a_1 = E / \left(\lambda^* V_R\right)$ and $a_2 = \lambda$. When the ACR is small, the increase curve approaches equilibrium only slowly, and if the measurement period is too short, it leads to a quasi-linear concentration curve.

Therefore, the ACR calculated from this curve can be associated with high uncertainty. For this reason, to obtain a more reliable estimate of the ACR, the increase in concentration should be recorded over a period of several hours and at intervals of several minutes. When metabolic CO_2 is used, measurements should preferably be performed while the occupants are sleeping. Because this is the closest thing we need to a constant source of CO_2. Figure 3.6 also shows the ACRs obtained from the analysis of the CO_2-related increase curves.

TABLE 3.4
Background Concentration and ACR with 95% Confidence Interval Were Measured in Six Rooms with Different Natural Ventilation (Closed Windows and Doors)

Curve	Time (h)	Air Change Λ (95% Confidence Interval)	Background Concentration (ppm) (95% Confidence Interval)	Fit (R^2)
1	14.5	0.942 (0.931–0.953)	393 (391–394)	0.998
2	23	0.456 (0.463–0.468)	392 (391–393)	0.999
3	48	0.154 (0.153–0.155)	380 (379–381)	0.998
4	60.3	0.091 (0.091–0.092)	367 (366–368)	0.999
5	60	0.067 (0.066–0.067)	406 (405–407)	0.999
6	57	0.026 (0.025–0.026)	357 (350–365)	0.999

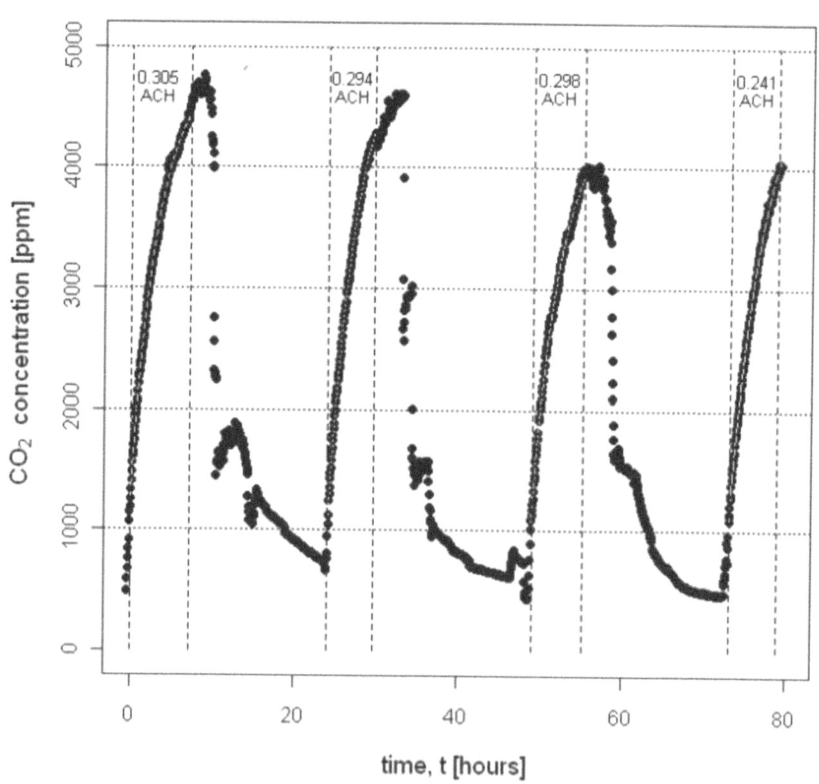

FIGURE 3.6 Curves of increasing CO_2 concentration in a bedroom equipped with double-glazed windows. Fixed injection method, room volume 30 m^3, residents: two adults and one child.

These CO_2 increase curves have been recorded in the same chamber as the drop curves in Figure 3.4. Measurement of CO_2 increase was done when the room was occupied by three people (two adults and one child). In both measurements, ACR curves for decreasing and increasing tracer gas concentrations differed only slightly, reflecting typical daily variations.

3.4.3.3 Comparison of Different Tracer Gases to Determine Air Change Rates

In certain conditions, the use of conventional tracer gases is not possible, and we must use the CO_2 method in these cases, especially for patients who do not accept the use of tracer gases in their homes due to the fear of health risks. In such cases, the CO_2 method is a valuable option. Therefore, it is necessary to examine the results obtained in this field.

CO_2 has different conditions compared to common tracer gases. In this context, several comparative studies using a common tracer gas such as hexafluorobenzene (C_6F_6) or sulfur hexafluoride (SF_6) in parallel with CO_2 are reviewed below.

3.5 DETERMINATION OF AIR CHANGE RATES USING C_6F_6 AND CO_2

In the following, the measurements preformed in two different rooms are examined: the first case, in a bedroom (effective volume 24 m^3) in a terraced house equipped with double-glazed windows and rubber insulation seals and measurements in a period of five days and the second case, in an office space (effective volume 72 m^3) in an old brick building equipped with two double-glazed windows. During the experiment, windows and doors were closed. In total, two measurement cycles were performed at two-month intervals. At the start of the ventilation phase, a diffusion tube containing hexafluorobenzene was placed in both chambers. The first sampling of the Tenax tubes was done about 72 h after closing the windows and doors. The next samples (two samples for each item) were gathered every 24 h. Room air was collected using sampling tubes (1 L per tube) with a blower pump, type Accuro Dräger. ACR was calculated by the specific emission rate of tubes containing C_6F_6 and equilibrium concentration and effective volume of the room and based on the fixed injection procedure described in the VDI standard.

In the bedroom, the CO_2 concentration was continuously measured during the entire study period by measuring 3-minute intervals using a Testo 400 device equipped with a CO_2 probe, and the measured values were recorded. In the office space, the CO_2 concentration was measured with the second device in a 24-h rhythm based on the American standard ASTM D 6245-98. Measurements were mainly made at night in an uninhabited office building. Before starting the measurement, the CO_2 concentration in the office space was set to approximately 600 ppm by the tester.

ACR was calculated by non-linear regression analysis using SPSS software, in which the measured concentration values were fitted to the exponential condition with consideration of the calculated background concentration.

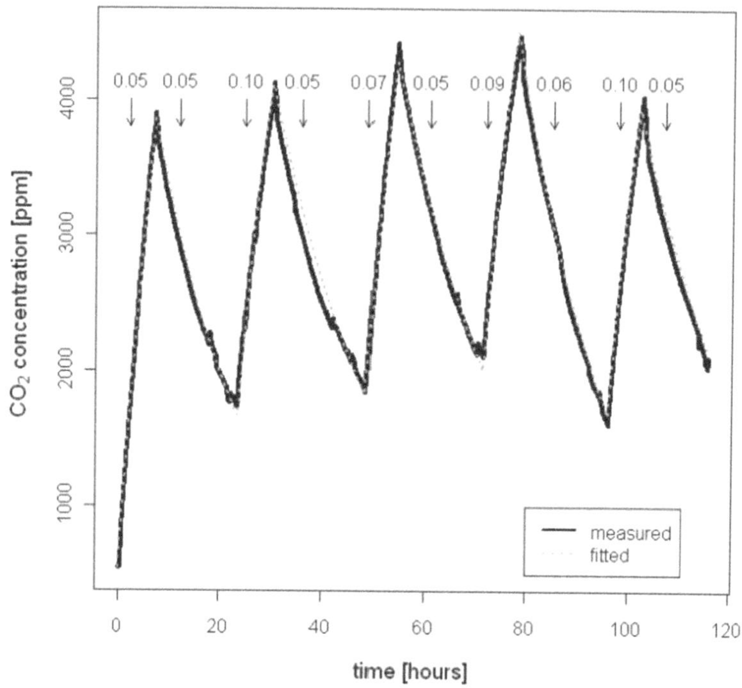

FIGURE 3.7 The trend of CO_2 concentration in a bedroom equipped with an insulated glass window with a rubber seal. Measured values (black) with air change rates at ACH, dotted curve: theoretical values. Room volume: 24 m^3; room conditions: presence of one adult.

3.5.1 CO_2 Measurement Results

The CO_2 concentration-time curve recorded for the bedroom is shown in Figure 3.7. The bold black line represents the measured values, while the gray line describes the plot obtained using cross-sectional nonlinear regression analysis. Above the curve segments, the calculated ACR is plotted. Given the ventilation in the bedroom, ACR h^{-1} was determined from 0.05 to 0.1. It is obvious that the concentration of CO_2 increases very rapidly during the night (an adult), which reaches levels greater than 4000 ppm, and decreases exponentially during the day, which reaches values of 2000–1500 ppm. Due to the low ACR, the equilibrium concentration of CO_2 during 7 h of sleep has not been considered. The statistical analysis of the CO_2 curves obtained for the office space equipped with double-glazed windows shows that the ACR is significantly high between 0.2 and 1.4 h^{-1}, which can be attributed to the characteristics of different window ratios.

3.5.2 Hexafluorobenzene Measurement Results

In Figure 3.8, hexafluorobenzene concentration values from a set of measurements in both rooms are shown along with the calculated curves. The continuous lines of the curve represent the concentration calculated based on the specified emission rate E,

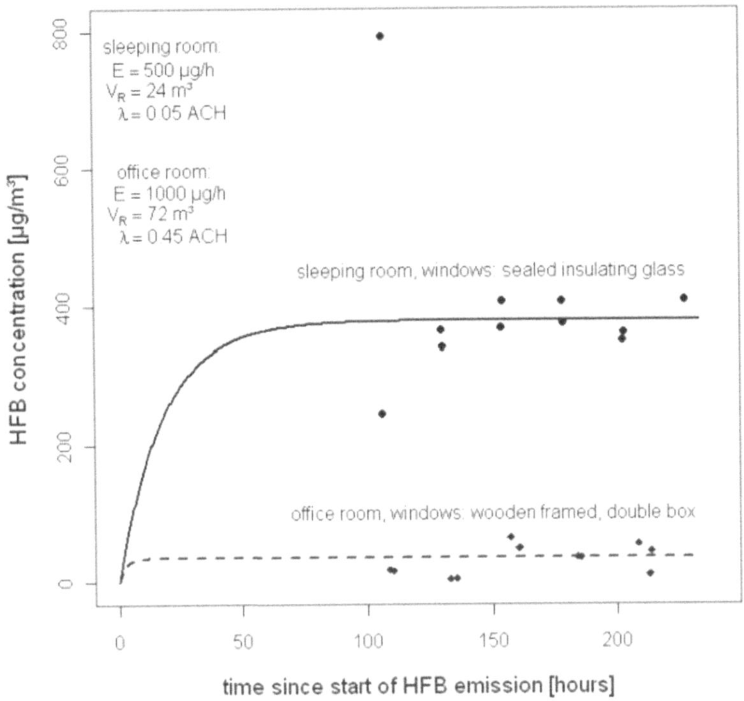

FIGURE 3.8 Hexafluorobenzene (HFB) concentrations in two rooms with different volumes and different types of windows, theoretical curves calculated from specific emission rates *E*, air change rate λ, and specific room volume V_R. Circle points: measured values.

the room volume V_R, and the specified air change rate λ; while the points represent the measured concentrations (bedroom: triangles; office space: squares). Bedroom measurements for ACR are expected to fluctuate around 0.05 h^{-1}. The large deviation of the first two measurements from the curve are probably due to the very different current resistance of the used Tenax tubes. Compared to the bedroom, the CO_2 values in the office (lower curve) show a lower level as opposed to the double emission rate. Also, the lower concentration increase is due to the larger volume of the room and the very high ACR around 0.45 h^{-1}. The calculation of ACR from the measured concentration values ranged from 0.05 to 0.14 h^{-1} for the bedroom and from 0.2 to 1.7 h^{-1} for the office space, respectively (Table 3.5).

3.5.3 Parallel Testing of CO_2 and SF_6

In another series of investigations, an office space with a volume of 48 m^3 and double-glazed windows was investigated for six consecutive days with CO_2 and SF_6, and the ACR was determined from the decline curves. The measurement of SF_6 was carried out in the same way as described in the section "Drop Curve." An example of the parallel concentration drop of both tracer gases is presented in

TABLE 3.5
ACR Results of Comparative Measurements with CO_2 and Hexafluorobenzene as Tracer Gases – Windows and Doors Are Closed

Building Type	Window Structure	CO_2 (ACH) Series 1	CO_2 (ACH) Series 2	(ACH) Hexafluorobenzene Series 1	(ACH) Hexafluorobenzene Series 2
New	Insulating glass with rubber seal	0.08–0.05 $n = 10$	0.1–0.05 $n = 10$	0.06–0.05 $n = 5$	0.14–0.09 $n = 6$
Old	Double glazing with wooden frame	0.3–0.5 $n = 10$	0.2–1.4 $n = 10$	0.2–0.9 $n = 5$	0.3–1.7 $n = 6$

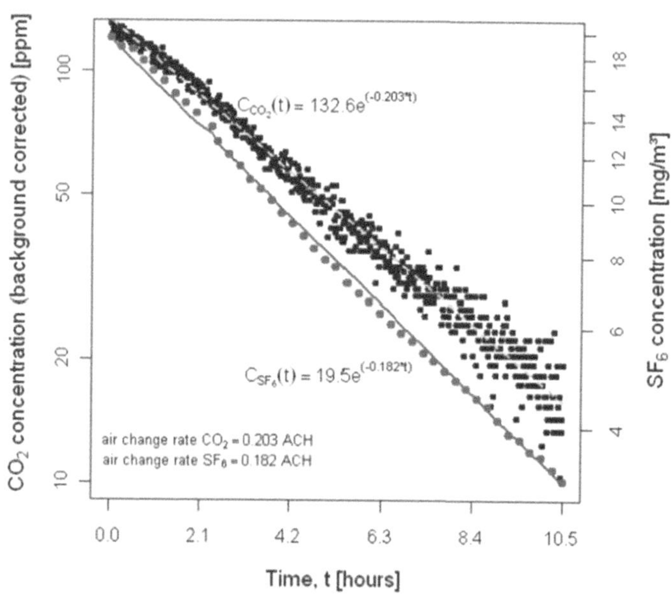

FIGURE 3.9 CO_2 and sulfur hexafluoride (SF_6) drop curves and fitting curve. Measurements at the same time in the same room.

Figure 3.9. The results of six experiments are presented in Table 3.6, which are only slightly different.

The average ACR of CO_2 was 0.304 h^{-1} (SD ± 0.09 h^{-1}) and 0.298 h^{-1} for SF_6 (SD ± 0.09 h^{-1}). A statistically significant difference was not observed between the compared tracer gases. Because the sample size was relatively small (only six samples for each tracer gas), the results of other studies were also examined for comparison. After considering these data, the difference was statistically significant.

TABLE 3.6
Comparative Measurements of CO_2 and SF_6 in the Office Space (Volume 48 m^3, Double-Glazed Window); (COV%): Coefficient of Variation = 100 * Standard Deviation (Standard Deviation)/Mean

	Air Change Rate		Air Change Coefficient
Measurement	CO_2	SF_6	ACR_{CO2}/ACR_{SF6}
1	0.211	0.240	0.879
2	0.477	0.438	1.021
3	0.330	0.327	1.009
4	0.203	0.182	1.115
5	0.330	0.320	1.031
6	0.300	0.280	1.071
Average	0.304	0.298	1.021
Standard deviation	0.090	0.087	0.080
COV (%)	29.7	29.2	8.0

3.6 EFFECTS OF WIND AND TEMPERATURE ON NATURAL AIRFLOW UNDER CO-VENTILATION CONDITIONS

Natural ventilation is produced by the air pressure difference between the indoor and outdoor environment, which is mostly seen in the opening parts of the room such as windows and doors. The air change rate is also affected by the impermeability of windows and doors against the seams and cracks of the building envelope. The air pressure difference is mainly caused by the temperature difference between the outside and inside atmosphere and local wind effects. On the windward side of a building, the wind is trapped when it reaches its highest dynamic power, and the flow angle is equal to 90°.

In the direction of a building that is parallel to the direction of the wind, negative air pressure is often formed. This air pressure difference leads to an increased air change rate at the edge of the wall openings, especially if there are leaks or windows open on both sides of the building because of wind suction and pressure (i.e., cross ventilation). The total air pressure in the facade depends on the wind speed, squalls, and the quality of the building materials (the impermeability of the facade) (Figure 3.10).

3.6.1 Wind Effects

To investigate the effects of weather conditions on air change rate (ACR), measurements were made in two different office buildings of the Robert Koch Institute in Berlin, Germany. Building No. 1 is an old brick building (built around 1900) surrounded by allotment gardens (north and south) and large open spaces. A three-story house is on the east side and a four-story office building is on the west side. Due to the location of the building, the unobstructed wind flow is limited to the southeast or southwest directions. All the five rooms selected in Building No. 1 are located on the

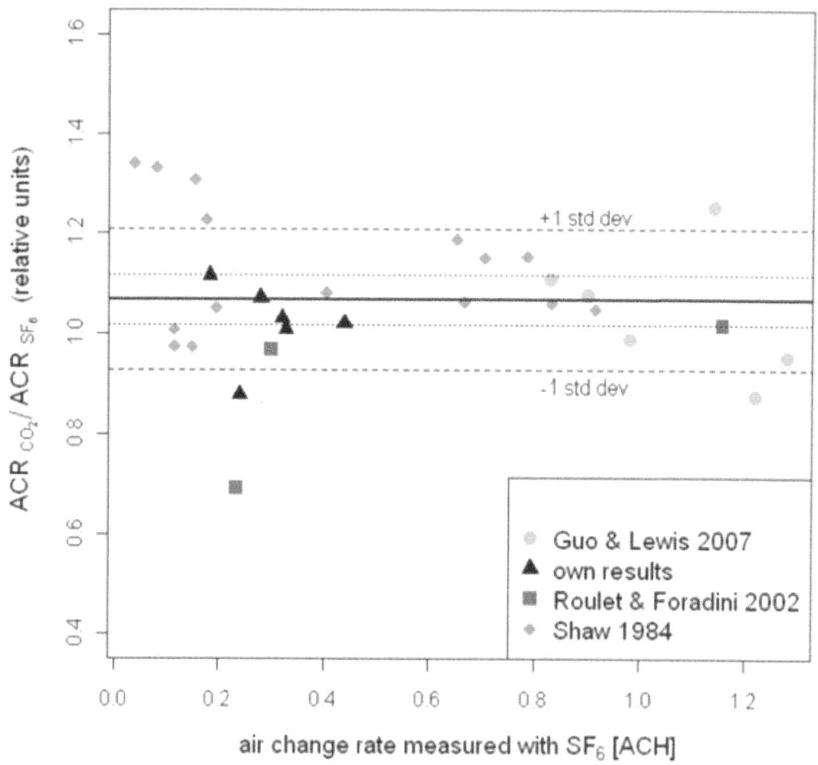

FIGURE 3.10 Comparative determination of air changes with CO_2 and SF_6 as tracer gases (the results of the studies). Solid line: mean; dotted lines: 95% confidence interval, and dashed lines: interval ± 1 standard deviation.

south side and are equipped with double-frame windows (rooms A to F; the volume of the rooms is between 30–90 m^3).

The rooms in Building No. 2, built in the 1970s, were equipped with strong windows with resistant glass and rubber insulation. The room chosen for measurement (room E) had a large windowsill with several windows in the northeast direction, and the volume of the room was 40 m^3.

In front of the windows was an open space with a length of approximately 150 meters that limited the wind flow only from the northeast to the southeast. On the southern side, the room E was surrounded by a building approximately 10 meters away. Likewise, on the northern side, the building was surrounded by a wing from northeast to southwest of the building (nearly 70 meters from room E). From January 1999 to December 2010, 611 measurements of air change in five rooms of Building No. 1 (rooms A to F) were done using the carbon dioxide drop method. The drop curves were recorded mainly at night or on weekends based on ASTMD 98-6245 standard, when the building was unoccupied.

In the same way, 388 measurements of air change were conducted between July 2002 and September 2008. During the measurement, all the doors and windows were closed. Meteorological conditions during the tests, such as outdoor air temperature, wind speed, and wind direction, were obtained from the websites related to regional weather stations in Berlin. In room E of Building No. 2, indoor air temperature was recorded during the measurement process. Table 3.7 provides an overview of the air change rate. For rooms with double-paned windows, the ACR values ranged from 1.8 h^{-1} to 0.05 h^{-1}, while the obtained medians (50th percentile) for rooms A to F varied between 0.6–0.3 h^{-1}. No significant difference was observed among the rooms.

The ACR rate for Room 3 in Building 2 equipped with double-glazed windows was significantly lower, averaging 0.08 hours. Their individual values ranged from 0.016–0.40 h^{-1}. It means that there was a big difference between the maximum and minimum values. The wind effects, especially wind speed, were the main cause of ACR variability in the two buildings (Figure 3.11). In Building No. 1 (rooms A–F), ACR was < 0.1 h^{-1} when the wind speed was less than 5 km/h, and ACR was 1/1–2/4 h^{-1} when the wind speed was between 30–40 km/h. Intense effects of wind led to significant and uncontrollable heat loss through leaky windows, which in turn caused harmful effects of cold wind flow.

Surprisingly, the results obtained for room E in Building No. 2 equipped with insulated windows and rubber seals also depended significantly on the wind effect, although the wind effects were much less than those in Building No. 1. For room E, ACR was reported by lower than or equal to 0.05 h^{-1} when the wind speed was less than 5 km/h, but it increased to 0.25 h^{-1} when the wind speed was higher than 40 km/h. Besides the wind speed, the wind direction had a significant effect on the ACR value. This is evident from the graphical presentation in Figure 3.12, where the ACRs are divided based on the prevailing wind direction during the test periods.

TABLE 3.7
Air Change Rate in Office Spaces with Different Windows: Rooms A to F with Double-Framed Wooden Windows – Room E with Insulated Glass Window and Plastic Frame and Rubber Seal

Room	Sample Size	Minimum [ACH]	Average (95% CI)	Maximum [ACH]
A	15	0.209	0.457 (0.715–0.369)	1.320
B	16	0.163	0.445 (0.351–0.552)	1.370
C	11	0.090	0.470 (0.126–0.696)	0.720
D	42	0.079	0.512 (0.317–0.374)	1.426
E	517	0.043	0.342 (0.317–0.374)	1.841
F	388	0.016	0.470 (0.126–0.696)	0.403

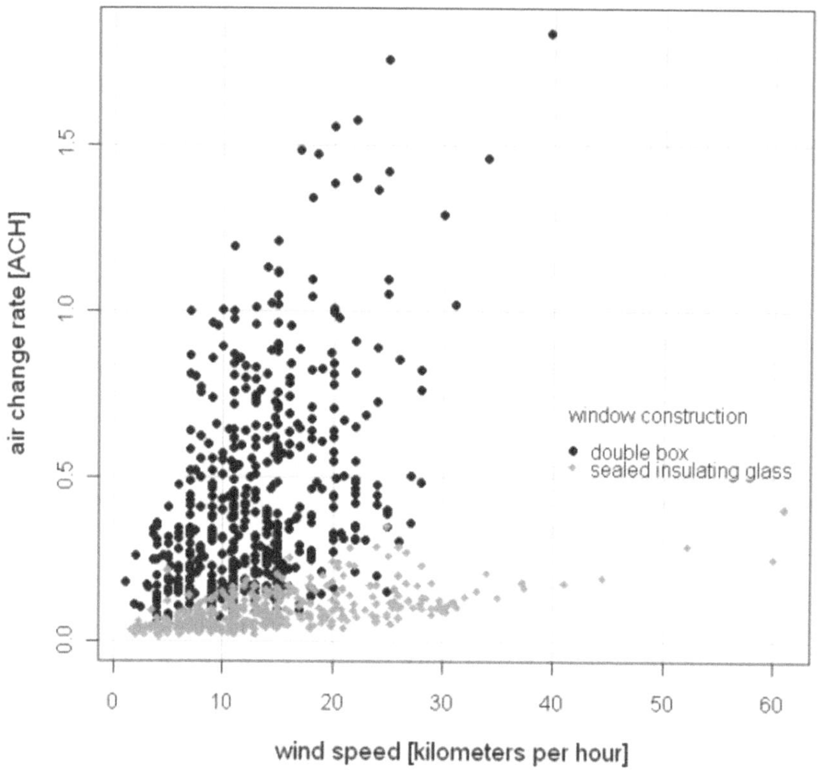

FIGURE 3.11 The effect of wind speed on ACR in office spaces equipped with different windows. Windows and doors were closed. Rooms with double-framed windows (A–D and F: $n = 611$) – Room with insulating glass window ($n = 388$).

Adverse wind flow in front of or parallel to the window has a greater effect than the wind flow from the back of the building, and this causes a lower ACR than the previous two cases (see Table 3.8).

In the case of rooms with double-paned windows, this leads to a threefold increase in air change, which is expressed as the ratio ac (luv): ac (lee) for wind directions from south to west. In a room with double glazed windows, for wind directions from north to east, we will have a twofold increase in the air change (70–100%). Overall air change ratios depending on the relative wind direction are not greatly affected by wind speed (Table 3.8). To summarize the wind effects, it should be stated that: regardless of wind strength and window type, it can be shown that airflow in the front of the window can increase the ACR by 1.5 to more than 2 times compared to airflow in the back of the window.

3.6.2 Effects of Temperature

The temperature difference between the inside and outside environment is one of the main causes of air conditioning. The effect of temperature gradient (room air

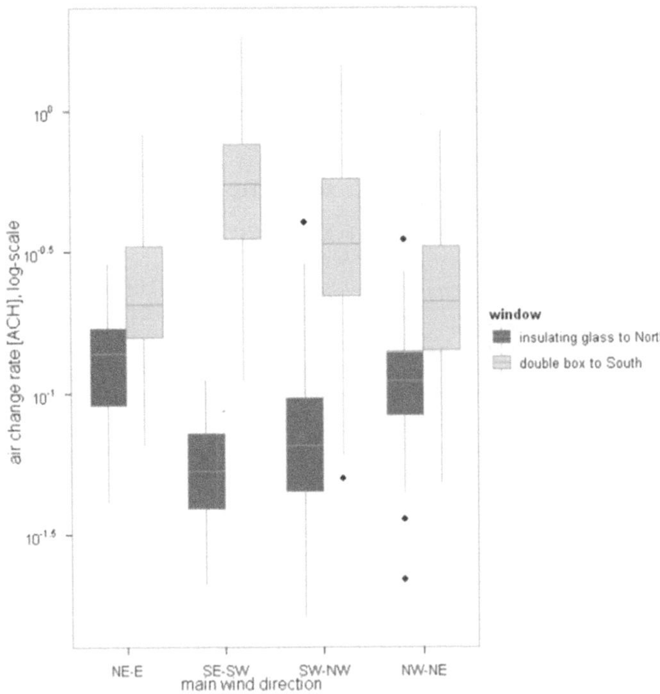

FIGURE 3.12 Wind direction and ACR in office spaces with different windows. Windows and doors are closed. Rooms with double-framed windows (A–D and F: $n = 601$) – Room with insulating glass window ($n = 388$).

temperature minus outside air temperature) on ACR in room E was investigated, and a statistically significant effect on air change rate was observed (Figure 3.14a), while the average air change rate (λ) was only 0.06 h^{-1} at zero-degree Kelvin temperature after correcting the wind effect. With the increase of the temperature gradient and the increase of air temperature to 30° Kelvin, the average air change rate (λ) rose up to 0.1 h^{-1}.

3.6.3 Air Change Rate during Window Ventilation

Wind and temperature gradients are important factors affecting the air change rate through building joints and seams. For indoor spaces in buildings with high leakage rates, window ventilation plays an important role in removing excess moisture, carbon dioxide, odors, and other volatile compounds. The effectiveness of ventilation through open windows depends not only on their number, size, and arrangement in the room, but also on the manner and frequency of active window ventilation. Half-open windows compared to fully open windows have lower ventilation. When the windows on both sides of the room are opened at the same time, cross-ventilation is formed. Cross ventilation effectively transfers moisture and pollutants from indoor air to outdoor air. The effectiveness of different modes of window ventilation expressed

TABLE 3.8
Dependency of ACR on Wind Direction, Natural and Adjusted ACR Coefficients, Rooms A–D and F, Windows to the South, Room E, Windows to the Northeast

Room	Wind Direction	Sample Size	ACR [ACH] avg (95% CI)	Natural Coefficient avg (95% CI)	Corrected Coefficient avg (95% CI)
A to D and F	Northeast to East	122	0.202 (0.171–0.243)	0.92 (0.171–0.243)	1.06 (0.941–1.189)
	Southeast to Southwest	242	0.55 (0.505–0.596)	2.50 (2.295–2.709)	2.89 (2.601–3.207)
	Southwest to Northwest	163	0.329 (0.294–0.395)	1.50 (1.340–1.801)	1.38 (1.239–1.547)
	Northwest to Northeast	74	0.22 (0.179–0.261)	-	-
E	Northeast to East	83	0.138 (0.109–0.153)	2.09 (1.651–2.317)	2.21 (2.038–2.399)
	Southeast to Southwest	81	0.054 (0.047–0.063)	0.82 (0.0–714.957)	1.10 (1.1–130.198)
	Southwest to Northeast	171	0.066 (0.056–0.072)	-	-
	Northwest to Northeast	53	0.111 (0.095–0.135)	1.68 (1.2–438.043)	1.92 (1.2–742.115)

in terms of ACR was investigated by several experiments in room E. Figure 3.13 shows the results related to effectiveness of these windows and the results related to joint ventilation.

ACR was determined by the drop method. The duration of the measurements varied from 0.5 to 3 hours in recording intervals of 12 seconds to 1 minute. After the emission of CO_2, a window in the room (0.8 m wide and 1.2 m long) was either half-open (i.e., upper opening with width of 10 cm) or fully open (i.e., open angle of 80 degrees).

A window was also fully open to achieve cross ventilation. For both ventilation modes, semi-open or fully open, ACR values were determined based on Table 3.9a.

A total of 239 measurements were performed with the half-open window. The average ACR (i.e., the median) with day-to-day changes (with range from 3.4 h^{-1} to 0.5 h^{-1}) was 1.15 h^{-1}. For the fully open window, the median ACR was 10 h^{-1}. The measurements were conducted with a minimum of 2.4 h^{-1} (179 measurements) and a maximum of 30 h^{-1} (321 measurements) for cross ventilation. ACR varied between 15 h^{-1} and 146 h^{-1} (median was 40 h^{-1}).

Based on these data, ventilation time was calculated to reduce the concentration of 1500 ppm CO_2 inside the room by 98% (it is related to the concentration of CO_2 that is about 30 ppm higher than the outside environment), which is 2 h long for the half-open window (range of 1–7 h), and 0.3 hours for the fully open window (range of 0.1–1.5 h), and 0.1 h for cross-ventilation (range of 1–14 min; see Tables 3.9a and 3.9b).

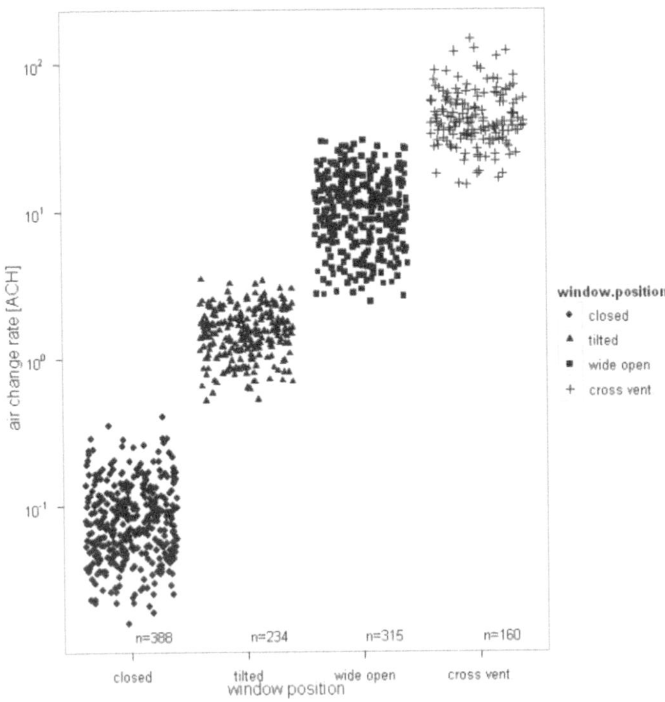

FIGURE 3.13 ACR and window situation, room E.

These results clearly confirm the enormous importance of ventilation with fully open windows and cross ventilation to effectively remove indoor pollutants. Wide variations in ACR can be explained by the effect of temperature gradient between indoor and outdoor environment (Figure 3.14a).

Compared to the temperature gradient, wind effects play a significant role, at least for the "half-open" and "fully open" windows tested here (Figure 3.14b).

The dependency of ACR on the temperature gradient can be estimated with a non-linear exponential curve for all window states (Figure 3.14a). From these findings, it can be concluded that in ventilation through a half-open window, with a temperature gradient of 10 Kelvin, an ACR equivalent to 1.4 h^{-1} (with a confidence level of 95%: 1 h^{-1}–2 h^{-1}) can be achieved, and with a temperature gradient of 20 Kelvin, an average ACR of ACR of 2.3 h^{-1} (with a 95% confidence level: 1.6 h^{-1}–3.1 h^{-1}) can be achieved. For ventilation with a fully open window with a gradient temperature 10 Kelvin, it is possible to achieve ACR of 9.2 h^{-1} (with a confidence level of 95%: 5.5 h^{-1}–15.3 h^{-1}) and with a temperature gradient of 20 Kelvin, it is possible to achieve an average ACR of 18.6 h^{-1} (with 95% confidence level: 11.1 h^{-1}–31.2 h^{-1}). In the cross ventilation, it is expected that with a temperature gradient of 10 Kelvin, ACR 39 h^{-1} can be achieved (with a confidence level of 95%: 23 h^{-1}–65.8 h^{-1}), and with a temperature gradient of 20 Kelvin, it is possible to achieve an average ACR of 50.2 h^{-1} (with 95% confidence level 29.6 h^{-1}–85.1 h^{-1}) achieved.

TABLE 3.9a
Dependency of ACR on Window Structure and Window Opening State

Room	Window Structure	Window Status	The Air Change Rate (ACR)					Sample Size
			Min	5th Percentile	Avg	95th Percentile	Max	
A to F	Double-glazed	Closed	0.043	0.111	0.36	1.030	1.841	602
E	Insulating glass	Closed	0.016	0.029	0.078	0.200	0.403	388
		Half open	0.52	0.70	1.50	2.60	3.43	239
		Fully open	2.4	3.5	10.4	23.4	29.7	321
		Cross ventilation	15.0	22.5	40.1	104.7	146.4	179

TABLE 3.9b
Ventilation Time Required to Reduce 98% Concentration of 1500 ppm Related to Carbon Dioxide

Room	Window Structure	Window Status	Ventilation Time [h] Max	95th Percentile	Avg	5th Percentile	Sample Size
A to F	Double-glazed	Closed	84	32.4	10	3.5	2.0
		Closed	225	124	46	18	8.9
		Half open	6.9	5.1	2.4	1.4	1.0
E	Insulating glass	Completely open	1.5	1.0	0.35	0.15	0.12
		Cross ventilation	0.24	0.16	0.09	0.03	0.02

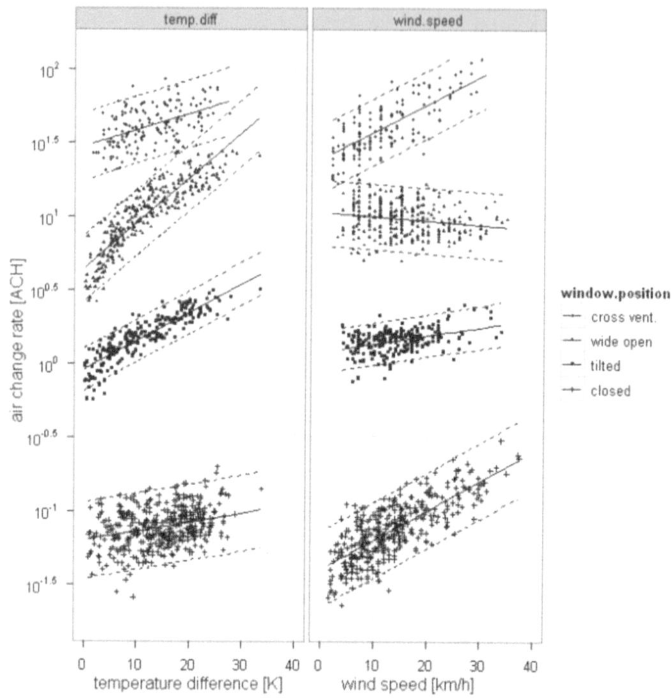

FIGURE 3.14 (a) Effect of temperature gradient (inside to outside) on ACR (corrected for wind effects). (b) Effect of wind speed on ACR (corrected for temperature effects) in an office place (room E) for different ventilation conditions: Natural ventilation: "closed," "half-open window," "fully open," and "cross ventilation" (from bottom to top: the room volume 40 m^3, fitted curve, the dashed lines (95% prediction intervals).

3.7 AIR CHANGE RATE

As part of place inspections and indoor air research during 1999–2005, ACR was determined for many buildings and rooms using the CO_2 concentration drop method. A total of 198 residential and workplaces—located in 152 buildings in and around Berlin, including 143 residential buildings and nine office buildings—were studied. In none of these rooms, there was no installed additional ventilation system. During the measurement of the tracer gas, the residents were asked to refrain from entrance to the rooms and close the windows and doors. The duration of tracer gas measurement was between 2 to 24 hours. To obtain reduction curves with a high number of nodes, time intervals of 1 to 3 minutes between CO_2 measurements were chosen. Two Testo 400 devices (air measurement devices of air conditioning system) with CO_2 detectors (infrared absorption) were used to record data. The results showed that under the worst conditions, the most values of ACR were between 1 h^{-1} and 0.5 h^{-1}. ACR of the studied rooms ranged from a minimum of 0.02 h^{-1} to a maximum of 1.98 h^{-1}. ACR in most of 198 rooms with natural ventilation (167 rooms) was 0.5 h^{-1} or < 0.5 h^{-1}. ACR in 12 rooms (6%) was only determined by 0.8 h^{-1} or higher.

Measurements were carried out in the same rooms using emission tubes and hexafluorobenzene, F_6C_6 or PFT as a tracer gas that relatively higher ACR values were reported for them compared to the CO_2 concentration drop method. This can be explained by the long-term exposures of the collection tubes from several days to several weeks. It means that there was one or more active ventilation phases influencing the air change during the measurement period (Table 3.10). However, even in these studies, the number of rooms with ACR < 0.5 h^{-1} was very high. Studies conducted in the United States also show that a high number of buildings (50% and more) have ACRs lower than 0.5 h^{-1}. Therefore, it is difficult to achieve an ACR of 0.5 h^{-1} under normal conditions. The classification of window types showed that ACR values for rooms with double-glazed windows or composite windows were significantly higher than those for rooms with insulating windows (double-glazed windows and rubber seals (Table 3.11)). However, for rooms with double-glazed windows or composite windows, most ACR values were lower than 0.5 h^{-1}.

The experimental frequency distribution of ACR for the entire sample has the usual shape of normal distribution with a steep slope on the left side and a long tail on the right side (Figure 3.15a). By calculating the logarithms of ACR, a good fit with the normal distribution can be obtained (Figure 3.15b).

3.8 VENTILATION AND CONCENTRATION OF POLLUTANTS

Proper ventilation is the cheapest and easiest way to effectively remove indoor pollutants. Previous studies have shown that inadequate air exchange is mainly due to renovations and other structural changes, mainly due to the installation of ventilation windows. In general, insufficient fresh air supply leads to a significant increase in indoor air pollutants, such as formaldehyde and other volatile compounds as well as radon gas.

Also, the studies presented in this section show a relationship between the concentration of volatile organic compounds and the amount of ventilation in rooms with natural ventilation.

Some known indoor air pollutants, including namely formaldehyde, acetaldehyde, acetone, 2-ethyl-1-hexanol, hexanal, d-limonene, alpha-pinene, decamethylcyclopentasiloxane (5D-siloxane), and toluene, were selected. To analyze these pollutants, air sampling was performed in the studied rooms (closed windows and doors). This sampling was conducted after the CO_2 detector gas measurement or directly before CO_2 loss record with air control samplers (model PCXR 822). A standard surface absorption medium was used for these pollutants. The air volume varied from 90 to 120 L, and the sampling airflow was between 1.5 to 2 L/min. Formaldehyde was somewhat special because it is semi quantitatively measured in most rooms. Interscan device (model 4000) was used for measurements.

Before determining the concentration of formaldehyde, zero balance was conducted by an air pump into an activated carbon filter located upstream for 10 minutes. After that, when the number was fixed, the concentration value was recorded. The difference of the measurements for 24 different rooms had a good agreement with the concentration levels obtained by the electrochemical method of formaldehyde with the Interscan device (model 4000) and the laboratory results obtained by the air sample analysis with the dinitrophenylhydrazine (DNPH: dinitrophenylhydrazine) cartridge.

TABLE 3.10
ACR Measurements in Germany and Scandinavia (Study Results)

Method	Gas Detector	Sample Size	Min [ACH]	Avg [ACH]	Max [ACH]	AM [ACH]	GM [ACH]	< 0.5 [ACH] (%)
Loss of concentration	N_2O	150	< 0.01	0.25	1.7	0.36	Unavailable	> 70
	CO_2	80	< 0.05	0.18	> 1.5	0.26	Unavailable	90
	CO_2	198	0.02	0.21	1.98	0.31	0.22	84
	TFE	114	< 0.01	0.28	> 1.5	Unavailable	Unavailable	72
Loss of concentration and accumulation	CO_2	300 items	-	0.44	-	0.62	0.46	57
		200 elements	-	0.42	-	0.62	0.46	
	C_6F_6	78	-	-	-	-	0.73[a]	-
			-	-	-	-	0.52[b]	-
Fixed injection (emission tubes)	PFT	117	0.16	Unavailable	0.96	0.37	0.34	82
	PFT	242	0.07	≈ 0.05	1.55	0.52	Unavailable	48
	PFT	38	0.15	Unavailable	1.4	Unavailable	Unavailable	58
	PFT	344	Unavailable	Unavailable	Unavailable	Unavailable	Unavailable	36
	PFT	320[c]	-	-	-	0.36[c]	-	80
		43[d]	-	-	-	0.48[d]	-	60

TFE: 1-1-2-Trichloro 2-2-1-Trifluoroethane, AM: Arithmetic Mean, GM: Geometric Mean: Geometric Mean, and < 0.5 H-1: Ratio (%) of Air Change Rate < 0.5 H-1.

a Before Reconstruction

b After Reconstruction

c Single-Family Houses

d Multi-Family Houses

TABLE 3.11
ACR Values in Rooms with Natural Ventilation (Closed Windows and Doors) Equipped with Different Types of Windows

Window Structure	Sample Size	Min	5th Percentile	25th Percentile	Avg (95%CI)	75th Percentile	95th Percentile	Max	Arithmetic Mean AM (95%CI)	Geometric Mean GM (95%CI)
Composite/double-glazed window	80	0.05	0.11	0.20	0.33 (0.27–0.44)	0.56	1.50	1.98	0.47 (0.37–0.57)	0.34 (0.28–0.41)
Insulating glass	118	0.02	0.05	0.10	0.16 (0.13–0.20)	0.25	0.54	1.00	0.22 (0.18–0.25)	0.16 (0.14–0.19)

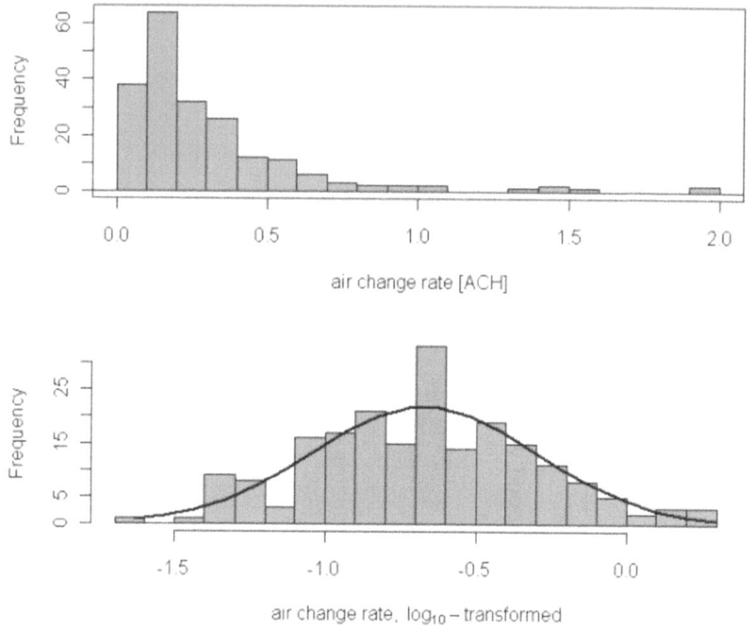

FIGURE 3.15 Experimental frequency distribution of ACR (natural ventilation) for 198 rooms in the city of Berlin: (a) determined values; (b) logarithmic distribution values. The continuous line represents the normal distribution.

The statistics of pollutants examined in this section are presented in Table 3.12. Acetone, formaldehyde, and acetaldehyde showed the highest concentrations, with average values of 75 μg/m³, 23 μg/m³, and 31 μg/m³, respectively. Because of their boiling points below 60°C, these pollutants are highly classified as very volatile organic compounds (VVOC) organic compounds. Therefore, these compounds, except formaldehyde, are not regularly found in indoor places.

As shown in Figure 3.16, the frequency of formaldehyde concentration higher than 100 μg/m³ seems to be associated with ACR values. In eight of 77 rooms in the first group with ACR < 0.2 h^{-1}, formaldehyde equal to and higher than 100 μg/m³ was estimated. Among the rooms in the second and third groups, in only one room, the amount of formaldehyde exceeded the permissible limit. The findings show that preventive measures to reduce emissions, including the use of low-emission products, can also be successful even in rooms equipped with ventilators.

Unlike formaldehyde, acetaldehyde and acetone concentrations show a clear dependency on the air change rate λ, which can be described by a power function $\left(C(\lambda) = a^{*}\lambda^{-k}\right)$. The air change rate of independent variable can explain nearly 50% of the total variance in acetaldehyde concentration and about 34% of this variance in acetone concentration. The curve fitting results are shown in Figure 3.17.

Given the obtained function parameters, it can be estimated that value of ACR increases from 0.1 h^{-1} to 0.5 h^{-1}, and the concentration of acetaldehyde increases from 40 μg/m³ to 15 μg/m³—and the concentration of acetone decreases from 145 μg/m³ to 60 μg/m³ inside the room. This reduction coefficient is equal to 2.4 to 2.7.

TABLE 3.12
Concentration (μg/m³) and 95% Confidence Interval for Selected Volatile Organic Compounds and Very Volatile Organic Compounds in the Air of Berlin City Rooms

Absorbent	Composition	Sample Size	Min	10th Percentile (95% CI)	Avg (95% CI)	90th Percentile (95% CI)	Max	Arithmetic Mean AM (95% CI)	Geometric Mean GM (95% CI)
				Very volatile material					
Discovery on site	Formaldehyde	171	< 20	< 20 (<20)	31 (25–38)	70 (63–96)	190	41 (36–45)	33 (30–37)
Silica	Slide	25	6	9 (n.d.)	23 (11–37)	53 (n.d.)	117	29 (19–39)	22 (16–30)
DNPH		25	18	39 (n.d.)	75 (53–128)	308 (n.d.)	481	118 (71–166)	84 (60–117)
					Volatile				
Anasorb 747/ Tenax absorbent	2- Ethyl-1-hexanal	148	< 1	1 (1–1.5)	4 (3–6)	12 (10–19)	65	7 (5–8)	4 (4–5)
Anasorb 747/ DNPH adsorbent	Hexanal	93	< 2	4 (3–6)	17 (14–21)	37 (30–45)	75	20 (17–22)	14 (12–17)
Activated carbon	Limonene	133	< 1	2 (1–3)	10 (8–13)	66 (36–89)	256	24 (17–31)	10 (8–13)
(NIOSH)	Alpha-pinene	133	< 1	2 (2–5)	15 (11–25)	102 (57–143)	623	45 (27–62)	16 (12–20)
Activated carbon	D 5-siloxane	99	< 1	< 1 (<1–1)	3 (2–4)	94 (39–321)	736	39 (61–17)	4 (3–6)
(NIOSH)	Toluene	133	2	4 (3–5)	14 (10–21)	91 (46–144)	396	34 (25–43)	16 (13–20)

N.D. 95% Confidence Interval Has Not Been Specified (Small Sample Size).

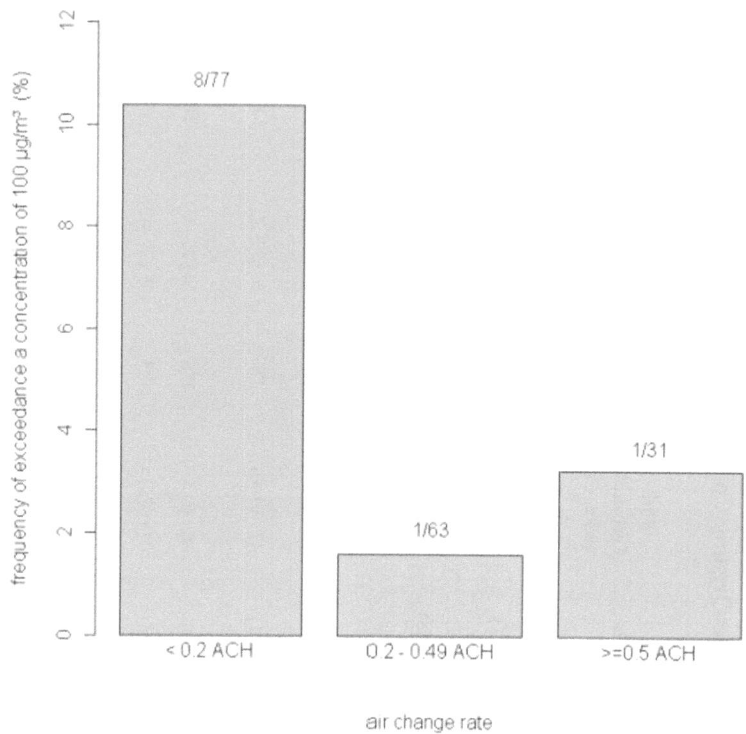

FIGURE 3.16 The frequency of formaldehyde concentration higher than 100 µg/m³ and ACR (natural ventilation) in the rooms of the Berlin city.

There are similar relationships for other volatile organic compounds (Table 3.12). Again, the dependency of indoor air concentration on ACR can be determined by power function (Figure 3.18). Regression analysis showed that increase of ACR from 0.1 h^{-1} to 0.5 h^{-1} decreases the concentration of 2-ethyl-1-hexanol and hexanal (or hexanaldehyde) with a coefficient of 2.3, because 2-ethyl-hexanol decreased from 6.2 µg/m³ to 2.6 µg/m³ and hexanal decreased from 21.2 µg/m³ to 9.1 µg/m³. For limonene and alpha-pinene, the reduction coefficients were 2.6 and 2.2, respectively. For these pollutants, the indoor air concentration decreased from 24 µg/m³ to 9 µg/m³ and from 30 µg/m³ to 14 µg/m³, respectively. For decamethylcyclopentasiloxane and toluene, the reduction coefficients are 2.7 and 2.4, respectively. The concentration reduction values were from 6.5 µg/m³ to 2.4 µg/m³ for decamethylcyclopentasiloxane and from 32 µg/m³ to 13 µg/m³ for toluene.

These results indicate that the concentration of volatile and very volatile organic compounds is reduced by half with a four-fold increase in ACR in indoor air. This means that the concentration is inversely related to the square root of the ACR, but it seems to contradict theoretical considerations if the emission rate of volatile compounds is assumed to be approximately constant. In this case, the concentration at equilibrium conditions has an inverse relationship with the ACR, which means that a fourfold increase in the ACR reduces the concentration to a quarter.

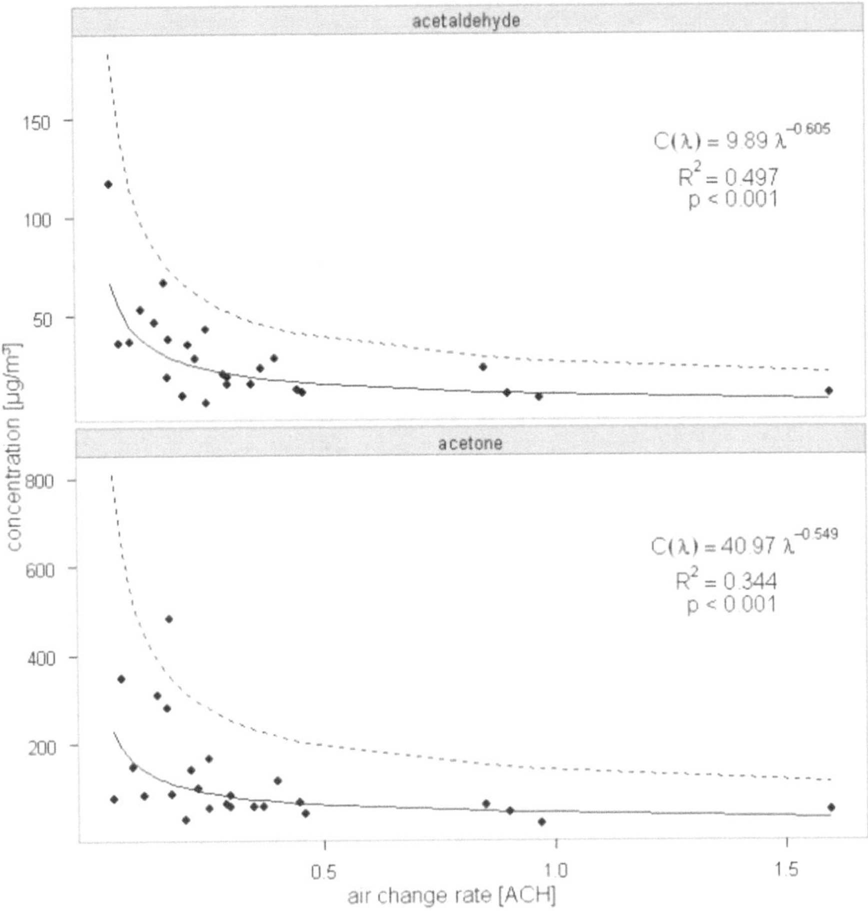

FIGURE 3.17 Acetaldehyde and acetone concentration and air change rates in rooms of Berlin city. Sampling: worst available conditions, solid line: mean concentration, dotted line: 90% predicted limit.

As a reason for the deviation of the results, many new buildings are built with high air sealing and thus, kinetic equilibrium cannot be achieved in normal sampling conditions (i.e., there is a time interval of several hours between the last ventilation through the window and the beginning of the measurement; see Figure 3.3b). This issue leads to two results, including low ACR ($<0.2\ h^{-1}$) and lower concentration record of volatile organic compounds compared to the actual value. Regression analysis shows a flatter curve and power lower than 1 for air change rate (λ). Another reason is the different interior design (furniture, wallpaper, other equipment) of the investigated places, which can cause different levels of air pollution. The VOC were sampled in rooms with different numbers of sources and types of sources. Therefore, it can be assumed that the air changes in the rooms are low, or there are no sources of volatile organic compounds, and or their emissions are low. As a result, only low

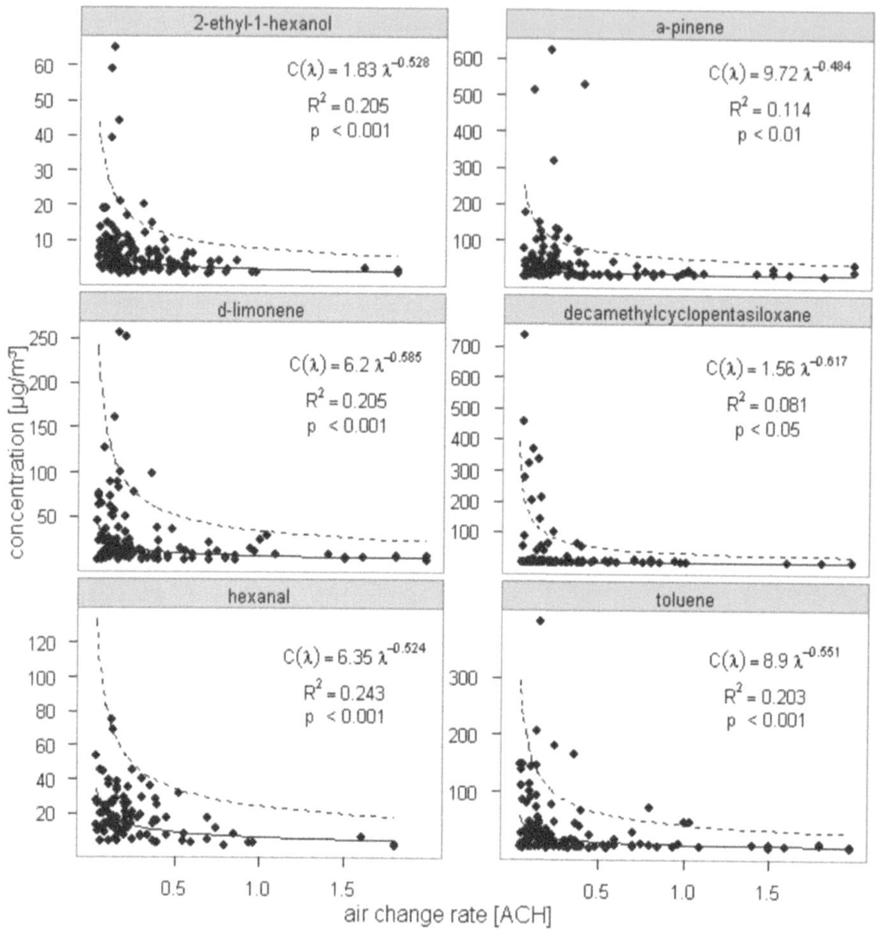

FIGURE 3.18 The concentration of selected volatile organic compounds and relevant ACR values in the rooms of the Berlin city. Worst-case sampling, solid line: mean concentration, dashed: 90% predicted limit.

concentrations can be measured in the air of these rooms. This issue also helps to flatten the air concentration change curve.

The effect of ACR on the concentration of volatile organic compounds is also quite correct when we consider a group of substances. To determine the frequency of excess concentrations, the measured values are separately summed. After that, the total values are divided based on the empirical distribution means.

The relationship between the concentration of selected substance groups in the ambient air and ACR was presented as follows (Figure 3.19): based on the type of substance, between 50% and 70% of the rooms with ACR < 0.2 h^{-1} showed concentrations higher than the average value. For ACR between 0.5 h^{-1} and 0.2 h^{-1}, this was correct for only 40 to 50 percent of the rooms. When the ACR was equal to or greater than 0.5 h^{-1}, this ratio decreased to 10 to 20%. This trend was statistically significant for all groups of studied materials. Even for siloxanes, this trend was significant.

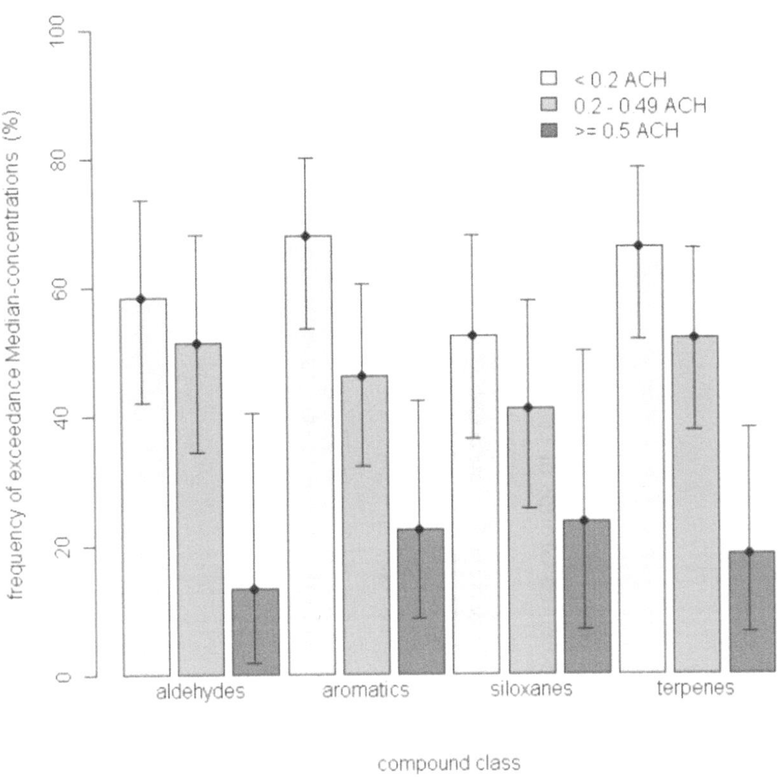

FIGURE 3.19 The frequency of concentrations higher than the median (50th percentile), for the groups of substances selected and classified by ACR. Active air sampling under worst-case conditions, error bars: 95% confidence intervals.

In general, it can be concluded that the probability of pollutants' concentrations higher than permissible values depends on the number of internal sources, power of their emission, and the ventilation of the rooms. However, no excess concentration is formed without the presence of internal sources. For this reason, the use of products with low emissions should be prioritized to avoid exposure to air pollutants. However, it is necessary to define the reference values related to the air change rate for the permissible and acceptable values of volatile pollutants in indoor air (Table 3.13).

In studies performed on residential air quality in Berlin, air change rates were low in many rooms. With worst case measurement, nearly 80% of the rooms had an air change rate lower than 0.5 h^{-1} and half of these rooms even had an air change rate lower than 0.2 h^{-1}. In general, rooms with newly installed windows with insulating glass and peripheral sealing strips had very low air change rates. Among these, about 50% of cases had an air change rate lower than 0.16 h^{-1}.

These values are completely different from rooms in older buildings with double-hung windows or composite windows, which recorded an average air change rate lower than 0.47 h^{-1}.

TABLE 3.13
Indoor Air Concentration and 95% Confidence Intervals [μg/m³] for Selected Groups of Chemical Compounds

Classification of Compounds	Sample Size	Avg (95% CI)	90th Percentile (95% CI)	Max	Average Mean AM (95% CI)	Geometric Mean GM (95% CI)	Purpose	Reference Rate
Aldehydes (C4-C10)	93	49 (36–59)	88 (76–127)	203	51 (43–58)	37 (30–45)	50	120
Aromatic	133	39 (30–51)	206 (149–248)	1140	82 (60–103)	45 (37–54)	50	200
Siloxane (D3-D5)	99	4 (2–6)	125 (51–338)	741	43 (21–65)	5 (4–8)	5	10
Terpenes and sesquiterpenes	133	51 (36–61)	230 (150–297)	1008	103 (73–133)	47 (37–58)	40	150

Low air change rates often lead to increased concentrations of volatile organic compounds in rooms with insulating windows, unless the production of gases and vapors is limited by products with low emissions. Improved ventilation, i.e., increased speed of air change, is an efficient solution to reduce the pollutant load in indoor places. Therefore, to decrease the unhealthy pollutants in the indoor environment, a combination of both measures is recommended.

3.9 AIR CHANGE METHOD

As mentioned, the air change method is the most common technique to calculate the air conditioning requirements based on the complete air changes in a room in a certain time. To determine the airflow required for proper ventilation, the following items are necessary:

1. Calculation of the room volume to be ventilated: width × length × height
2. Calculation of the required air volume by multiplying the room volume by the air change rate (ft^3/h). The following equation is used to compute the ventilation air (cubic feet per minute):

$$Q\left(\mathrm{ft}^3/\mathrm{min}\right)=\frac{(\mathrm{ACH})\times \text{Room size}}{60\,\mathrm{min/h}} \tag{3.16}$$

In this equation, Q is the calculated air volume flow rate, and ACH is the number of expected air changes per hour, based on the type of structure (solid, medium, or loose) under certain conditions.

As a general rule, divide the room volume (cubic feet) by the "appropriate air change minutes." As presented in Table 3.14, note that air changes per hour = 60 min/air changes per minute.

TABLE 3.14
Suggested Air Change Rate for Proper Ventilation

Type of Space	Change the Weather in Minutes	Type of Space	Change the Weather in Minutes
Assembly halls	3–10	Plating rooms	1–2
Attic room	2–4	Hospital	4–6
Lecture halls	4–15	Kitchens	2–4
Bakeries	1–3	Laboratory	2–5
Banks	3–10	Laundries	2–5
Reception halls	3–4	Locker rooms	2–5
Lounge	2–5	Lodge rooms	3–5
Farm barn	10–15	Machine repair workshops	3–5
Beauty salons	2–5	Meeting rooms	4–6
Heating houses	2–5	Mill	3–8

(Continued)

TABLE 3.14 (Continued)

Type of Space	Change the Weather in Minutes	Type of Space	Change the Weather in Minutes
Bowling halls	2–8	Paper factories	2–3
Self-services	3–5	Textile factory	5–15
Church/Mosques	4–10	Administrative office	2–8
Schools/Classrooms	4–6	Canning factory	2–5
Sports clubs	3–7	Plating rooms	1–5
Halls/Corridors	6–20	Printing houses	3–8
Dairy factory	2–5	Show room	1–2
Aerobic hall	3–7	Game rooms	2–8
Dining halls	3–6	Residences	3–6
Dormitories	5–8	Restaurants	3–7
Dry cleaners	2–5	Sanitary facilities	3–6
Engine houses	1–2	Retail sales	3–8
Factories (style)	5–10	Ship storage areas	8–10
Factories (heavy)	2–6	Shops	3–10
Blacksmith workshops	1–3	Theaters	3–8
Foundries	1–4	Material transfer room	1–5
Garages	2–10	Transformer rooms	1–5
Power generator room	2–5	The tunnels	6–10
Glass factories	1–2	Turbine rooms	2–6
Gymnasiums	3–8	Warehouses	3–10

3.10 CONCLUSION

Measurement with tracer gas facilitates the determination of air change values between indoor and outdoor environments under different ventilation conditions. In this chapter, methods were described to determine the air change rate by the tracer gas measurement.

In this chapter, special attention was paid to carbon dioxide as a tracer gas, because this tracer gas is used to evaluate the indoor air quality and room ventilation. Methods to determine the air change rate with carbon dioxide were presented, which are compared by the results of other tracer gases.

4 Local Exhaust Ventilation

A. H. Khoshakhlagh, H. M. Moda, S. Yazdanirad, and A. Gruszecka-Kosowska

4.1 OBJECTIVES OF THE LESSON

Upon completing this lesson, readers should be able to:

- Explain the general principles of local ventilation
- Identify the types of hoods
- State special hood requirements
- Explain the speed types in ventilation
- State the agents affecting the design of local ventilation
- Design the local ventilation system

4.2 LOCAL EXHAUST VENTILATION

Local exhaust systems are designed to capture and remove process emissions before they escape from the workplace. The local exhaust hood is the entry point to the exhaust system, and it includes all the exhaust openings regardless of their physical configuration. The main function of the hood is the creation of an airflow field to effectively capture the pollutant and transit it inside the hood. Figure 4.1 provides the nomenclature associated with local exhaust hoods.

4.3 POLLUTANT PROPERTIES

Inertia effects: Gases, vapors, and fumes do not show significant inertial effects. Also, fine dust particles, 20 μm or less in diameter (including respirable particles), do not exhibit significant inertial effects. These materials will only mix considering the moving air. In such cases, the hood must create an airflow pattern and sufficient capture rate to control the movement of air with pollutants plus additional airflows due to room cross draughts, etc.

Effective specific gravity: Normally, the place of exhaust hoods is wrongly based on the hypothesis that the pollutant is "heavier than air" or "lighter than air." In most health risk programs, this criterion has little importance (Figures 4.2a and 4.2b). Fine dust particles, vapors, fumes, and gases follow air streams as airborne particles and are not subject to significant upward or downward movement due to their density. The natural air movement is due to an equal mixture of these pollutants. Exceptions

DOI: 10.1201/9781003501091-4

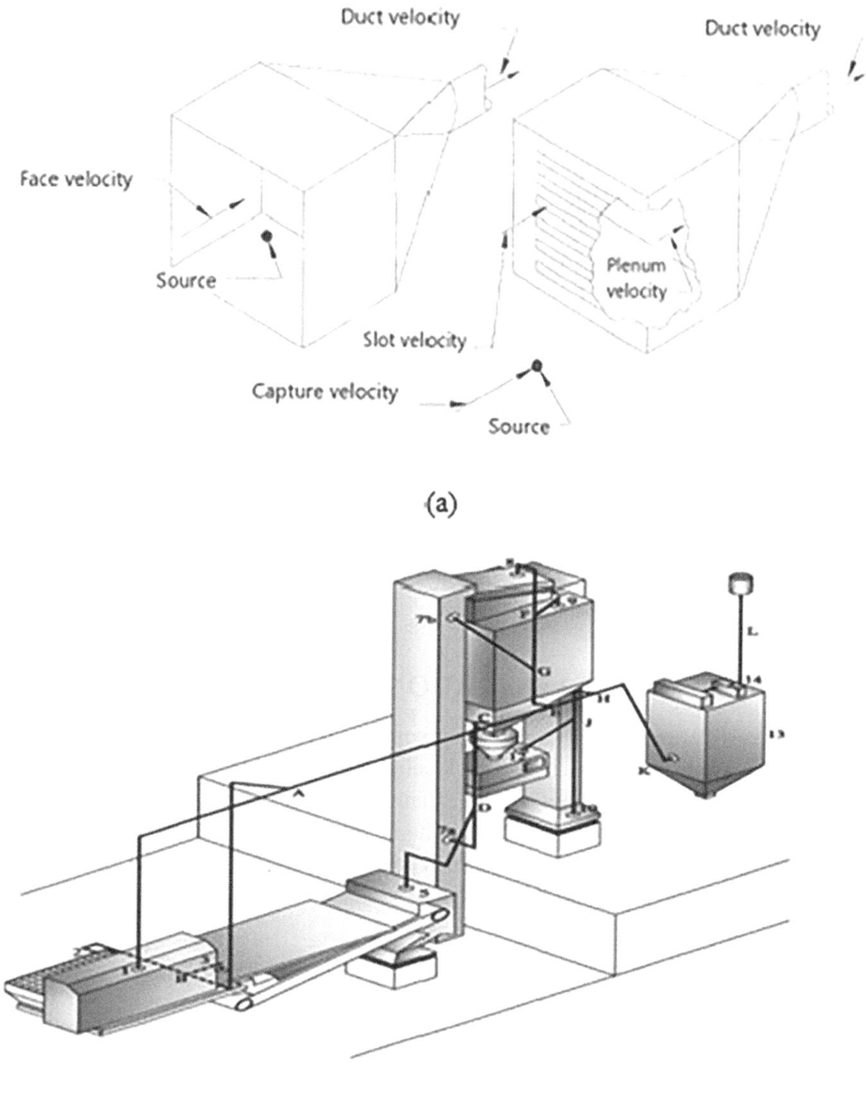

FIGURE 4.1 (a) Local exhaust hood; (b) Isometric plan of local exhaust ventilation system.

may occur in very hot or very cold operations, or where the pollutant is produced in very large quantities and is controlled before the pollutant is diluted.

Turbulent path effects: When air flows around an object, a phenomenon known as "boundary layer separation" occurs. This phenomenon leads to the formation of a turbulent path downstream of the object, similar to the conditions observed when a

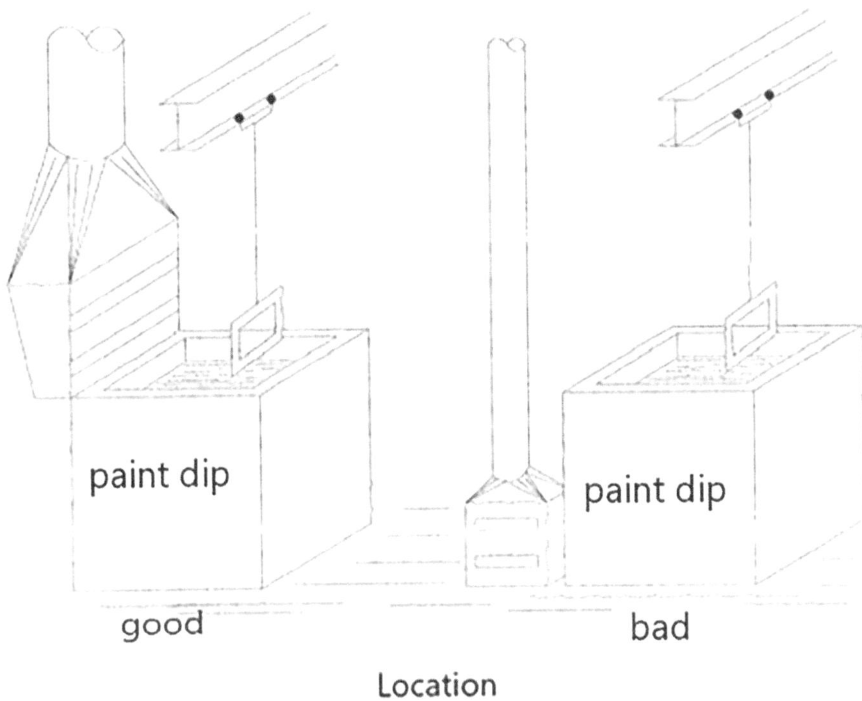

FIGURE 4.2a Effects of specific weight.

ship moves. A turbulent path is a region of intense mixing and recirculation. If a person is working with or near a pollutant-producing source, the recirculation of pollutants in the respiratory area can occur. It is important that this turbulent path around the human body is minimized, and these pollution sources should be placed outside of these recirculation areas as much as possible.

Most control devices are placed at a short distance from control emission sources. This separation may be required for several reasons. There may not be enough space to install the control device close to the source, or the device may collect many emissions from different sources in the facility. Therefore, those should be located at a place with suitable and equal distance, or the auxiliary connections required by the control device may be available in some remote places. Regardless of this reason, the contaminated gas stream must be transmitted from the source to the control device and from there to a stack before it can be released into the atmosphere.

The type of equipment needed to transfer contaminated gas is the same for most types of control devices. These include: (1) hoods, (2) ductwork, (3) stacks, and (4) fans, which form a ventilation system. A hood captures emissions from a source. The ductwork is used to transfer air to the control device. A stack is used to disperse the air, and a fan is applied to provide energy required for air movement through the control system.

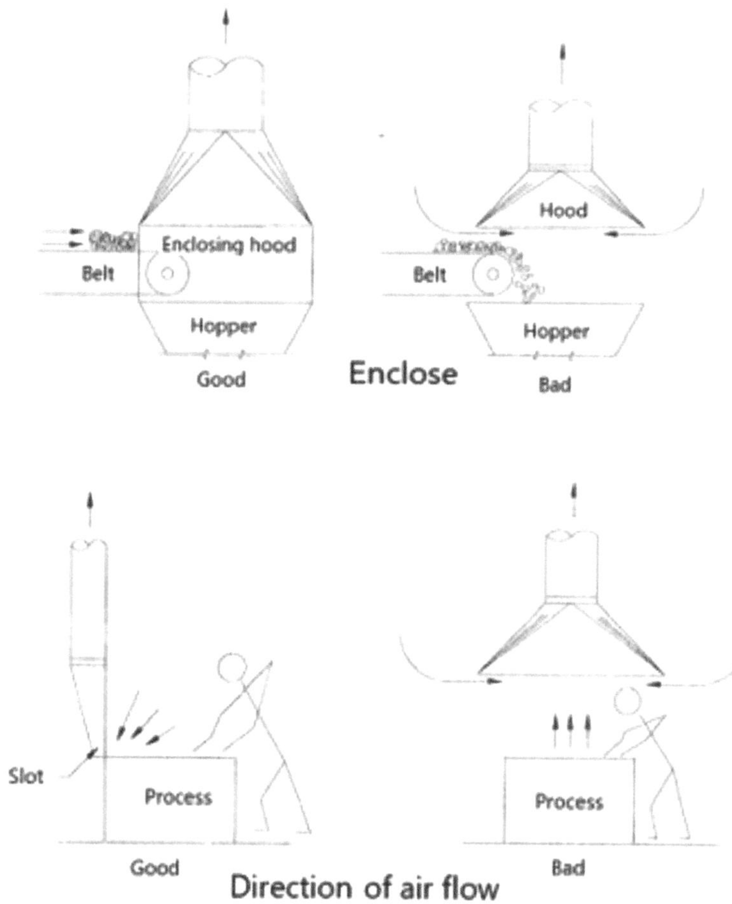

FIGURE 4.2b Relationship between equipment, operator, and enclosure hood.

4.4 HOODS

Among several components of the air pollution control system, the capture device is the most important part. Because if the emissions are not effectively captured at the source, those cannot be transferred to the control device and removed from the air. There are two general categories of capture devices: (1) direct exhaust connections (DEC) and (2) hoods. The DEC is the part of the channel (specifically the elbow) in which emissions flow directly. These connections are typically used when the emission source is itself a channel or vent, such as a process vent in a chemical plant or oil refinery.

Hoods are a wide category of DECs. Hoods are used to capture particulates, gases, and/or mists emitted from various sources, such as basic oxygen steelmaking furnaces, welding operations, and plating tanks. Hood processes are generally classified as either "hot" or "cold." This characteristic influences hood selection, place, and design.

The source characteristics also affect the materials used to construct the hood.

Mild (carbon) steel (medium carbon steel or common structural steel) is a good choice for outlet streams that is non-corrosive and with moderate temperature. However, if there are corrosive materials (for example, acid gases) with high concentrations, stainless steel or plastic (e.g., fiberglass-reinforced plastic (FRP)) is required.

4.4.1 Types of Hoods

Although some hoods have different names, there is a general agreement on their classification depending on the ventilation source. The four hood types are: (1) enclosures, (2) booths, (3) capturing hoods, and (4) receiving hoods.

Enclosure hoods completely or partially enclose the process or pollutant production point. A complete enclosure can be a laboratory glove box or similar enclosure with minimum number of openings. The partial (relative) enclosure can be a laboratory hood or spray booths.

Enclosure hoods are preferred if operation and process are allowed. If it is not possible to use the full enclosure, the partial enclosure should be applied as much as possible (Figure 4.3).

Enclosures include two types: (1) enclosures completely closed to the outside environment and (2) enclosures with openings for material entry/exhaust. The first type is used only for the transition of radioactive materials. They are impervious to dust and gas. These types of enclosures are rarely applied in air pollution control. The second type is used in applications such as control of pollutant emissions from electric arc furnaces. They are equipped with small wall openings (NDO: natural draft openings) that allow materials to be moved in or out and to be ventilated. However, the area of these pores should be compared to the total area of the enclosure walls (usually 5% or less).

Another use of enclosures is measurement of the capture efficiency related to volatile organic compound control devices. Capture efficiency is the fraction of all volatile organic compounds produced and emitted from a source directed to a control device. In this application, a complete enclosure is a temporary structure that completely encloses a diffusing process. So, it traps all emissions of volatile organic compounds for discharge through ducts or stacks. The airflow rate through a complete enclosure must be high enough so that the concentration of the VOC mixture within it is lower than the Occupational Safety and Health Administration (OSHA) standard limits and the vapor explosion limits (25% of the lower explosive limit (LEL)) for the mixture of volatile organic compounds. In addition, the overall apparent speed of the air through the enclosure should be at least 200 ft/min.

The surfaces of the temporary enclosure are usually manufactured of plastic film or hard materials such as insulating panels or plywood. The plastic layer has advantages such as being light, transparent, cheap, and easy to use. However, its weakness points include weak strength, flammability, and a relatively low melting point. In addition, the plastic should be hung on a frame made of wood, plastic piping, or scaffolding.

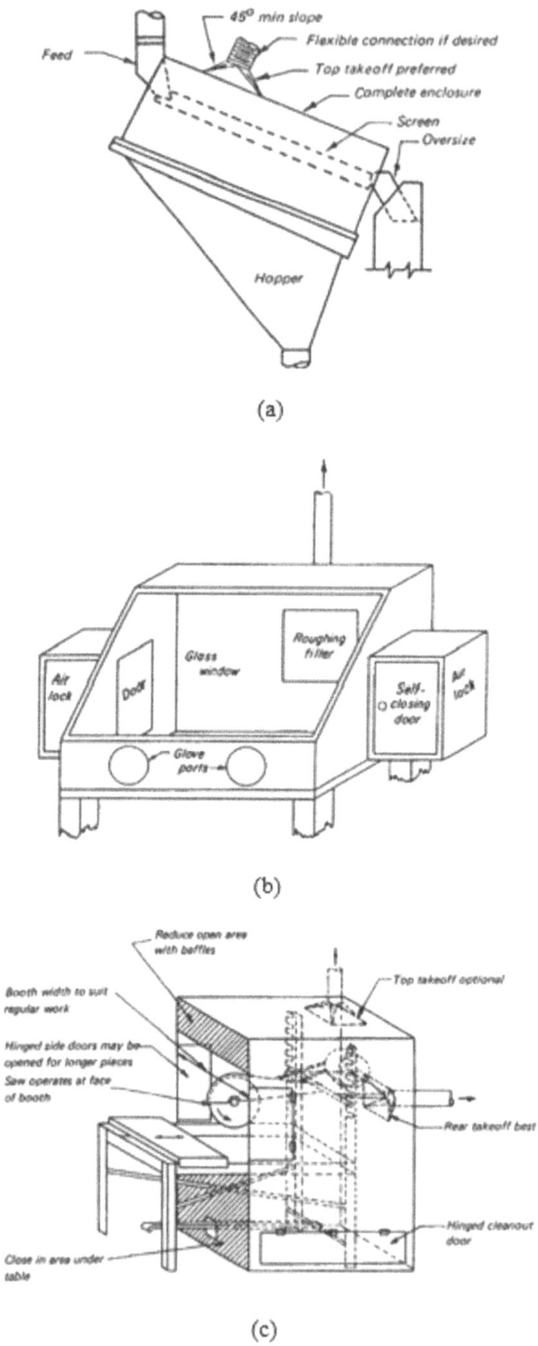

FIGURE 4.3 Examples of enclosure hoods: (a) glove box; (b) flat wall-mounted hood; (c) abrasive cutting saw chamber enclosure hood; (d) spray painting chamber enclosure hood; (e) tunnel-type enclosure hood with several straight-line rollers.

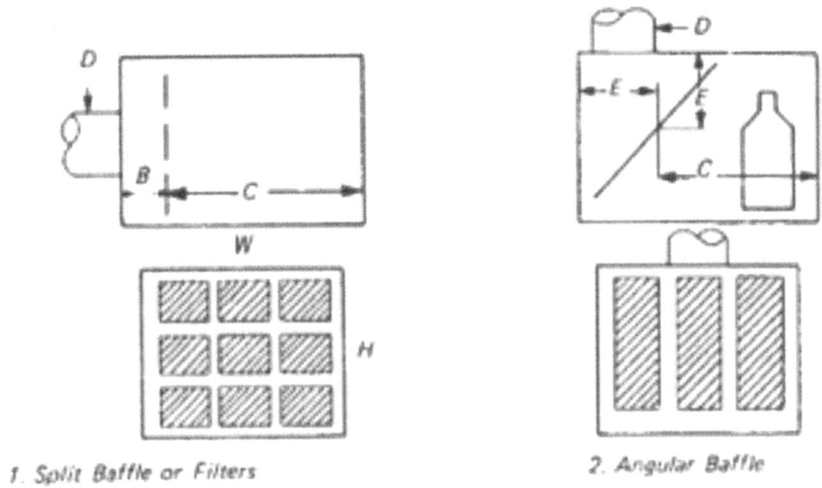

(d)

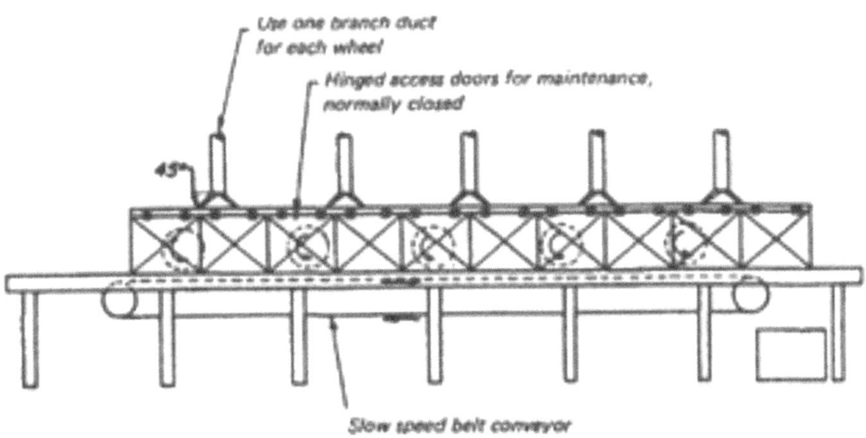

(e)

FIGURE 4.3 (Continued)

Although rigid materials are more expensive and less suitable than plastic, those are more durable and can withstand greater pressure differences between the interior and exterior spaces of the chamber.

Booths are similar to enclosures because those surround the emission source, except that a wall (or part of it) is removed to allow access to operators and equipment. Like enclosures, booths should be large enough to prevent the penetration of particles to the inner walls. Those are used in operations (and emission sources) such as painting and mobile grinding, polishing, and buffing operations.

Capturing hoods (also known as active or external hoods) do not enclose the source in any way.

External hoods are placed in the vicinity of the emission source without enclosure. Examples of external hoods are slots along a tank edge or a rectangular opening on a welding table. Where the pollutant is a gas, vapor, or fine particle and does not diffuse at any significant speed, hood orientation is not necessary. However, if the pollutant contains large particles that diffuse at a significant speed, the hood should be placed in the diffusion path (for example, the grinding operation).

If a process emits hot polluted air, it will rise to the top due to the thermal buoyancy force. The use of external side hoods (placed horizontally to the hot process) may not provide satisfactory capture because of the hood airflow inability to overcome the heat-induced airflow. This is especially true for very high temperature processes such as a melting furnace. In such cases, a canopy hood located above the process may be suggested.

Another example of an external hood is a push-pull system. In this case, a jet blows air along a pollutant source toward the hood flow field. Pollutant control is primarily provided by jets. As the function of the exhaust hood, blown air is received and discharged. The advantage of the blow-suction system is that the blow jet can move air in a controlled manner over much greater distances than the air sucked by an exhaust hood. The push-pull system is proposed for some operations of open-surface containers and plating. However, the exhaust portion of the system has the potential to increase operator exposure if it is not properly designed, installed, or used.

External hoods have one to three corners, and those are placed at a low distance from the source. Those exhaust emissions through vents. Capturing hoods are categorized into types of side-draft/backdraft, slot, downdraft, and high-velocity and low-volume (HVLV) hoods.

Side-draft/backdraft hoods are typically placed next to/behind the emission source, but as close to it as possible, because air velocity decreases inversely with distance. Examples of this hood include snorkel-type welding hoods and side shake-out hoods.

Slot hoods work in a similar way to side-draft/backdraft hoods. However, the inlet opening is very small, narrow, and long. Additionally, a slot hood is placed in the peripheral portion of an emission source such as a narrow and open tank. This type of hood is also used in bench welding operations.

While slot and side-draft/backdraft hoods are placed next to or behind a source, a bottom-draft hood is located below the source. Contaminated air is directed from the bottom of the exhaust source to a control device. Some applications of this hood are the casting vibration part, soldering operations on the worktable, and torch cutting.

In HVLV hoods, very high velocities (capture velocities) are used to collect pollutants at the source and distribute the velocities throughout the hood optimally. To maintain a low volumetric flow rate, these hoods are placed as close to the source as possible.

The last group is receiving hoods (inactive or canopy hoods). A receiving hood is typically placed above or next to a source to collect emissions accelerated by the source. For example, a canopy hood may be placed directly above an open tank containing a hot liquid (a floating source). With air entrance, the vapors emitted from the liquid are gathered by the hood. Here, the canopy hood acts like a passive collector as gases are forced into the hood by natural forces (Figure 4.4).

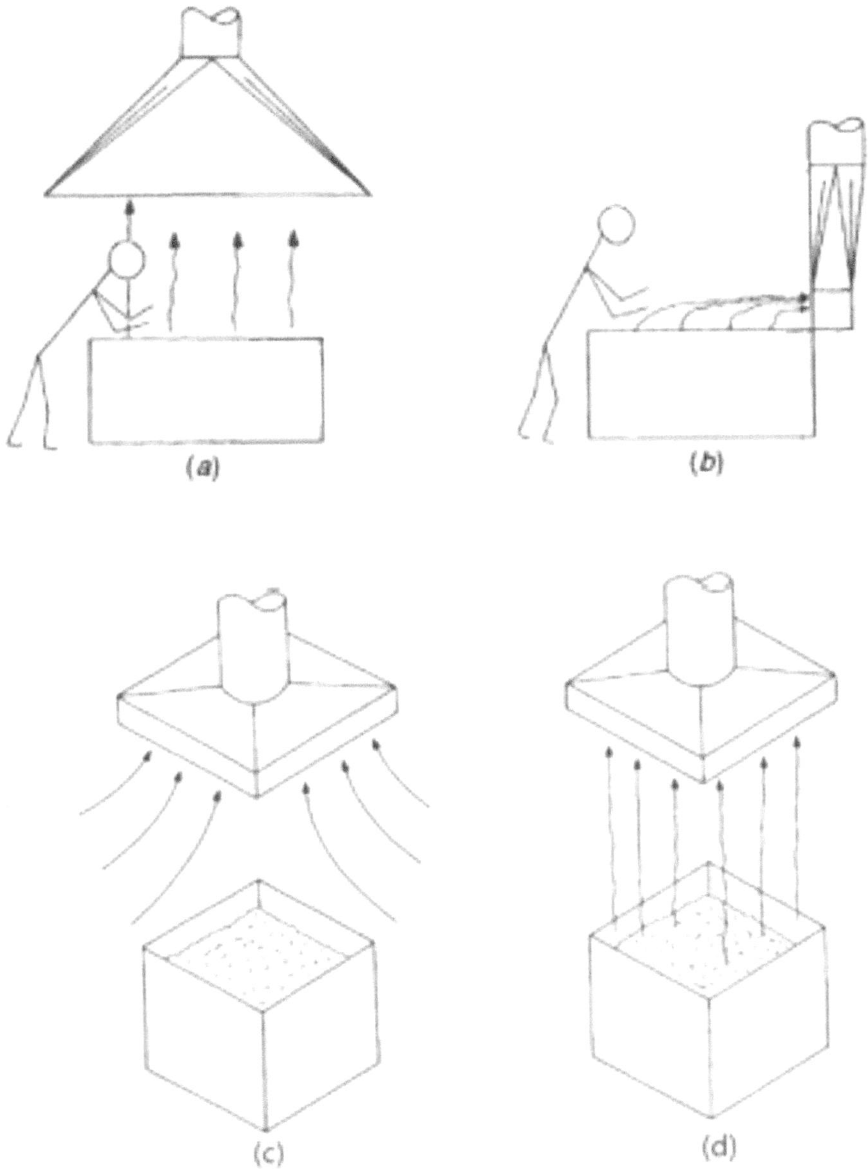

FIGURE 4.4 Airflow patterns used in the open surface tank: (a) canopy hood; (b) slot hood; (c) the airflow in the canopy hood of the process equal to the ambient temperature; (d) airflow in the canopy hood in the thermal process.

4.5 TYPES OF SPEEDS IN THE VENTILATION SYSTEM

Capture velocity: it is defined as the air velocity in front of the hood or in the opening of the hood, which is necessary to counteract opposing air streams and capture the polluted air at that point.

Face velocity: it is described as air velocity at the hood opening.

Slot velocity: it is defined as air velocity through the openings in a slot hood. It is used to achieve uniform distribution of air along the opening of the hood.

Velocity in the equalizing chamber: It is described as air velocity in the homogenizer chamber. For good distribution of air in all types of slot hoods, the maximum air velocity in the equalizing chamber should be 1.2 or less than the velocity in the slot.

Speed in the duct: It is defined as air velocity in the cross-section of the duct. When solid materials are present in the air stream, the duct velocity should be equal to or greater than the minimum air velocity required for the movement of particles in the air stream.

Solvent vapors are not considered heavier than air in terms of health hazards. Exhaust from the floor is usually done only for fire-related issues.

4.6 DESIGN PROCEDURES

As mentioned in the above section, the hood is one of the main elements in any air pollution control system. Before the design procedures for hoods are described, the general principles of ventilation are presented. These basic principles cover basic concepts of fluid flow, and those explain the application of these principles to air pollution control ventilation systems. However, these concepts will be presented as simply as possible to calculate the design parameters.

4.6.1 Design Principles

4.6.1.1 Bernoulli's Equation

The fluid flow in the hood, duct, tube, stack, or other chamber is computed by the Bernoulli equation. This equation states that the total mechanical energy of an element in a moving fluid is constant in the whole system, which includes potential energy, kinetic energy, and pressure energy. However, since no system is ideal, Bernoulli's equation must be adjusted to consider frictional losses. Added items resulting from energy from ventilators, pumps, etc., should also be calculated. For a pound mass (lb_m) of fluid flowing in a steady state system, the adjusted Bernoulli equation is expressed as:

$$\int_1^2 v\,dp + \Delta z\left(\frac{g}{g_c}\right) + \frac{\Delta\left(u^2\right)}{2g_c} = W - F \tag{4.1}$$

In which:

v = specific volume of fluid (ft^3/lb_m)
p = static pressure-gauge (lb_f/ft^2)
z = height of the fluid above the reference point (ft)
u = fluid velocity in the channel, hood, etc. (ft/sec)
g = gravitational acceleration (ft/sec^2)

g_c = gravitational constant (32.174 [lb_m-ft/sec^2]/bl_f)
W = work added by fan, etc. (ft-lb_f/lb_m)
F = energy loss due to friction (ft-lb_f/lb_m)

Each term on the left side of Equation 4.1 represents the energy change between points 1 and 2 in a pound of liquid. The terms of work (W) and friction (F) represent the energy amount added/lost between points 1 and 2.

The units of each term in Equation 4.1 are "$ft - lb_f / lb_m$" energy per mass unit. The units in the English system are "lb_f" "and "lb_m", which are numerically equivalent for all purposes because the ratio of the term gravitational acceleration (g) to the gravitational constant (g_c) is very close to 1. In fact, the equation unit is liquid feet or the corresponding height of the liquid in feet. In air pollution control conditions, the liquid often has air properties because the pollutants present in the polluted gas stream have small quantities. So, the stream physical properties are approximately the same as those of pure air.

Because air is a compressible fluid (where the density cannot be assumed to be constant under static conditions, a fluid is considered compressible), its specific volume compared to the specific volume of the incompressible liquids such as water is much more sensitive to changes in pressure and temperature.

Therefore, the term "*vdp*" in Equation 4.1 must be integrated between points 1 and 2. However, in most air pollution control ventilation systems, neither the pressure nor the temperature can be changed from the emission point to the control device inlet place.

Consequently, the specific volume is constant throughout the ventilation system for all practical purposes, and there is no need to integrate the term *vdp*. With this assumption, the first term in Equation 4.2 is expressed as:

$$\int_1^2 v\,dp = v\int_1^2 dp = v\Delta p \tag{4.2}$$

Example: A volatile organic compound emitted by an open surface tank is captured by a hood and transmitted by a fan through a 150-ft, 12-in-diameter duct to an outdoor refrigeration condenser. The fan is located immediately before the condenser inlet place. Therefore, the entire ventilation system is under vacuum. The flow temperature and absolute pressure are 100°F and approximately 1 atmosphere (14.696 lb_f/in^2), respectively.

The inlet height of the refrigeration condenser is 30 feet below the tank surface. The air velocity at the source is zero, while the duct transition velocity is 2000 ft/min. The static pressure varies from −0.5 inches of water column at the source to 4.5 inches of water column at the fan outlet place. Finally, the calculated friction loss through the duct and hood is 1.25 inches of water column.

In this example, calculate the amount of mechanical energy that the fan adds to the gas flow. Assume that the gas temperature remains completely constant.

Solution: First convert "inches of water" to "feet of air":

$$\text{Feet of air} = (\text{Inches of water})\left(\frac{1\,\text{ft}}{12\,\text{in}}\right)\left(\frac{v_{a100}}{v_{w100}}\right) \tag{4.3}$$

Where:

v_{w100} = water specific volume in 100°F = 0.01613 ft^3/lb_m

v_{a100} = air specific volume in 100°F and 1 atmosphere

Since the absolute pressure of the system is close to atmospheric pressure, the polluted gas behaves like an ideal gas. Therefore, the specific volume can be calculated by the ideal gas law:

$$v_a = \frac{RT}{pM} \tag{4.4}$$

Where:

R = ideal gas constant 1545 ft-lb_f/(lb_m-mole)(ER)

T = gas absolute temperature 100 + 460 = 560 R

M = gas molecular weight (air) 28.85 lb_m/lb_m-mole

P = absolute pressure 2116 lb_f/ft^2

By replacing, the following equation is obtained:

$$v_a = 14 \cdot 17 \frac{\text{ft}^3}{\text{lb}_\text{m}}$$

Finally, by replacing these values for v_a and v_w in Equation 4.3, the following expression is obtained:

$$\text{Feet of air } 100°\text{F}.1\,\text{atm} = 73 \cdot 207 \times \text{inches of water}$$

Calculate the changes in mechanical energy and friction loss between the hood inlet (point 1) and the blower outlet/condenser inlet (point 2):

$$\text{Pressure}: v\Delta p = \left(4 \cdot 5 - \left[-0 \cdot 50\right]\text{in} \cdot \text{wc.}\right)\left(\frac{73 \cdot 207\,\text{ft air}}{\text{in} \cdot \text{w c}}\right) = 336\,\text{ft air}$$

(Point 2 is lower than point 1) Δz = –30 ft air (potential):

$$\text{Kinetic}: \frac{\Delta u^2}{2g_c} = \left(\frac{\left[2000\frac{\text{ft}}{\text{min}}\right]}{\left[\frac{60\frac{\text{ft}}{\text{min}}}{1\frac{\text{ft}}{\text{sec}}}\right]}\right) \times \left(\frac{1}{2}\right)\left(\frac{32 \cdot 174\left[\text{lb}_\text{m} - \frac{\text{ft}}{\text{sec}^2}\right]}{\text{lb}_\text{f}}\right)^{-1} = 17.3\,\text{ft air}$$

Friction loss: $F = 1.25$ in. wc $\times 73 \cdot 207 = 915\,\text{ft air}$

Replace the above items into Equation 4.1 and calculate the added energy of the fan for W.

$$W = 44 \cdot 6 \frac{\text{ft} - \text{lb}_\text{f}}{\text{lb}_\text{m}\ \text{air}} = 6 \cdot 07\,\text{in} \cdot \text{wc}$$

To convert fan energy (W) to horsepower (hp_f), it must be multiplied by the air mass flow rate (lb_m/sec), and the result must be divided by the horsepower conversion factor (550 ft-lb_f/s-hp). Mass flow rate is the volumetric flow rate $(Q, \text{ft}^3/\text{s})$ divided by the specific volume

$$hp_f = W\left(\frac{Q}{v_a}\right)\left(\frac{1}{550}\right) = 0 \cdot 001818 \frac{WQ}{v_a} \tag{4.5}$$

Accordingly, Q is a function of the velocity in the channel $(u_t, \text{ft/s})$ and the channel diameter (D_d, ft).

$$Q = u_t\left(\frac{\pi D_d^2}{4}\right) \tag{4.6}$$

Equation 4.6 is only applied for circular channels.

If we combine Equations 4.5 and 4.6 and replace the inputs for this example, it will be:

$$hp_f = 444 \cdot 6\left(\frac{2000}{60}\right)\left(\frac{\pi}{4}\right)(1)^2\left(\frac{1}{14 \cdot 17}\right)\left(\frac{1}{550}\right) = 1 \cdot 49\,\text{hp}$$

Some notes on this example include:

- The units of W and other terms in Equation 4.1 were "ft-lb_f per b_m air," which are expressed as "ft air"; those express energy, not length
- Potential and kinetic energy conditions compared to pressure and friction energy are limited; if those are ignored, the results will not change significantly
- The pressure and friction terms clearly show the importance of units' directness—as shown in step (1), one inch of water is equivalent to more than 73 feet of air; however, as shown in Equation 4.3, the pressure related to air and water column equivalent heights will be the same
- The fan input power depends not only on the required corresponding head (feet of air), but also on the gas flow rate—also note that the horsepower calculated from Equation 4.5 is a theoretical value; those must be adjusted to consider the fan and fan motor efficiencies—fan efficiencies range from 40 to 70%, while fan motor efficiencies are typically 90%; these efficiencies are usually combined with a unit efficiency (ε, fraction) divided by the theoretical horsepower power to calculate the actual need for horsepower

4.6.2 Pressure: Static, Velocity, and Total

Although it is more accurate and consistent to express Bernoulli's equation in terms of feet of air (or precisely, ft-lb_f/lb_m of air), industrial ventilation engineers prefer to use the units of "inches of water column" (in. wc). These units were chosen because the results expressed in "feet of air" often have a wide range. Furthermore, the total pressure changes in ventilation systems are relatively small compared to those in fluid flow systems. Thus, "inches of water" are a special case, because the values expressed in this measurement unit are usually in the range of 1 to 10. Consequently, pressure changes are measured with water-filled manometers.

In the previous paragraph, a new quantity was mentioned, called total pressure (TP). Total pressure, known as impact pressure, is the sum of the static pressure (SP) and the velocity pressure (VP) at any point of a duct, hood, etc., which are expressed in terms of in. w.c.

$$TP = SP + VP \tag{4.7}$$

Where:

SP = (cf) vp
VP = (cf) $u^2/2g_c$

The cf in the terms of SP and TP are the coefficients for converting the energy expressions from ft air to in. wc. Those are at standard temperature and absolute pressure (70° Fahrenheit and one atmosphere). This conversion factor (cf) must be calculated using Equation 4.8:

$$cf = \frac{\text{in}\cdot\text{wc}}{\text{ft.air}} = 12\left(\frac{v_{w70}}{v_{a70}}\right) \tag{4.8}$$

Where:

v_{w70} = specific volume of water at 70°F = 0.01605 (ft^3/lb_m)
v_{a70} = specific volume of air at 70°F = 13.41 (ft^3/lb_m)

Therefore:
$cf = 0.01436$ in. wc/ft air

The conversion factor varies as a function of temperature and pressure. For example, at 100°F and 1 atmosphere, $cf = 1 \div 73.207 = 0.01366$

The omitted term from Equation 4.7 is the potential energy term z (g/g_c). This omission is not accidental. In ventilation systems, potential energy (P.E.) is usually less useful than other terms. Of course, P.E. is a function of the vertical distance of the desired measurement point relative to the datum level, which is usually the ground. This distance will not be more than 20 or 30 feet. Given that P.E. is approximately 0.3 to 0.4 in. wc, we can usually ignore the contribution of P.E. in ventilation systems without a significant error.

The static pressure in a duct is the same in all directions, while the velocity pressure, which is a function of the gas velocity, varies throughout the duct. The duct velocity is the highest value at the duct center and lowest value at the duct walls. However, for airflow in a long and straight duct, the average velocity along the duct (u_t) approximates the centerline velocity (ucl). This is an important point because the average velocity in the duct is often measured by a Pitot tube placed in the center of the duct.

By replacing c_f into Equation 4.7, we can obtain a simple equation relating velocity to velocity pressure (V_P) at standard conditions:

$$V_P = \frac{0 \cdot 143 u_t^2}{2 g_c} \quad (4.9)$$

$$u_t \left(\frac{\text{ft}}{\text{sec}} \right) = 66 \cdot 94 \left(V_P \right)^{1/2} \quad (4.10)$$

Or

$$u_t \left(\frac{\text{ft}}{\text{sec}} \right) = 66 \cdot 94 \left(V_P \right)^{1/2} \quad (4.11)$$

These equations are applicable for any duct, regardless of its shape.

As Burton describes, static pressure can be considered as "stored" energy in a ventilation system. This stored energy is converted into kinetic energy, velocity, and friction losses (primarily heat, vibration, and sound). Frictional losses have several types:

- Losses through straight ducts
- Losses through duct connections—elbows, tees, converters, etc.
- Losses at branch inlets and control device
- Losses in hoods due to turbulence, shock, and throat constriction
- Losses in fans
- Losses in stacks

In general, most of the static pressure energy is lost because of friction, which converts to velocity pressure energy. Friction losses (ΔSP_f) are typically expressed in terms of velocity pressure:

$$F = \Delta SP_f = kVP \quad (4.12)$$

where, k = empirically determined loss coefficient (unitless)

Alternatively, Equations 4.11 and 4.12 should be combined to express F (in.wc) in terms of the average velocity in the duct u_t (ft/min).

$$F = \left(6.200 \times 10^{-8} \right) k u_t^2 \quad (4.13)$$

4.6.3 Temperature and Pressure Adjustments

Equations 4.8 to 4.13 are developed by assuming standard temperature and pressure of the polluted gas stream. In fact, these conditions are 70°F and 1 atmosphere (14.696 in^2/ lb$_f$). While approximately 1 atmosphere is considered as the standard pressure, several different standard temperatures are used in scientific and engineering calculations: 32°F, 68°F, and 77°F, as well as 70°F. The chosen standard temperature varies considering the industry type or engineering field. For example, occupational hygienists and air-conditioning engineers prefer 70°F as the standard temperature, while combustion engineers prefer 77°F.

To use these equations for polluted gas streams at other air pressures and temperatures, their variables must be adjusted. As mentioned in the section above, the flow of pollutant gases in air pollution control programs follows the ideal gas law.

From this law, the following adaptive equation can be obtained:

$$Q_2 = Q_1 \frac{[T_2]}{[T_1]} \frac{[P_1]}{[P_2]} \tag{4.14}$$

Where:

Q_2, Q_1 = Gas flow rates in conditions 1 and 2, respectively (ft^3/min)
T_2, T_1 = Absolute temperatures in conditions 1 and 2, respectively (°R)
P_2, P_1 = Absolute pressures in conditions 1 and 2, respectively (atm)

However, based on Equation 4.6:

$$Q = u_t \left(\frac{\pi D_d^2}{4} \right)$$

D_d: cross-sectional area (ft$_2$)
D_1, D_2: cross-sectional area (ft$_2$) in conditions 1 and 2
D_f=D_c=0.8$_y$
y: reference height or depth parameter (feet)

If Equations 4.6 and 4.14 are combined, the following equation is obtained:

$$u_{t2} = u_{t1} \frac{[T_2]}{[T_1]} \frac{[P_1]}{[P_2]} \frac{[D_{d2}^2]}{[D_{d1}^2]} \tag{4.15}$$

If the gas flow is in circular ducts, the last term can be used to fit u_t into any equation.

4.7 HOOD DESIGN PROCEDURE

4.7.1 Hood Design Factors

When designing a hood, several factors must be considered:

- Hood shape
- Volumetric flow rate

- Capture velocity
- Friction

Each of these factors and their interrelationships are explained in this section.

The hood shape depends on the source nature, which includes factors such as the temperature and composition of the emission, as well as the dimensions and configuration of the emission stream. Also, environmental factors, such as the velocity and temperature of adjacent air streams, are also important.

The hood shape determines the volumetric flow rate required to capture the pollutant emission. Since the hood is under negative pressure, air is drawn into it from all directions. Consider the simplest type of hood, which is a flat channel with an open end. Then, imagine a sphere around the duct opening. The center of this sphere is at the center of the duct opening, while the radius of the sphere is the distance from the duct end to the point where the emission is captured. Air is drawn from this imaginary sphere into the hood duct. The volume of air sucked through the sphere is a result of multiplying the cross-sectional area of the sphere and the capture velocity speed of the hood, u_c as follows.

$$Q = u_c\left(4\pi x^2\right) = 12.57\,u_c\,x^2 \tag{4.16}$$

Where:

Q = airflow at the exhaust point, cfm
u_c = velocity at distance x, fpm
x = imaginary sphere radius, ft

Equation 4.16 is applied to a duct in which its diameter is small relative to the sphere radius. However, if the duct diameter is larger, the capture area must be reduced by the duct cross-sectional area $\left(D_d\right)$, or:

$$Q = u_c\left(4\pi x^2 - \left(\frac{\pi D_d^2}{4}\right)\right) \tag{4.17a}$$

Similarly, if an infinite linear source is considered, the surface will be a cylinder, and the flow rate (with ignoring final effects) will be:

$$Q = u_c\left(2\pi \times L^2\right) = 6.28\,u_c \times L \tag{4.17b}$$

Where:

L = length of linear source, ft

In practice, exhaust sources are not as point or linear but those have physical dimensions that cause the flow surface deviation from the standard geometric shape. Velocity lines are determined empirically. The flow for round hoods and rectangular hoods, which are essentially square, can be calculated by the following equation:

$$Q = u_c\left(10x^2 + A\right) \tag{4.17c}$$

Where:

Q = airflow, cfm
u_c = centerline velocity at distance X from hood, fpm
X = distance outside the axis, ft (Note: The equation is only valid for a finite distance X, where X is 1/5 D)
A = hood opening area, ft^2
D = diameter of round hoods or square hoods, ft

If distances X are greater than 1/5, the flow rate increases at a slower rate than that in Equation 4.17c.

Equation 4.17c shows that the velocity decreases inversely with the distance square from the hood (Figures 4.5a–4.5f).

Figures 4.6 and 4.7 show the flow and level lines for flanged and planar circular hood openings. The level lines are lines with equal velocity in front of the hood. The

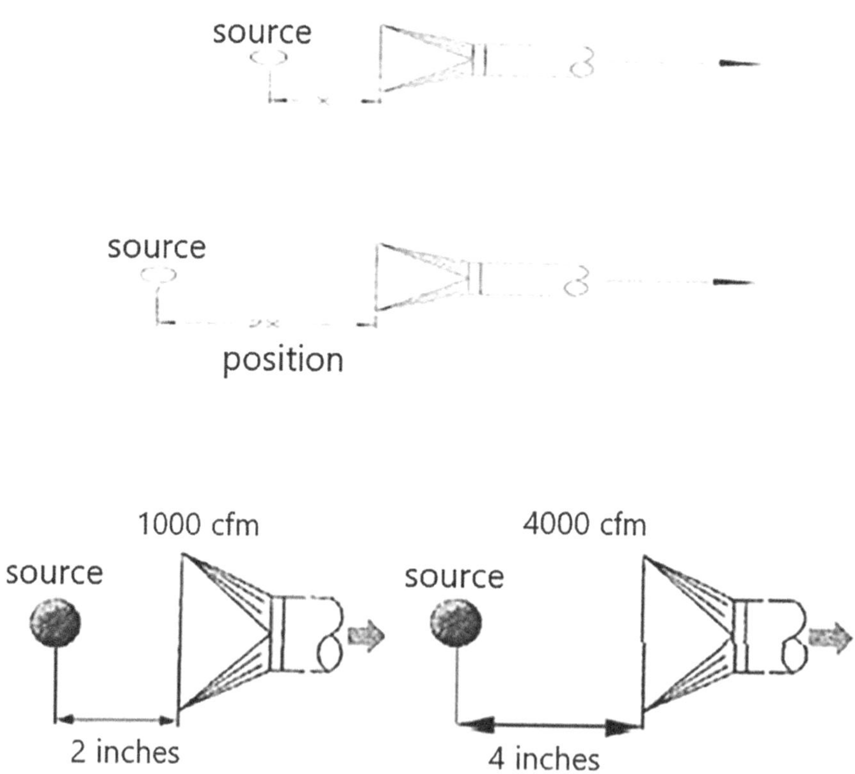

FIGURE 4.5a Flow rate considering distance from hood. Place hood as close as possible to the pollutant source. Required velocity values are related to the distance square from the source.

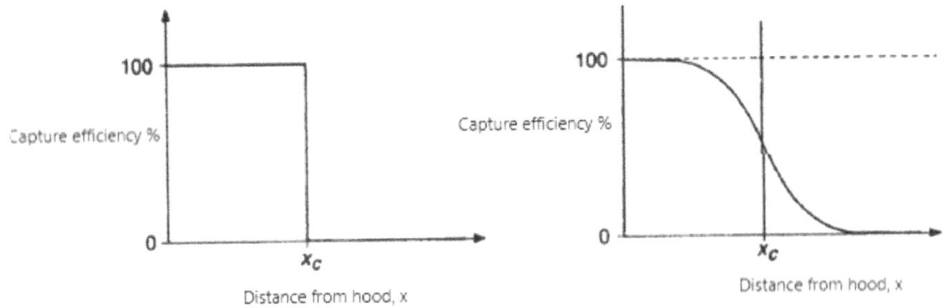

(b) (c)

FIGURE 4.5b Hood capture efficiency versus distance from hood opening: (b) theoretical, without any turbulence and (c) actual, with turbulence.

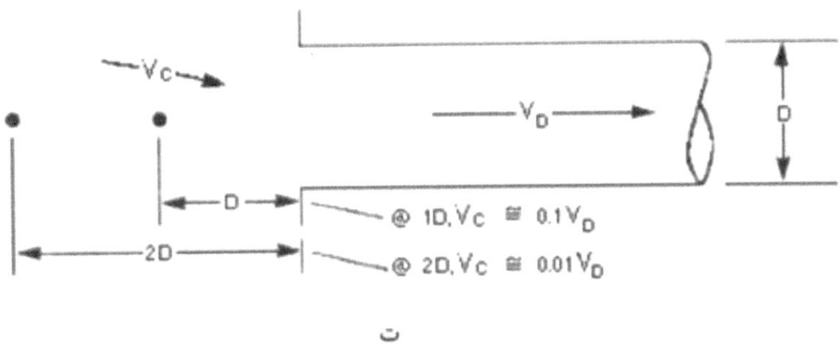

(d)

FIGURE 4.5d Relationship between capture velocity (V_c) and duct velocity (V_d).

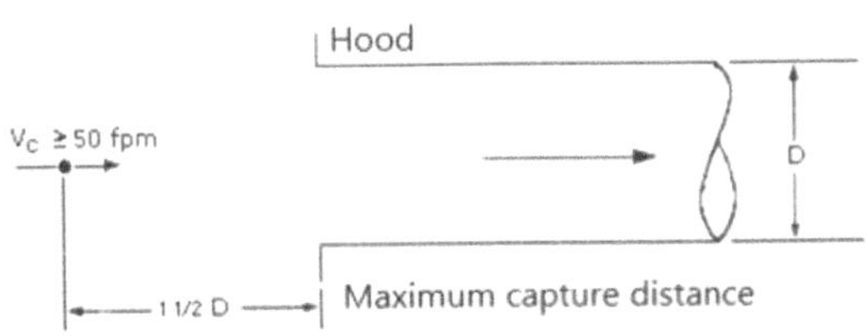

(e)

FIGURE 4.5e Maximum capture distance.

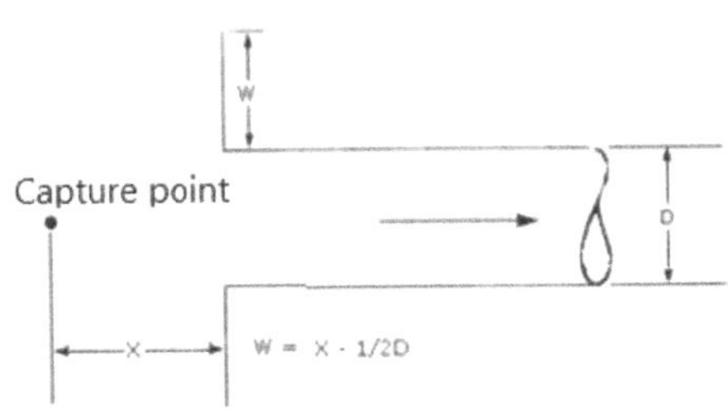

(f)

FIGURE 4.5f Determination of the effective width of the flange.

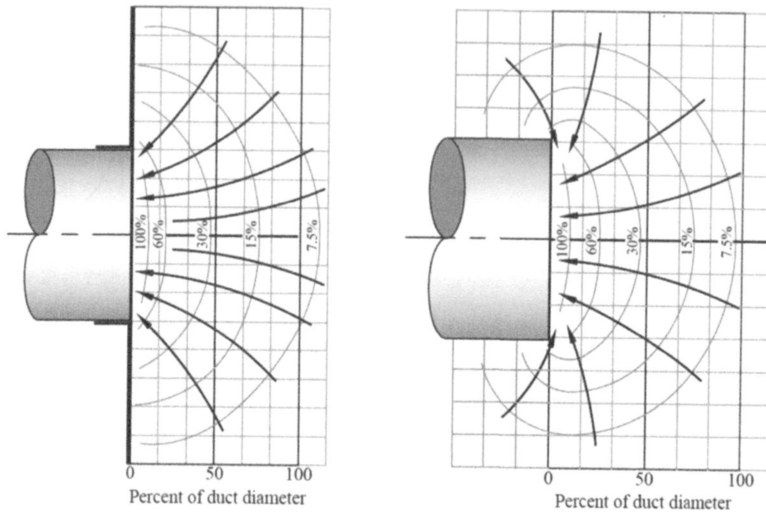

FIGURE 4.6 (a) Velocity level lines - flanged circular opening; (b) Velocity level lines - simple circular opening.

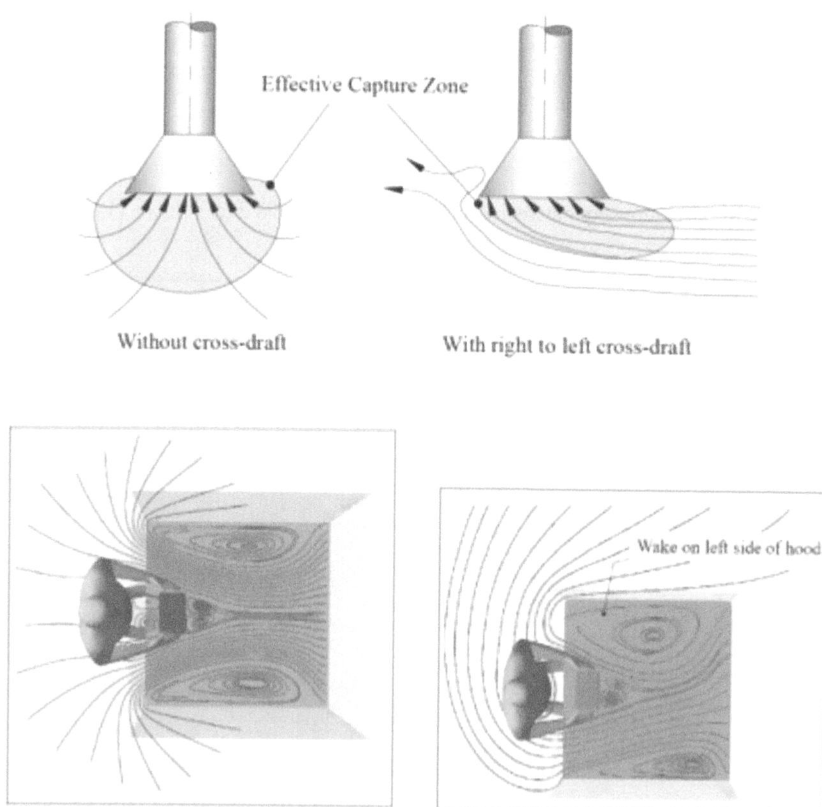

FIGURE 4.7 Effective capture area and the effect of cross-draft flows.

streamlines are perpendicular to the velocity contour lines (tangential to a streamline at any point indicating the direction of airflow at that point).

The equations of flow capture velocity for various hood configurations are presented in Figures 4.8–4.11a.

4.7.2 Relationship between Capture Velocity (V_C) and Duct Velocity (V_D)

To operate effectively, the hood should be placed close to the emission source. The maximum distance from the emission source should not exceed 1.5 times the duct

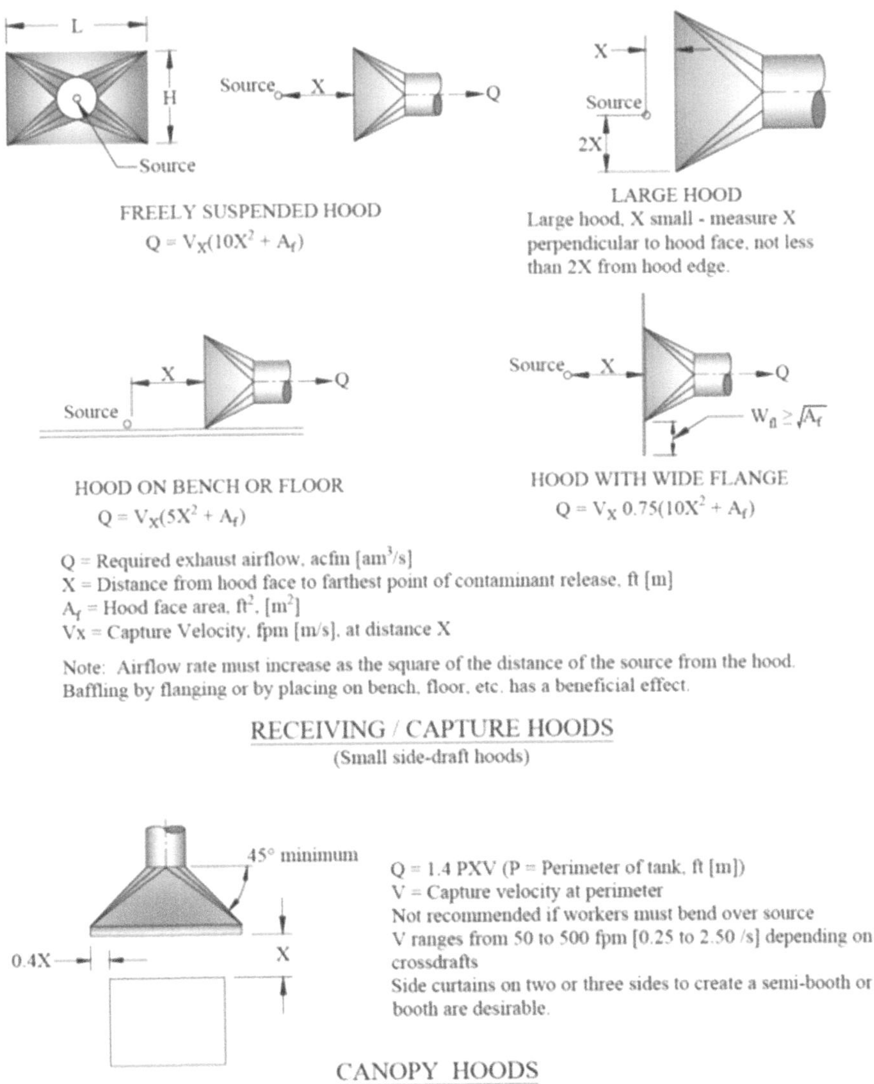

FIGURE 4.8 Capture velocity/flow rate of hoods.

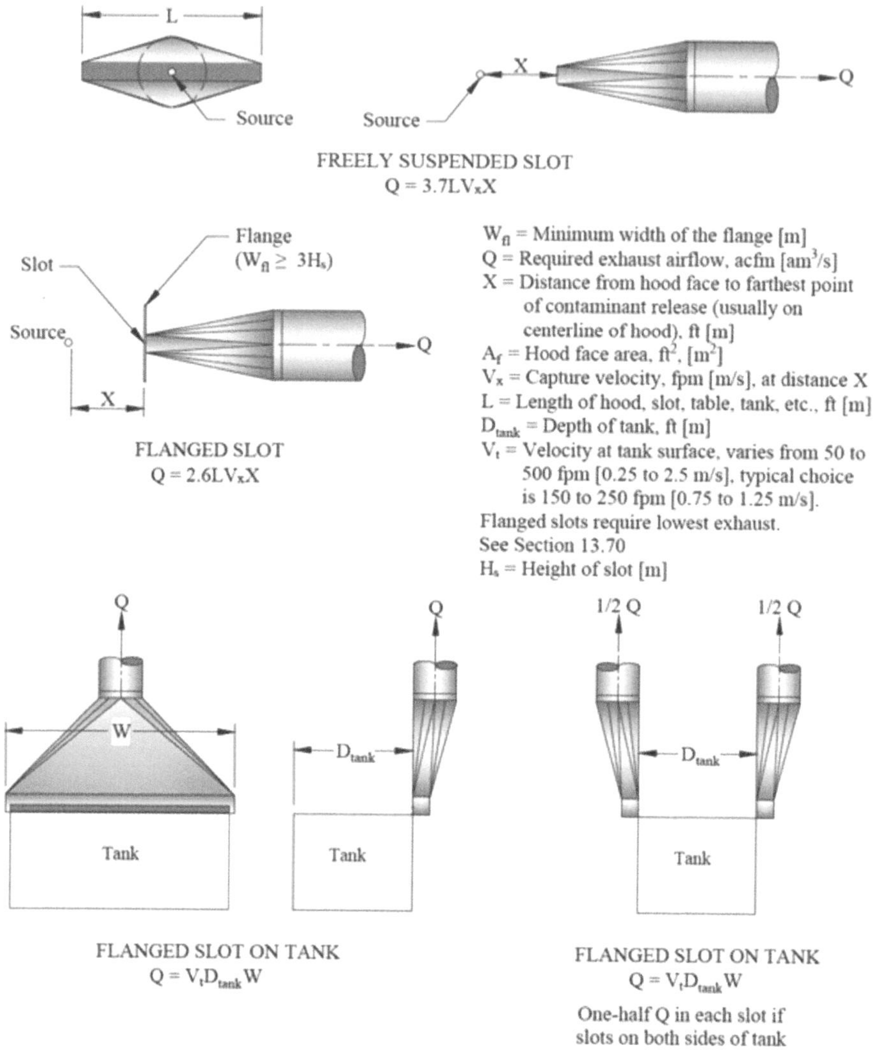

FIGURE 4.9 Capture velocity/flow rate of slotted hoods.

diameter. An approximate relationship between the capture velocity (V_c) and the duct velocity (V_d) for a simple or narrow flat flanged hood is shown in Figure 4.5d. For example, if the position of an emission source is equal to one duct diameter in front of the hood and the duct velocity is $(V_d) = 3000\,\text{fpm}$, then the expected capture velocity (V_c) will be 3000 fpm. At twice the duct diameter from the hood opening, the capture velocity decreases by a coefficient of 10 to 30 fpm.

The maximum capture distance should not exceed 1.5 times the duct diameter. Figure 4.5e shows a general rule that can be used for simple capture hoods. If the duct diameter (D) is 6 inches, the maximum distance from the emission source to the hood should not exceed 9 inches. Similarly, the minimum capture velocity should not be less than 50 fpm.

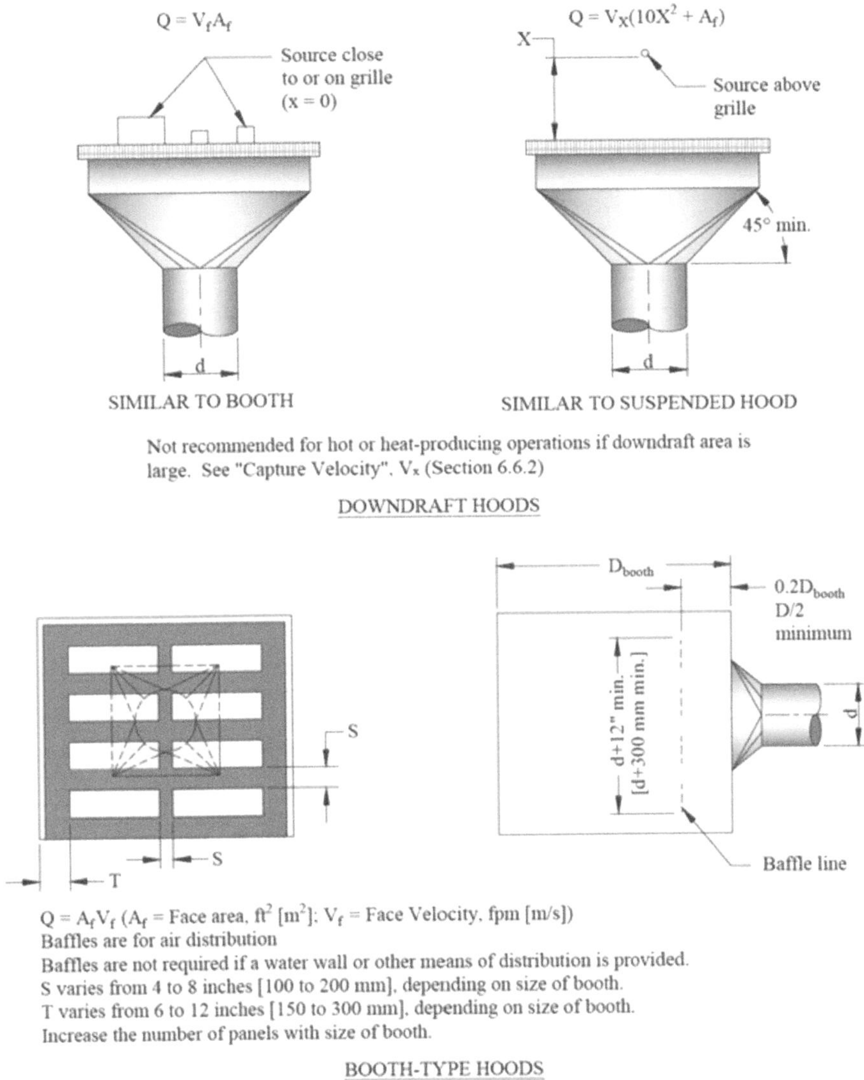

FIGURE 4.10 Capture velocity/flow rate of chamber type and downflow hoods.

Figure 4.5f provides guidance for determination of the effective flange width (W).

4.7.3 Effects of Flanges and Baffles

A flange is a surface parallel to the hood opening that provides a barrier to unwanted airflow in the back of the hood. A baffle is a surface that provides a barrier to unwanted airflow in the front or sides of the hood.

If the exhaust source is placed on a plane, the flow area is reduced (1/2 in both cases), resulting in a reduction in the flow rate required to achieve the same required velocity.

Hood type	Description	Aspect ratio (aspect ratio) W/L	Airflow rate
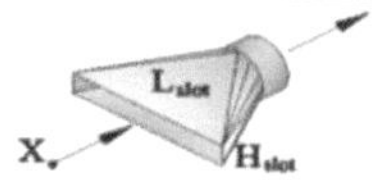	Slotted	2.0 or less	$Q = 3.7\,LV_xX$
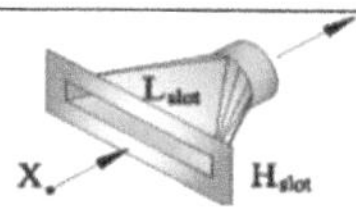	Flanged Slotted	2.0 or less	$Q = 2.6\,LV_xX$
	Simple Opening	2.0 or more and round	$Q = V_x(10X^2 + A_f)$ $A_f = WH$
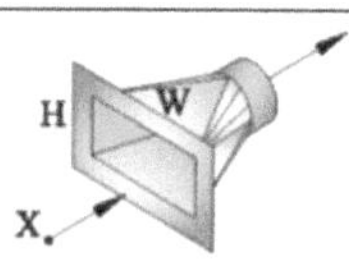	Flanged Opening $W_f \geq \sqrt{(A_f)}$	2.0 or more and round	$Q = 0.75\,V_x(10X^2 + A_f)$ $A_f = WH$
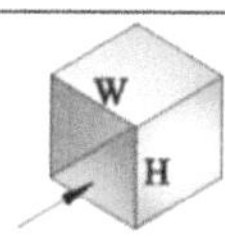	Chamber	Based on work type	$Q = VA = V_xWH$
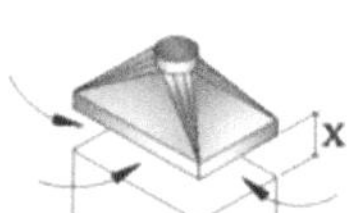	Canopy	Based on work type	$Q = 1.4\,PVD$ P = workplace or tank D = height above work
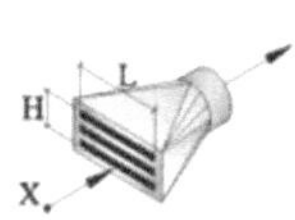	Simple multiple slotted opening with 2 or more slots	2.0 or more	$Q = V_x(10X^2 + A_s)$ $A_s = HL$
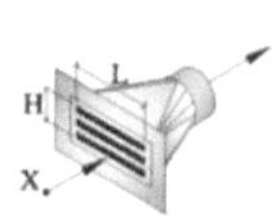	Flanged multiple slotted opening with 2 or more slots	2.0 or more	$Q = 0.75\,V_x(10X^2 + A_s)$ $A_s = HL$

FIGURE 4.11a Types of hoods.

A flange around the hood opening has the same effect of reducing flow rate required to achieve a certain capture velocity. In practice, a flange can reduce the flow rate (or increase the velocity) by approximately 25% (see Figures 4.6a, b and 4.11a). In most applications, the flange width should be equal to the square root of the hood area $\left(\sqrt{A}\right)$.

Baffles can make a similar effect. The extent of this effect depends on the place and size of the baffle.

Figure 4.11a shows several hood types and the velocity/flow formulas.

Similarly, if a flange is installed around the outer end of the duct, the cross-sectional area, through which air is exhausted, reduces the volumetric flow rate by half. So, the flange blocks the airflow from the points behind it. Hence:

$$Q = u_c\left(2\pi x^2\right) \tag{4.18}$$

From these examples, the hood shape has a direct relationship with the gas flow rate. However, Equations 4.16 to 4.18 are only applicable to hoods with spherical flow patterns. For other hood types, other flow patterns, such as cylindrical, planar, etc., are applied. This relationship can be generalized between volumetric flow rate and hood design parameters as follows:

$$Q = f\left(u_i, x, Sh\right) \tag{4.19}$$

Where:

$f(\ldots)$ denotes the function …
Sh denotes the factors affecting the hood shape
u_i = design-capture velocity, at the opening and the slot

Table 4.1 presents the design equations for several commonly used hood shapes. As shown in this table, Q is a function of x, the hood shape, and, in general, the entrainment velocity $\left(u_c\right)^1$. In the case of a chamber hood, the design velocity is the hood opening velocity $\left(u_f\right)^2$. For slotted backflow and side flow hoods, the slot velocity $\left(u_s\right)^3$ is the design velocity. In practice, both the opening velocity and the slot velocity are the same, since those measure the velocity of the gas passing through the hood inlet opening(s). Figure 4.7 shows the effective capture area and the effect of cross-draft flows.

When the gas enters the hood, mechanical energy is lost due to friction. This loss due to friction is calculated using Equations 4.1 and 4.2, if the potential energy contribution from gravity $\left(\Delta z\left(g/g_c\right)\right)$ and the work added to the system $\left(W\right)$ are both zero. Thus:

$$vp2 - vp1 + \frac{u_2^2}{2g_c} - \frac{u_1^2}{2g_c} = F \tag{4.20}$$

By replacing these terms with the expressions in Equations 4.7 and 4.12, Equation 4.21 is obtained:

$$SP_2 - SP_1 + VP_2 - VP_1 = -Hc = -K_h VP_2 \tag{4.21}$$

Where:

i = Static pressure at point i (in. w.c.)
VP_i = Velocity pressure at point i (in. w.c.)
H_c = Hood entry loss (in. w.c.)
K_h = Hood loss coefficient (unitless)

In this equation, subscript 1 refers to a point outside the hood surface. Subscript 2 represents a point in the duct at downstream of the hood where the duct static pressure $(SP_2 \text{ or } SP_h)$ and the duct transfer velocity $(u_2 \text{ or } u_t)$ are measured. At point 1, the velocity pressure in the hood (VP_1) is zero because the air velocity is negligible in this place. Furthermore, the static pressure (SP_1) will be zero because the absolute pressure, i.e., the reference pressure at point 1, is assumed to be one atmosphere. After these simplifications, Equation 4.21 can be adjusted to resolve the hood loss coefficient (k_h):

$$K_h = -\frac{[-SP_h]}{[VP_2]} - 1 \tag{4.22}$$

Firstly, it seems that k_h can be negative, because VP is always positive. However, since the hood entry air is under a vacuum condition produced by a downstream fan

TABLE 4.1
Design Equations, Loss Coefficients, and Inlet Coefficients for Selection of Hood Types

Hood Type	Design Equation	Loss Coefficient K_h	Input Coefficient C_e
Duct end (round)	$Q = 4x^2u_c$	0.93	0.72
Flanged duct end (round)	$Q = 2x^2u_c$	0.50	0.82
Free standing slotted hood	$Q = 2xLu_c$	1.78	0.55
Slotted/side, back hood	$Q = 0.5xLu_c$	1.78	Not applicable
Cone hood	$Q = 2xu_c$	0.06[a]	0.97
Cone updraft/chamber hood (round)	$Q = u_f A_h$	0.25	0.89
Canopy hood	$Q = 1.4Pxu_c$	0.25	0.89
Canopy/vent hood	$Q = 1.4Pxu_c$	0.1	0.71
Immersion tank hood (slotted)	$Q = 125A_t$	1.78	Not applicable
Paint booth hood	$Q = 100A_b$	0.25	Not applicable

[a] Both k_h and C_c are related to round ducts and hoods with a 45° cone. At other angles, k_h and C_c are different. Q = flow rate in hood (ft³/min); x = distance from hood to source (ft); u_c = hood capture velocity (ft/min); u_f = velocity at hood opening (ft/min); u_s = velocity at hood slot (ft/min); A_h = hood chamber area (ft²); P = source perimeter (ft); L = hood slot width (ft); A_t = tank + discharge chamber cross-sectional area (ft²); A_b = chamber cross-sectional area (ft²).

(SP_h) must be negative. Therefore, the term "SP_h / VP_2" is used because the absolute value of SP_h is greater than VP_2, $k_h > 0$, which must be positive.

The hood loss coefficient varies based on the hood shape. For horn-shaped hoods, it can range from 0.04 to 1.78. A parameter related to the hood loss coefficient is the entry coefficient (c_e), which is expressed as:

$$c_e\left[\frac{1}{(1+K_h)}\right]^{1/2} \tag{4.23}$$

c_e depends only on the hood shape, and it may be used to calculate k_h and related parameters. The values of k_h and c_e are given in Table 4.1.

Example: The static pressure (SP_h) is −1.75 inch-water. The duct velocity (u_t) is 3500 ft/min. Calculate the hood drop coefficient and entry coefficient. Assume standard temperature and pressure.

Solution: First calculate the duct velocity pressure. By readjusting Equation 4.11 and replacing it for u_t, it will be:

$$VP = \left[\frac{u_t}{4016}\right]^2 = \left[\frac{3500}{4016}\right]^2 = 0.76\,\text{in.w.c}$$

Next step, replace VP into Equation 4.22 and calculate:

$$k_h = \left[\frac{-SP_h}{VP}\right] - 1 = -\left[\frac{-1.75}{0.76}\right] - 1 = 1.30$$

Finally, use this value and Equation 4.23 to calculate the entry coefficient:

$$c_e = \left[\frac{1}{(1+1\cdot30)}\right]^{1/2} = 0.66$$

The hood design velocities are represented in Table 4.2. Three types of velocities are given: (1) capture, (2) in the opening, and (3) in the slot. As stated in the previous section, capture velocity is the air velocity induced by the hood to capture pollutants emitted at some distance from the hood entry. The opening velocity is the average air velocity to pass through the hood entry. A similar parameter is the slot velocity, which is the average air velocity at the hood's slot openings that their areas are only a fraction of the total hood surface area. As a result, the slot velocity is usually much higher than the opening velocity.

Note that these velocities range from 50–100 ft/min to 2000 ft/min for the velocity at the slot in backflow/side flow hoods with slots (tank and degreasing hoods). As a reference, the air velocity in industrial operations with thermal mixing should be 50 ft/min.

Two other velocities are also discussed in the occupational health literature, although those are not as influential in hood design as the capture velocity at the

TABLE 4.2
Hood Design Velocities

Hood Type/Operation	Type of Speed	Speed Range (ft/min)
Tanks, degreasing	Capture	50–100
Drying oven	In the opening	75–125
Paint booth	Capture	100–200
Canopy hood	Capture	200–500
Grinding, blow cleaning	Capture	500–2000
Slotted hood	In the slot	2000

TABLE 4.3
Capture Velocity Ranges

Pollution Emission Conditions	Examples	Capture Speed, fpm
Emission without initial velocity in calm air	Evaporation from tanks, and washing	50–100
Emission at low velocity in relatively calm air	Spray chambers, intermittent filling of containers, low speed conveying via conveyor, electroplating, and pickling	100–200
Active production of pollutants in an area with fast air movement	Spray painting in shallow chambers, barrel filling, conveyor loading, and shredders	200–500
Emission at high initial velocity in an area with very fast air movement	Grinding, abrasing, and tumbling	500–2000

opening or slot. These two velocities are the plenum velocity (a plenum is defined as an enclosed space that may be in the ceiling and acts as an air duct). It may be built onto the room floor or installed in the walls. This enclosed space is used to pass the ducts inside the building. The velocity in the plenum chamber is the gas flow velocity as it passes through the conical section of a hood (the plenum chamber) between the hood opening and the duct connection. This homogenizer chamber is a transition zone between the hood opening and the duct. As a result, the velocity in the plenum chamber is higher than that in the hood opening, but lower than the velocity in the duct (transfer). The transfer velocity is the gas velocity through the duct, which varies with the composition of the polluted gas.

This is an important parameter in determination of the duct diameter, static pressure loss, and fan size and fan motor systems.

Table 4.3 provides the capture velocities based on the pollution emission conditions.

In each group above, a range of capture velocity is presented. The appropriate selection of values depends on several factors: At the low limit: (1) minimum desirable room airflows for capture, (2) contaminants with low toxicity or nuisance, (3) low and intermittent production, (4) large hood with high air volume.

At the high limit: (1) disturbing airflows in the room, (2) high toxicity contaminants, (3) high production and heavy use, (4) local control by a small hood.

4.7.4 Procedures of Hood Size Determination

Like many auxiliary and control devices, there are various methods for hood size determination.

Some of these approaches are quite complex and involve a series of complex calculations. For example, as one method, the dimensions of the hood (length and width for rectangular hoods and diameter for circular hoods) are firstly determined. In the next step, the amount of sheet metal (ft^2) needed to build a hood with these sizes are estimated using parametric curves (no curves are provided for non-metal hoods). This sheet area is the input to an equation that includes a "pricing factor" and the price per pound of metal. The labor cost required to manufacture the hood is estimated from equations similar to the sheet area relationships. Finally, the metal and labor costs are computed to obtain the total cost of the hood.

A simpler method of size determination is the calculation method of the hood opening area $\left(A_f\right)$. This area is equal to the hood entrance area, which is correlated with the cost of hood construction. To calculate A_f, the following information is required:

- Hood type
- Hood opening distance from the source (x)
- Velocity at the slot (u_s), capture (u_c), or at the opening (u_f)
- Source dimensions (for some hood types)

As the equations in Table 4.1 show, these are the same parameters used to determine the volumetric flow rate (Q) of the hood and duct network. For most control and auxiliary devices with a certain size, a value of Q is given. For hoods, Q must be usually calculated.

Example: A circular canopy hood is used to extract emissions from a chrome plating tank.

The hood surface is 6 feet above the tank in a circular chamber 8 feet in diameter. The capture velocity for this example is 200 ft/min. With assumption of standard tank ambient conditions, calculate the required volumetric flow rate directed into the hood, the velocity at the hood opening, and the area of the hood opening.

Solution: Use the canopy hood equation from Table 4.1.

$$Q = 1.4\,Pxu_c \tag{4.24}$$

Where:

P = tank perimeter (ft)
x = hood distance from top of tank (ft)
u_c = capture velocity (ft/min)

Because the tank is circular, $P = \pi$ (8 ft) = 25.1 ft.
So:

$$Q = (1.4\,\text{ft})(6\,\text{ft})\left[200\frac{\text{ft}}{\text{min}}\right] = 42200\frac{\text{ft}^3}{\text{min}}$$

For this type of canopy hood, the hood diameter is 40% larger than the tank diameter (the "1/4" coefficient in Equation 4.24). Thus:

$$A_f = \left[\frac{\pi}{4}\right]([1.4][8])^2 = 98.5\,\text{ft}^2$$

Finally, the velocity at the hood opening (u_f) will be:

$$u_f = \frac{Q}{A_f} = \frac{42200}{98 \cdot 5} = 428\frac{\text{ft}}{\text{min}}$$

In this example, note that the velocity at the hood opening is greater than the capture velocity. It seems logical, given the fact that the hood entry area is smaller than the area into which the tank fumes are directed. In some hoods, the velocity at the opening is even greater. For example, for slotted hoods, the minimum velocity is 1000 ft/min. In fact, one manufacturer designs the openings of their slotted hoods so that the velocity at the opening is equal to the transfer velocity in the duct.

4.7.4.1 Air Distribution

Slotted hoods are hoods with a width to length ratio (W/L) of 0.2 or less. Slotted hoods are typically used to provide a uniform exhaust airflow and adequate capture velocity within the area of contaminant generation, for example, an open tank or over the opening of a large hood such as a side suction design. The slot plays a role to uniformly distribute the air. The velocity in the slot does not contribute to the capture velocity. High velocity in the slot results in high pressure loss. Note that the capture velocity equation (Figure 4.11b) shows that the capture velocity is related to the suction volume and the slot length, not to the velocity in the slot.

Slotted hoods usually consist of a narrow suction opening and a plenum chamber. Uniform distribution of the exhaust air along the slot is obtained by determination sizes of the slot width and the plenum chamber depth, where the velocity in the slot is much higher than that in the plenum chamber. Splitter vanes may be used in the plenum chamber. However, in most industrial suction systems, the vanes are subject to corrosion and/or erosion, and material accumulates in these places. Adjustable slots can be used but those may be subject to tampering and misalignment. The most practical hood

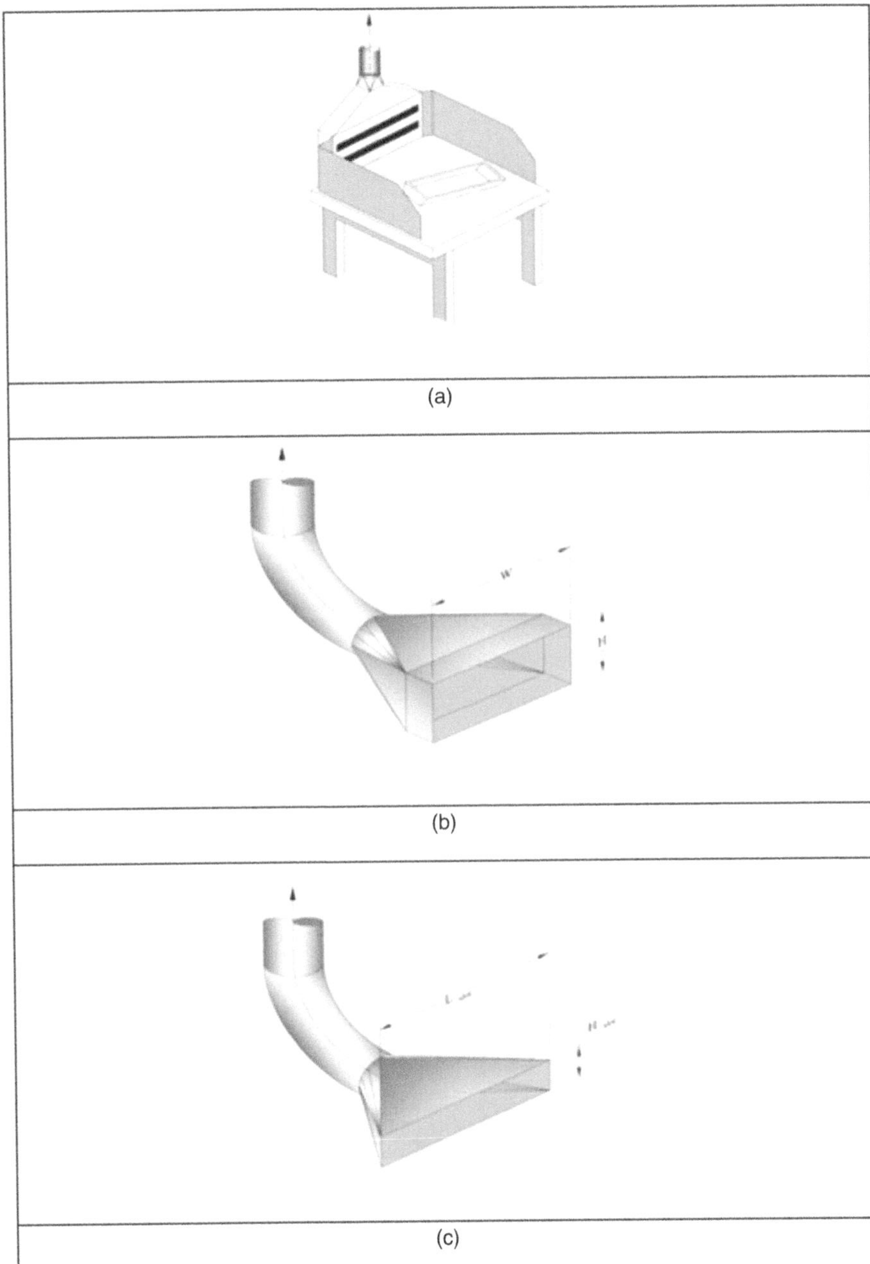

FIGURE 4.11b (a) Slotted hood with baffles; (b) Simple opening hood; (c) Slotted hood; (d) Slotted hood-plenum chamber; (e) Multiple slotted hood; (f) Overhead canopy hoods.

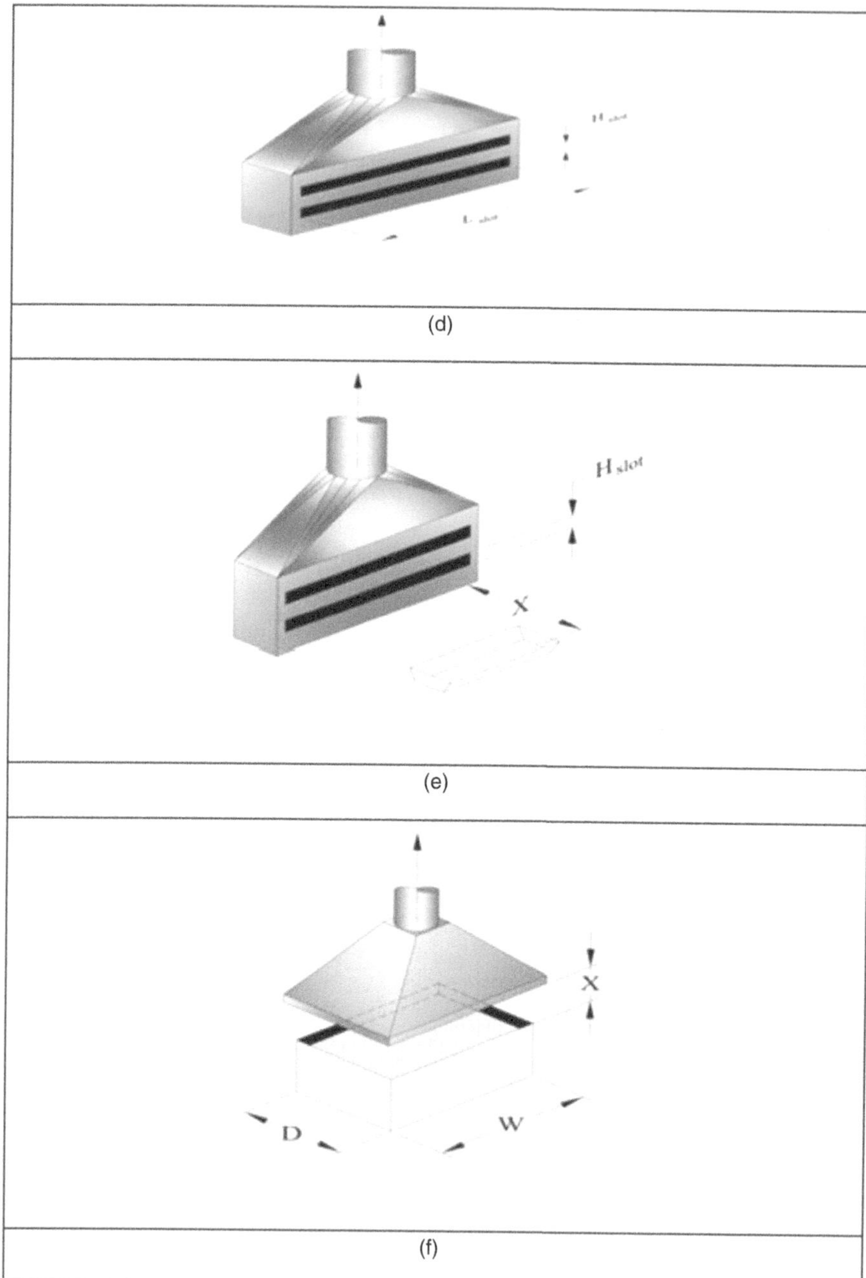

FIGURE 4.11b (Continued)

is the fixed slot type with an unobstructed plenum chamber. The design of the slot and plenum chamber is such that the pressure loss across the slot is higher compared to that across the plenum chamber. Therefore, all parts of the slot are essentially affected by equal suction, and the velocity in the slot will be essentially uniform.

There is no direct method for calculating pressure loss at two ends of a combined structure of plenum-slot. As a very useful approximation used for most hoods, the maximum velocity in the plenum can be designed by half the velocity in the slot. For most slot hoods, the velocity in the slot is 2000 fpm, and the velocity in the plenum is 1000 fpm, which is a reasonable choice for flow uniformity and moderate pressure loss. The design of the central suction duct connection results in a smaller plenum chamber because the air approaches to the duct from both directions. In industries requiring large plenum chambers, such as foundry vibration section hoods, the velocity in the slot may be as high as 1000 fpm and the velocity in the plenum chamber may be as high as 500 fpm.

4.7.4.2 Rectangular and Round Hoods

Air distribution for rectangular and round hoods is determined by the airflow in the hood rather than by the pressure loss. The slotted hood (the length of the hood from the opening to the duct connection) should be as long as possible. The hood duct connection should include a cone angle of 60 to 90 degree. For long hoods, multiple take-offs (passages where air moves from the main duct tube to a new branch) may be required. Configuration of end-of-duct connection requires large plenum chamber sizes because the air must pass in one direction. Figures 4.12 and 4.13 show air distribution methods.

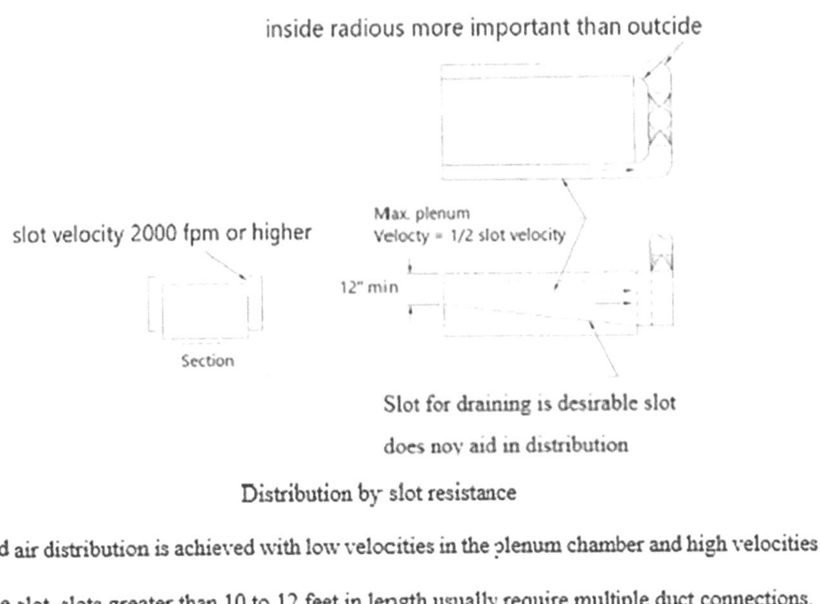

FIGURE 4.12 Air distribution methods.

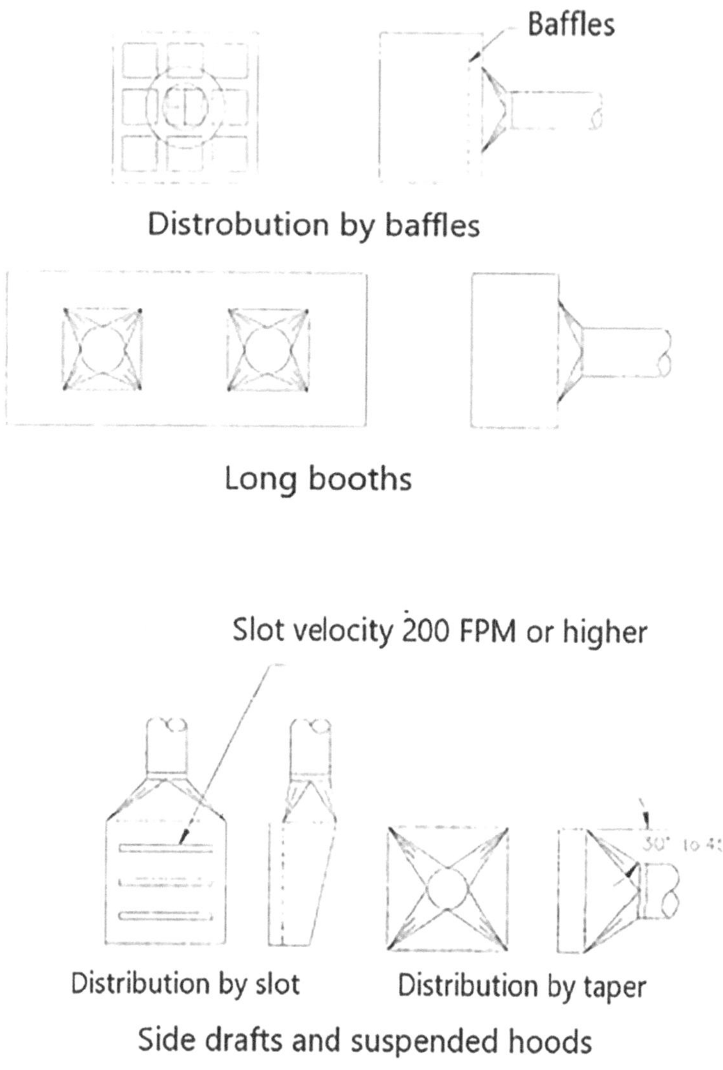

FIGURE 4.13 Air distribution methods.

4.7.4.3 Worker Position Effect

The goal of industrial ventilation is to decrease worker exposure to airborne toxic pollutants in a safe and reliable manner. One of the primary engineering controls is local exhaust ventilation near the point of the pollutant production. Usually, the position of the worker with respect to the airflow direction is not considered. Studies have shown that this factor is an important parameter to determine the concentration in the breathing zone. In Figure 4.14, Position 2 shows a worker with his back to the airflow. Immediately downstream of the worker, a zone of reverse flow and turbulent mixing is produced due to boundary layer separation. The pollutant is released in this zone (for example, from a source near the source or manually), it is mixed in the

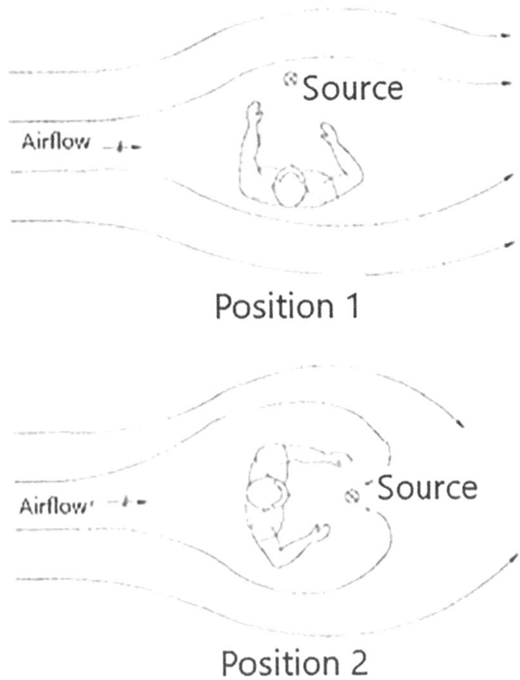

FIGURE 4.14 Effect of worker position.

breathing zone, and it ultimately results in exposure. In Figure 4.14, Position 1 shows a worker positioned at a 90° angle to the flow direction. Here, a reverse flow zone is formed, and there is little chance for emission of the pollutant into the worker's breathing zone.

The results of the studies show that this phenomenon is important when large chamber-type hoods are used or there is a uniform airflow. In a study, exposure by a sewing mannequin to simulate an operator in a chamber-type hood was applied to show powder transfer. In all cases, the exposure for position 1 was lower than position 2 by at least a coefficient of 2000.

In the second case study, use of a spray and brush for a chloroform-based adhesive was investigated among females. The results showed that there was significant exposure despite the use of a chamber-type hood. When workers were positioned on the side of the air stream (position 1), worker exposure was reduced by 50%. Subsequent modifications to the spray method resulted in a 30° angle to the air stream. So, the nozzle placed downstream of the stream provides optimal conditions.

In this situation, it is assumed that the worker is not in the path of an upstream source, and the pollution source has negligible momentum. In cases where the pollution source has significant momentum (e.g., high-pressure compressed air in paint spray operations), the effect of position on exposure may be reversed; for example, position 1 in Figure 4.15 may result in greater exposure. This is made by the spray deflection in the upstream of the worker and subsequent recirculation through the

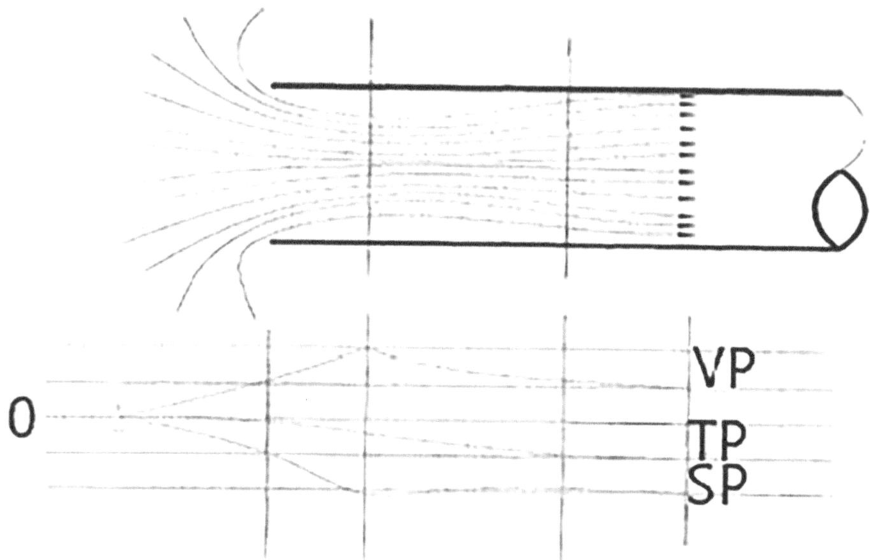

FIGURE 4.15 Airflow in a narrowed throat.

breathing zone. Further research and field studies are needed to evaluate the reverse flow direction in more complex situations. Although the effects of boundary layer separation using smaller local exhaust hoods have not been fully investigated, three studies have suggested that a 90° orientation is appropriate in this case. It has been suggested that a lateral orientation (i.e., position 1) is the preferred orientation in situations. Bottom suction may have similar advantages under certain conditions. It is recommended that a lateral orientation (i.e., position 1) be considered as a preferred working method where possible. Exposure assessment through individual sampling is essential.

4.8 HOOD LOSSES

Simple openings, flanged duct openings, canopies, and similar hoods have only one significant energy loss. A bottleneck forms when air enters the duct, along with a small energy loss as static pressure converts to velocity pressure (Figure 4.15). As the air passes through the bottleneck, the flow area increases to fill the duct, and the velocity pressure is converted to static pressure. At this point, the uncontrolled reduction in air velocity from the bottleneck to duct downstream causes a major portion of the entry loss. The more prominent role of the bottleneck can be associated with the greater energy loss and static pressure of the hood.

Compound hoods have two or more significant energy loss points and must be considered separately. Those must be combined to calculate the total hood drop. Common examples of hoods with two entry losses include: slotted hoods and multi-entry hoods; side suction hoods used in electroplating, paint dipping, and degreasing tanks; and side suction ventilation in foundry vibration sections. The hood entry loss

(h_e) is obtained from the hood loss coefficients (F_f) when multiplied by the slot or channel velocity pressure (VP), in inches of water. The hood static pressure is equal to the hood entry loss plus the velocity pressure in the channel. The hood entry loss represents the energy required to overcome the drops caused by air movement through or into the duct. The velocity pressure expresses the energy required to accelerate the air from a stationary state to a velocity in the channel, which is expressed as follows:

$$SP_h = h_e + VP_d \tag{4.25}$$

$$SP_h = (F_s)(VP_s) + (F_d)(VP_d) + VP_d \tag{4.26}$$

Where:

h_e = total hood inlet drop = $h_d + h_s$, wg”
h_s = loss at the opening or loss = $(F_s)(VP_s)$, wg”
h_d = entry loss in the duct = $(F_d)(VP_d)$, wg”
F_s = slot loss coefficient
F_d = entry duct loss coefficient
VP_s = velocity pressure in the opening or slot, wg”
VP_d = velocity pressure in the duct, wg”

A special case can occur when the velocity at the slot (or the entry velocity in other parts of the hood) is greater than that in the duct. In such a case, the velocity pressure acceleration used in determination of SP is higher than the velocity pressure in the opening or slot.

In Figure 4.16, the hood entry loss coefficients for several common hood types are shown.

4.8.1 Simple Hoods

A simple hood is depicted in Figure 4.17. If the velocity in the hood opening for a simple hood is less than 1000 fpm, h_s will be negligible and the loss will depend only on h_d. If the velocity in the hood opening is greater than 1000 fpm, both h_s and h_d must be considered. Velocities greater than 1000 fpm in the hood opening usually occur only in relatively small areas of the hood opening (0.25 to 0.5 ft^2) (Figure 4.17).

4.8.2 Compound Hoods

Figure 4.18 shows a hood with two entry losses, which is a simple slotted hood with a plenum chamber and a transition from the plenum chamber to the duct. The purpose of the plenum chamber is to provide a uniform velocity across the slot opening. Air enters the slot, in which case a sharp-edged orifice is used, and the energy at this point decreases due to the bottleneck. The air then continues to flow through the plenum chamber, with a greater portion of the velocity maintained in the slot. Since the airflow along the plenum chamber is similar to the blowing flow shown in Figure 4.7 (maintaining velocity in the plenum chamber due to the short length of the plenum

Hood type	Hood type	Hood type
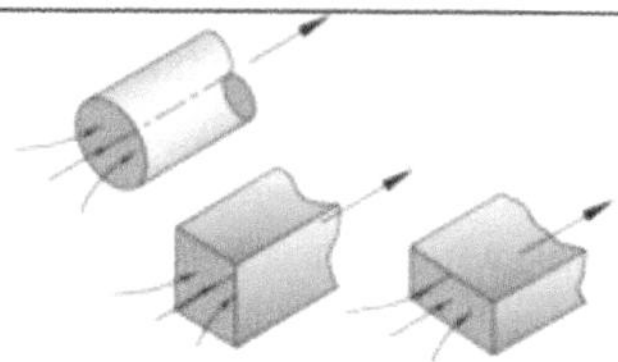	Plain opening end	0.93
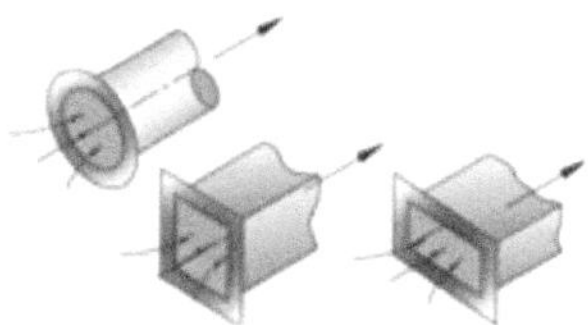	Flanged opening	0.6
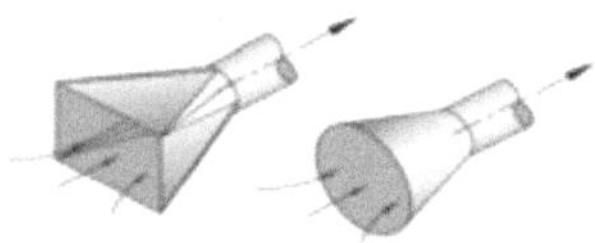	Funnel or conical hood	-
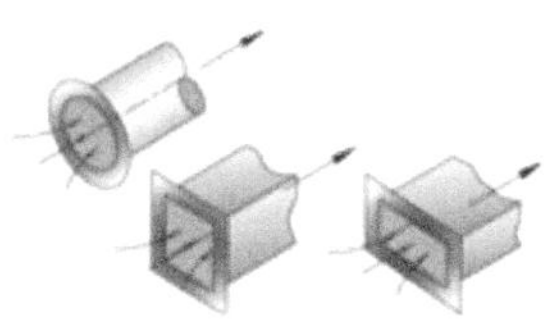	Thorn entry opening	0.04
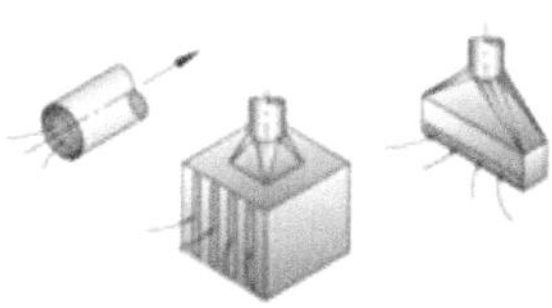	Orifice	-
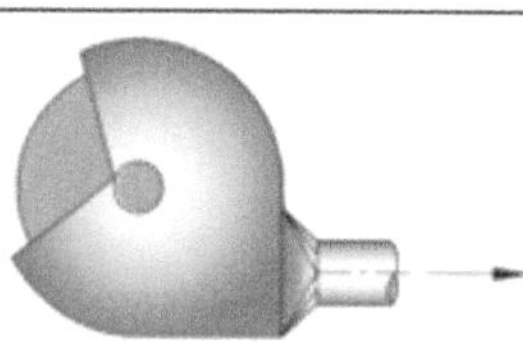	Typical grinding hood	0.65 (Direct duct connection) 0.40 (45 degree duct connection)

FIGURE 4.16 Hood loss coefficients.

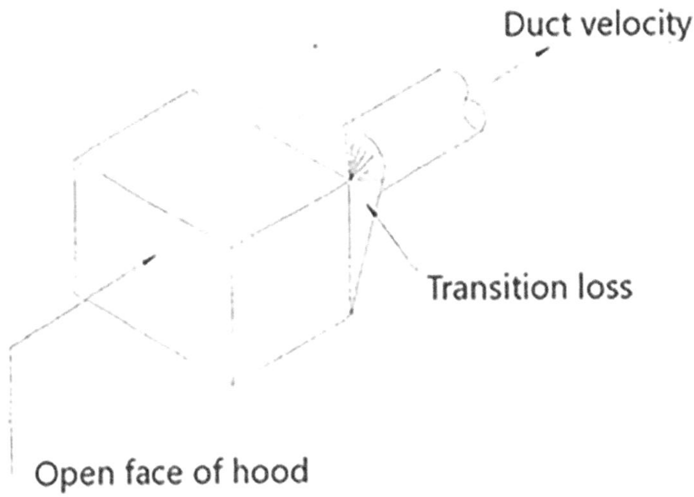

FIGURE 4.17 Simple hood.

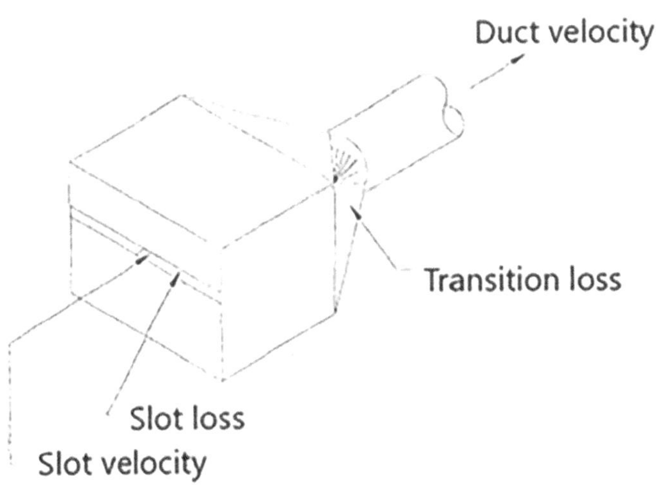

FIGURE 4.18 Composite hood.

chamber is a feature of most local exhaust hoods). In the case of very large hoods or enclosed exhaust rooms, velocity loss must be considered. Finally, the air enters the duct through a transition where the second most significant energy loss occurs. For this type of hood, both h_s and h_h must be considered. The static pressure drop of a slotted compound hood is computed by the following equation:

$$SP_h = h_s + h_d + VP_d \tag{4.27}$$

where, *hs* = gap drop, wg". The velocity in the channel is greater than that in the slot.

4.9 MINIMUM DUCT VELOCITY

The pressure velocity, VP_d, is used to determine hood losses. In the previous sections, the air velocity in the duct was measured immediately downstream of the hood to the duct interface. This velocity is determined by the type of material transferred in the duct.

For particle control systems, a minimum design velocity is required to prevent settling and duct blockage. On the other hand, excessively high velocities cause power loss and may cause rapid wear in the duct. The suggested minimum design velocities are higher than the experimental and theoretical values to avoid the following:

1. Blockage of one or more branches will reduce the total flow rate in the system and thus, the velocity will decrease, at least in some parts of the duct system
2. Damage to the ducts, e.g., denting, will increase the resistance and reduce the flow rate and velocity in the damaged part of the system
3. Leakage in the duct increases the flow rate and velocity in its downstream but reduces the airflow rate in its upstream and other parts of the system
4. Corrosion or wear of the fan wheel or slipping of the fan belt reduces the speed velocity flow rate
5. The velocity should be sufficient to remove or re-enter dust settled due to improper operation of the suction system

The designer should note that in some conditions, such as the presence of sticky materials, the presence of dust in humid conditions, effects of strong electrostatic, etc., speed alone cannot prevent blockages and other special cases.

Some typical velocities in the duct are represented in Table 4.4.

4.10 SPECIAL HOOD REQUIREMENTS

4.10.1 Ventilation for Highly Toxic and Radioactive Processes

Ventilation for highly toxic and radioactive processes requires knowledge of the hazards, the use of highly effective control methods, and adequate maintenance and repair, including supervision. This section covers only the general principles. For radioactive processes, the standards and regulations of the nuclear regulatory agencies should be referenced. The enclosed type of local exhaust hoods with the maximum possible coverage should be applied. Where the use of a complete or nearly complete enclosure is not possible, control velocities of 50 to 100% higher than the minimum available standards should be used. If the enclosure is not complete and the operator must be in an opening, for example, in front of a laboratory hood, the maximum control velocities should not exceed 125 fpm. Air velocities above this value in front of the operator will make vortices that may direct pollution from the hood into the operator's breathing zone. Air replacement should be done at a low velocity and in a direction that does not disrupt cross-ventilation at the hood opening.

TABLE 4.4
Minimum Duct Design Velocity Ranges

Nature of the Pollutant	Examples	Design Speed
Vapor, gas, and smoke	All fumes, gases and fumes	Any velocity desired (economically optimal velocity is usually 1000–2000 fpm)
Fumes	Welding	2000–2500
Very fine and light dust	Cotton fibers and stone powder	2500–3000
Powders and dry dust	Fine rubber dust, Bakelite molding powder dust (plastic made from formaldehyde and phenol), hemp fibers, cotton dust, shavings (light), soap dust, and leather shavings	3000–4000
Medium industrial dust	Dust from grinding, fiber sieving (dry), woolen hemp dust (vibration residue), coffee grounds, shoe dust, granite dust, silica dust, brick cutting, clay dust, casting (common), limestone dust, and packaging and weighing asbestos dust in textile industries	3500–4000
Heavy dusts	Sawdust (heavy and wet), metals grinding, foundry turning and vibrating parts, sandblasting dust, wooden molds, brass grinding, cast iron drilling dust, and lead dust	4000–4500
Heavy or wet dust	Lead dust with small chips, wet cement dust, Asbestos pieces from pipe cutting machines, fiber binders (adhesives), and lime dust	4500 and higher

4.10.2 Laboratory Activities

Glove boxes should be used for high-activity alpha or beta emitters and for highly toxic and biological materials. Suction type of airlocks should be used in glove boxes if those are connected directly to the room. Typically, the suction flow rate for each glove box is 35–50 cfm.

For laboratory work with low-activity radioactive materials, a laboratory fume hood may be acceptable. For such hoods, an average opening velocity of 80–100 fpm is recommended.

For new buildings, it is usually necessary to estimate the ventilation before detailed design and equipment specifications.

4.11 SUPPLY-EXHAUST VENTILATION

Supply-exhaust ventilation consists of a blower nozzle, an exhaust hood to gather the blown air, and jet pressure. Supply-exhaust ventilation is most used in open surface containers such as plating tanks but may effectively be applied in other situations.

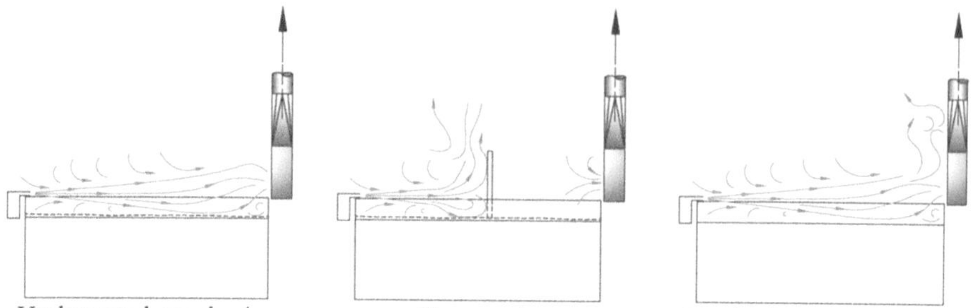

FIGURE 4.19 Supply - exhaust ventilation for immersion tanks.

The advantage of supply-exhaust ventilation is that a blower jet maintains velocity over long distances, 20–30 feet or more, whereas the velocity in front of a fume hood decreases rapidly with increasing distance from the hood. When proper use of blower jet, it picks up polluted air and transports this air to the exhaust hood over relatively long distance. Thus, this design provides adequate control. Otherwise, collection of pollutants becomes difficult or impossible (Figure 4.19).

Blower jet: Ambient air enters into the blower jet, and it is blown through the nozzle at high velocity. Then, this air is sucked in several times by the exhaust hood. The jet velocity decreases with distance from the nozzle. The entrainment ratio for a long and narrow slotted nozzle is calculated using the following equation:

$$\frac{Q_x}{Q_o} = 2\sqrt{\left(\frac{ax}{b_o}\right) + 0 \cdot 41} \tag{4.28}$$

The speed ratio is computed using the following equation:

$$\frac{V_x}{V_o} = \frac{12}{\sqrt{\left(\frac{ax}{b_o}\right) + 0 \cdot 41}} \tag{4.29}$$

where Q_o is the supply flow of the blower nozzle, Q_x is the jet flow rate at a distance x from the nozzle, V_o is the air outlet velocity from the blower nozzle, and V_x is the maximum velocity of the blower jet at distance x.

A is a nozzle characteristic coefficient (0.13 for slots and tubes, x is the distance from the nozzle, b_o is the width of the slot* [*If the nozzle is freely suspended (free plane jet), b_o is equal to half the total width of the slot. If the nozzle is at or very close to the plane surface (wall jet), b_o is equal to the full width of the slot. For perforated tubes, b_o is the width of the slot with an equivalent area]. Typical jet velocity profiles are shown in Figure 4.20.

Obstructions in the jet path should be minimized. Objects with small cross-sections, such as hangers, will cause serious problems. However, the use of flat surfaces in large objects should be avoided. If the jet is spread out at great distances from the nozzle, larger objects are acceptable if they are placed inside the jet.

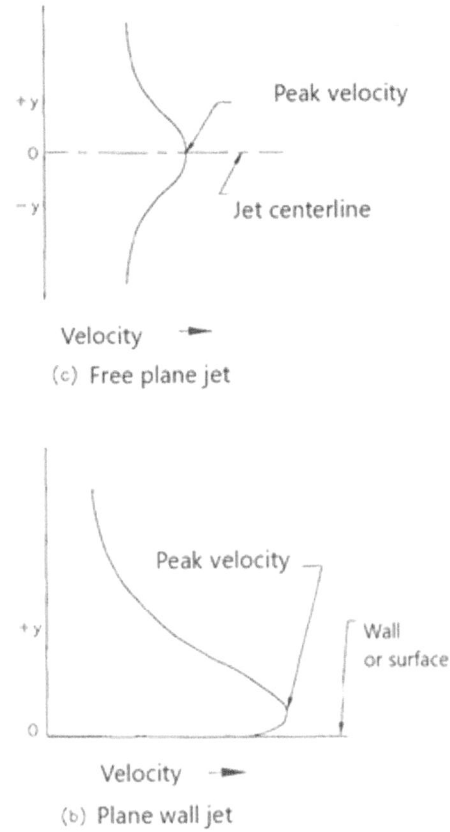

FIGURE 4.20 Jet velocity profile.

The nozzle may be a long and thin slot, a tube with holes, or a nozzle with a hole. The total area of the nozzle outlet should not exceed 50% of the nozzle cross-section of the plenum chamber to ensure uniform flow distribution. The slot width can be in the range of 0.25–0.125 inches for short supply lengths such as plating tanks (4–8 feet). The hole size should be 0.25 inches at distances with diameters of 3–8 inches. The nozzle movement size coefficient is proportional to the output flow of nozzle per foot of nozzle length multiplied by the velocity of nozzle output $\left(Q_o \times V_o\right)$, which should be sufficient to achieve an effective jet, but it should not be so high that the exhaust hood receives too much pressure. The range of $Q_o V_o$ should be approximately 50,000–75,000 per foot of nozzle length for short distances of 4–8 feet.

4.12 EXHAUST HOOD

The exhaust hood gathers and expels the blower jet flow. The same design considerations for flow distribution, hood entry losses, etc., apply to the exhaust hood. The hood exhaust flow should be approximately 1.5–2 times the blower flow directed to the hood. If design criteria specifying the exhaust flow rate are not available, Equation 4.28 can be used.

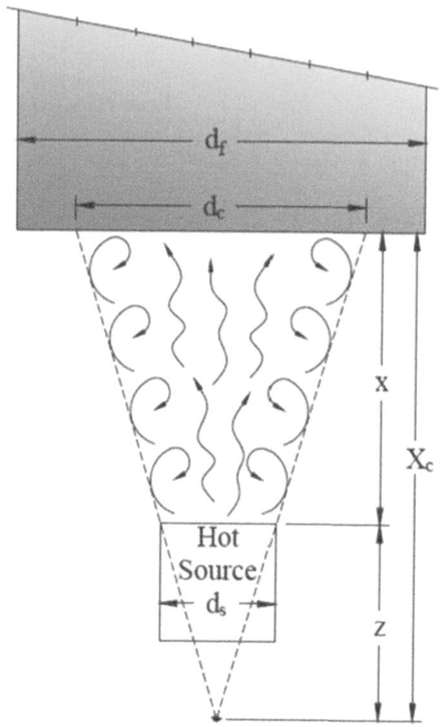

FIGURE 4.21 Dimensions used in the design of long canopy hoods for hot sources.

The hood opening height should be equal to jet width. However, smaller opening heights are acceptable if the hood flow rate is 1.5–2 times the jet flow.

The exhaust jet velocity $\left(V_x\right)$ at the hood opening should be 150–200 fpm (Figure 4.21).

4.13 HOT PROCESSES

Hood design for hot processes requires different considerations than that for cold processes. When significant amounts of heat are transferred to the air above and around the process by conduction and convection, a thermal suction is created that causes the upward airflow at air velocities up to 400 fpm. This thermal suction must be considered in the hood design and exhaust flow rate.

Figure 4.22 shows an ineffective pollution hood for a hot process. Figure 4.23 shows a suitable pollution hood for hot sources.

Long circular canopy hoods: As the hot air rises, it is mixed with the surrounding air in a turbulent manner. This leads to an increase in the diameter of the air column and the volumetric flow rate. The column diameter (Figure 4.21) can be calculated by the following equation:

$$D_c = 0.5\,X_c^{\,0.88} \tag{4.30}$$

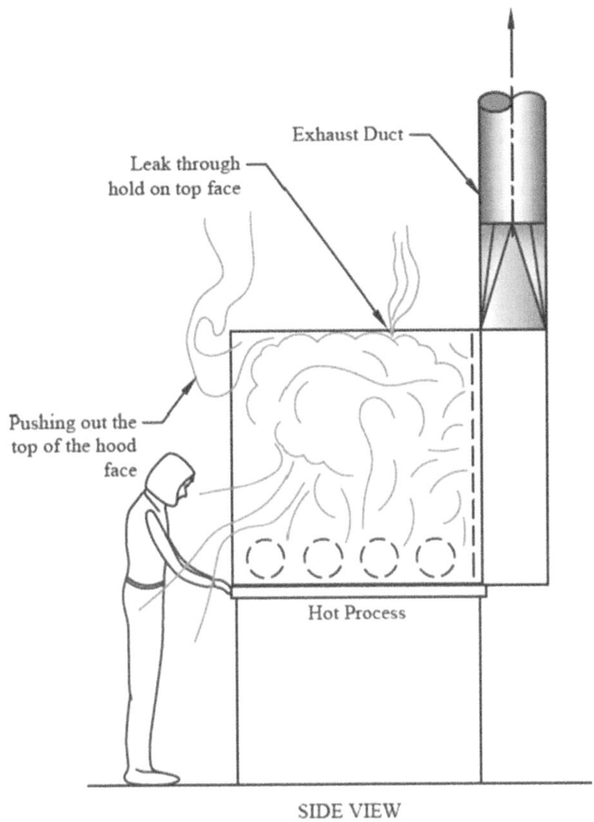

FIGURE 4.22 Ineffective pollution hood in hot process.

where D_c is the diameter of the column at hood opening (ft), X_c is $y+z$ distance from the hypothetical point source to hood opening (ft), y is distance from the process surface to hood opening (ft), z is the distance from the process surface to hypothetical point source (ft).

z can be calculated from the following equation:

$$z=\left(2\Delta_s\right)^{1.138} \tag{4.31}$$

where Δ_s is the diameter of hot source (ft).

The velocity of the hot air column can be computed by the following equation:

$$Vf=8\left(As\right)0.33\frac{\left(\Delta t\right)^{0.42}}{X_c^{0.25}} \tag{4.32}$$

where V_f is hot air column velocity at hood opening (fpm), A_s is hot source area (ft^2), Δt is temperature difference between hot source and ambient temperature (F°), and $X_c=y+z$ is distance from the hypothetical point source to hood opening (ft).

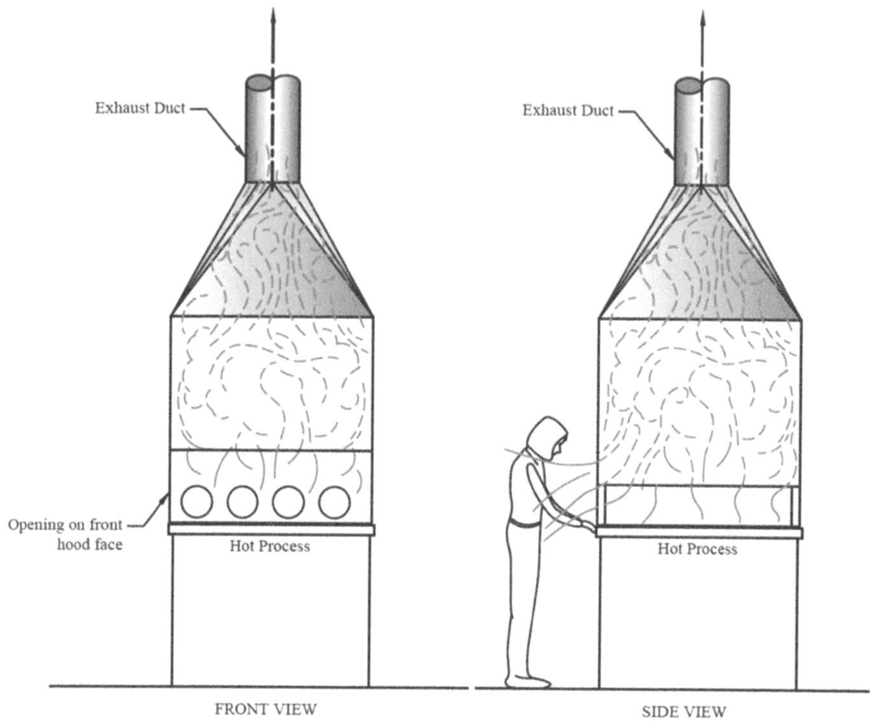

FIGURE 4.23 Enclosure hood designed for hot source.

The diameter of the hood opening must be larger than the diameter of the hot air column to ensure complete extraction.

The hood diameter can be calculated by the following equation:

$$D_f = D_c = 0.8\text{y} \tag{4.33}$$

where, D_f = Diameter of hood opening, feet

The total airflow rate in the hood will be:

$$Q_t = V_f\, A_c + V_r \left(A_f - A_c\right) \tag{4.34}$$

where, Q_t is total entry volume to the hood (cfm), V_f is hot air column velocity at hood opening (fpm), A_c is hot air column area at hood opening (ft^2), V_r is air velocity required to pass other hood surfaces (fpm), and A_c is total hood opening area (ft^2).

4.13.1 Long Rectangular Canopy Hoods

Hot air columns from non-circular sources may be better controlled by a rectangular canopy hood. Airflow calculations in the hood are made in the same way as for circular hoods except the dimensions of the hot air column in the hood (and the dimensions of the hood). It is determined by the length and width of the source.

Equations 4.30, 4.31, and 4.33 are used separately to estimate the length and width of the hot air column and those of the hood. The remaining values are calculated in the same way as for circular hoods.

4.13.2 Short Canopy Hoods

If the distance between the hood and the hot source does not exceed approximately the source diameter or 3 feet, the hood may be considered a short canopy hood. In such a situation, the diameter or cross-sectional area of the hot air column will be approximately equal to the diameter and cross-sectional area of the source. The diameter or lateral dimensions (length and width) of the hood need only be 1 foot greater than the source. The total flow rate for a circular short canopy hood will be:

$$Q_t = 4.7\left(D_f\right)^{2.33}\left(\Delta t\right)^{0.41} \tag{4.35}$$

where Q_t is total airflow through the hood (cfm), D_f is hood diameter (ft), and Δt is the temperature difference between the heat source and the ambient temperature (F°).

The total flow rate for a short rectangular hood will be:

$$\frac{Q_t}{L} = 6.2\,b^{1.33}\Delta t^{0.42} \tag{4.36}$$

where Q_t is total airflow through the hood (cfm), L is length of the rectangular hood (ft), b is width of the rectangular hood (ft), and Δt is temperature difference between the hot source and the ambient temperature (F°).

Table 4.5 presents the advantages and disadvantages of dilution and local ventilation.

TABLE 4.5
Advantages and Disadvantages of Dilution and Local Ventilation

Dilution Ventilation		Local Ventilation	
Advantages	**Disadvantages**	**Advantages**	**Disadvantages**
Typically, low installation and equipment costs.	It cannot be used for highly toxic chemicals.	It is a single option for highly toxic airborne chemicals.	It requires regular cleaning, inspection, repairment, and maintenance.
Less maintenance and repairment requirements.	It is not effective for metal dust or fumes or large amounts of gases or vapors.	It can control all types of pollutants including dust and metal fumes.	-
Effective control for combustible or flammable gases or vapors.	It requires large amounts of cooled or heated replacement air.	It requires less replacement air since less air amount is sucked.	-
Best ventilation for limited dispersion or moving sources of pollution.	It is not effective for controlling irregular emissions and rapid flows of gases and vapors.	It requires lower energy costs since there is less replacement air to heat or cool.	-

5 Hood Design for Specific Activities

A. H. Khoshakhlagh, H. M. Moda, S. Yazdanirad, and A. Gruszecka-Kosowska

5.1 OBJECTIVES OF THE LESSON

Upon completing this lesson, readers should be able to:

- State the general principles of hood design for specific activities
- Identify the hood types for different specific activities
- State the hood requirements for different specific activities
- Design different hoods for various specific activities

5.2 HOOD DESIGN FOR SPECIFIC ACTIVITIES

5.2.1 Plating

Many metalworking and coating operations are associated with the plating industry. Typically, these processes are carried out in open-topped tanks, and the toxic air pollutants released by these processes must be controlled by local exhaust ventilation methods. In the early 20th century, open-topped tanks were often placed on the exterior wall of the plant and were suctioned by a wall fan. Open-topped canopy hoods were favored in early ventilation designs; however, the scarcity of these hoods limited their use to modern industries. Two types of hoods, enclosure hoods and side-exhaust hoods, provide effective control of pollution emitted from open-surface tanks. The design of both hood types is presented in the following section.

5.2.1.1 Hood Design

Enclosure hoods are preferred for many industrial operations where efficient control of airborne toxic pollutants is possible with minimal airflow. The main design problem is that the hood may restrict access to the operation. An enclosure hood with access through a slot and open end of a conveyor (Figure 5.1a) has found widespread use in automated plating operations. In a shop floor operation where parts are manually transferred from one tank to another, an enclosure hood with an open front opening may be acceptable (Figure 5.1b). The open area of an enclosure hood should be precisely specified at the design stage so that the airflow can be calculated production management should provide optimum conditions for the airflow enclosure in the operation. Typically, enclosure hood systems are properly designed and installed,

 DOI: 10.1201/9781003501091-5

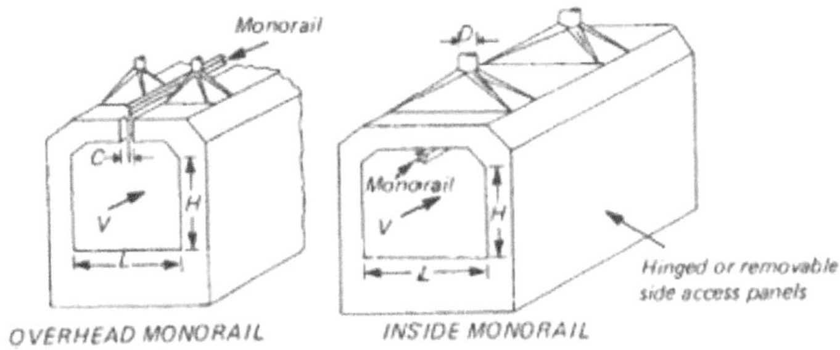

Q is exhaust volume (cfm) equal to AV, H is height of rectangular opening (ft), L is length of opening (ft), A is area of openings in enclosure equal to area of monorail slot + 2HL (ft^2), V is minimum velocity at opening (fpm), velocity in duct equal to minimum 1500 fpm (7.62 m/s), and entry loss equal to entry loss coefficient for conical hood × pressure velocity in duct. L, C, and H must be large enough to allow equipment to pass easily.

(a)

FIGURE 5.1a Enclosure hood in a plating shop with two open ends for material transition by conveyor.

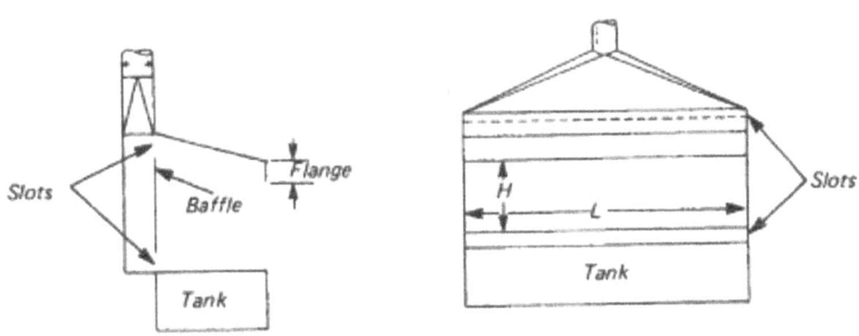

Q is exhaust volume (cfm) equal to AV, H is height of rectangular opening (ft), L is height of opening (ft), A is area of openings in enclosure equal to area of monorail slot + 2HL (ft^2), V is minimum velocity at opening (fpm), duct velocity is minimum 1500 fpm (7.62 m/s), slot velocity is minimum 2000 fpm (10.16 m/s), and entry loss is 1.78 velocity pressure at slot + entry loss coefficient for conical hood × duct velocity pressure.

(b)

FIGURE 5.1b Enclosure hood with one opening acceptable for manual transfer of parts.

but when the workers remove part of the enclosure to access the transfer tank, their performance is severely reduced.

The three types of exhaust hoods shown in Figures 5.2a, 5.2b, and 5.3 are side-exhaust hoods, which are designed to create a horizontal movement of air along the tank surface for transition of airborne pollutants to the hood. The most popular type of enclosure hood has access to open-surface tanks in plating shops. The operator has access to the tank contents from three or four sides, which allows easy transfer of parts from one tank to another. An important feature of hoods with an upward plenum chamber (Figure 5.2a and 5.2b) is that the fishtail taper from the hood to the duct provides a barrier to protect the tank from room suction that can interfere with the hood's capture efficiency. At the edge or perimeter of the slotted exhaust hood (Figure 5.3), external barriers or tanks should be installed against a wall to provide equal protection from interfering suction.

5.2.1.2 Airflow

The initial layout for a plating shop presented to the ventilation designer includes the tank size, approximate place, tank contents, operating temperature, and structure materials. As mentioned in the previous section, if the facility is automated, semi-enclosed hoods (Figures 5.1a and 5.1b) may be selected by the designer to provide effective control of airborne toxic pollutants at a minimum airflow. In shop operations with need to direct access to the tank, one of the hoods shown in Figures 5.2 or 5.3 may be suitable. The width to length ratio of the tank has a significant effect on the minimum airflow required for effective control of airborne toxic pollutants. If the tank is less than 3 feet wide, a side-exhaust hood with suction on one side of the tank will be adequate. A wider tank may require side-slotted hoods on both sides of the tank, centerline suction, or a supply-exhaust system.

The choice of hood type will have a significant effect on the airflow required to control airborne toxic pollutants released from the tank. A rational approach for calculation of the airflow required for operations of open-surface tanks was developed in the 1930s. Since the 1940s, the extraction airflow for plating tanks has been based

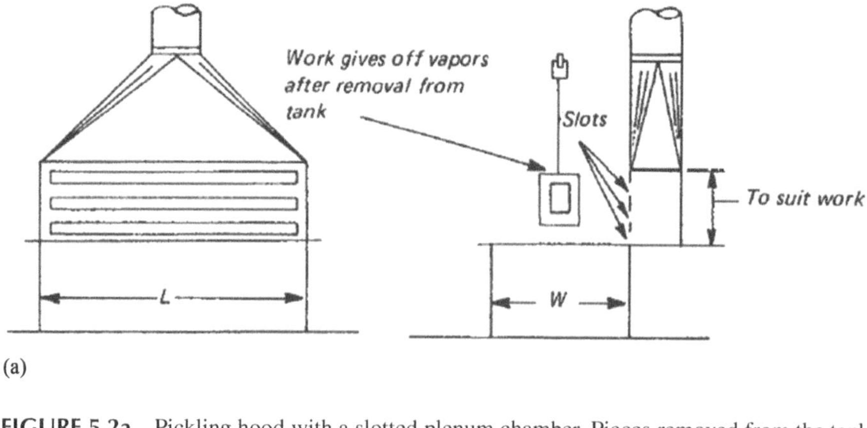

(a)

FIGURE 5.2a Pickling hood with a slotted plenum chamber. Pieces removed from the tank can be placed in front of this hood to gather toxic air pollutants.

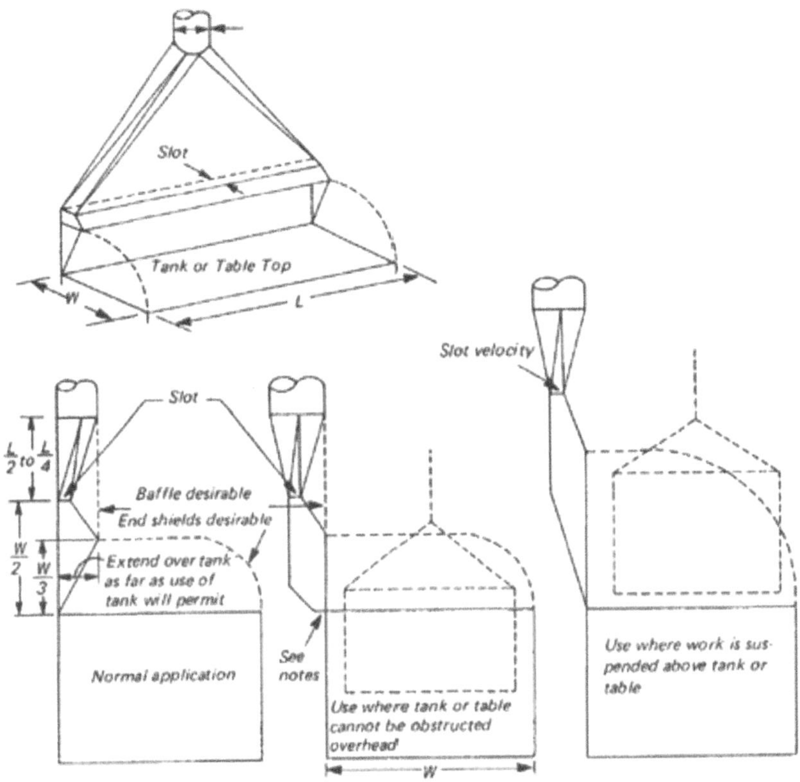

Q is exhaust volume (cfm or m^3/s) equal to qLW, W is tank width (ft or meters), L is tank length (ft), q is tank surface to maintain minimum capture velocity (cfm/ ft^2 or $m^3/s/m^2$), duct velocity is minimum 1500 fpm (7.62 m/s), slot velocity is minimum 2000 fpm (10.16 m/s), and entry loss is 1.78 slot velocity pressure + entry loss coefficient for cone hood × duct velocity pressure. W/L not to exceed 2 to 1 and maximum length of transition piece is 4 ft (1.22 m). Use greater than one for long hoods. Hood may be extended up to 6 in (0.15 m) so that tubes are placed over the tank edge. The gap should be covered with baffles.

(b)

FIGURE 5.2b Fishtail hood in tanks with a width to length ratio of less than 0.5.

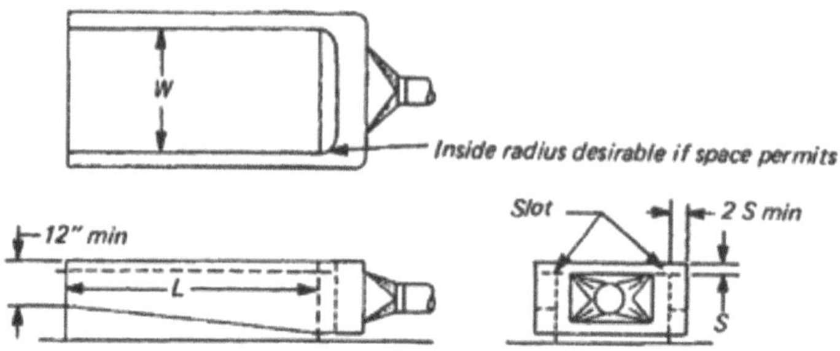

Q is exhaust volume (cfm or m^3/s) equals to qLW, W is tank width (ft or meters), L is tank length (ft), q is tank surface area to maintain minimum capture velocity (cfm/ft^2 or $m^3/s/m^2$), duct velocity is minimum 1500 fpm (7.62 m/s), slot velocity is minimum 2000 fpm (10.16 m/s), and entry loss is 1.78 slot velocity pressure + entry loss coefficient for cone hood × duct velocity pressure.

FIGURE 5.3 Side exhaust hood with one end connection in duct.

on a constant airflow per unit tank area (cfm/ft^2) ($m^3/s/m^2$). During this period, the New York City Department of Industrial Hygiene specified 120 cfm/ft^2 (0.06 $m^3/s/m^2$) for plating tanks controlling toxic chemicals and operating at operating temperatures of 160°F (71°C).

The first step toward a design approach consistent with the American National Standards Institute (ANSI) Z9.1 Emissions Standard was taken in 1957. This standard proposes that the airflow per unit tank area is determined based on a two-part code to determine the hazard potential of the tank contents and the likelihood of the contents' emission into the work environment. The first part of the code is the hazard potential classification (A through D); This is a measure based on the toxicity and fire and explosion characteristics of the bath contents (Table 5.1). Classifications are selected based on accepted standards for the most toxic components of the bath and the flash point of the bath. Typically, the greater of the two hazard potential groups are used in design.

Second part of the code (Groups 1 through 4) describes the potential for the emission of toxic air pollutants into the workplace based on the operating temperature of the tank, the rate of evaporation, and the rate of gassing (Table 5.7). The rate of gassing in metalworking water baths is similar to that in alkaline cleaning and electroplating. In a metalworking water bath, not all the energy is spent on the metal transportation from the bath to the part. Some of the energy, depending on the efficiency of the bath, is used to separate the water from the hydrogen gas released at the cathode and the oxygen at the anode. The gas bubbles produced in this process increase and appear on the surface of the bath, which produces the mist of the bath contents. These mists may be inhaled by the worker. The volatile bath contents may also be emitted as the bath heats up. Finally, air agitation is used to improve product quality, which may produce dense mist.

TABLE 5.1
Determination of the Potential Hazard Group

Risk Potential	Health Standards		Flash Point
	Gas or Vapor (ppm)	Mist (mg/m³)	
A	0–10	0.1–0	-
B	11–100	1–11.0	Below (40°C) 100°F
C	101–500	1.1–10	(40–94°C) 100–200°F
D	Above 500	Above 10	Above (95°C) 200°F

TABLE 5.2
Vapor, Gas, and Mist Emission Rate Classification

Amount	Liquid Temperature [°F (°C)]	Degrees Below Boiling Point [°F (°C)]	Relative Evaporation[a]	Gassing
1	Above 200 (95)	0–20 (0–10)	Fast	High
2	150–200 (65–95)	21–50 (11–25)	Medium	Medium
3	94–149 (35–65)	51–100 (25–50)	Slow	Low
4	Below 94 (35)	Above 100 (50)	Zero	Zero

[a] The relative evaporation rate under 100% evaporation conditions are as follows: fast, 0–3 h; medium, 3–12 h; slow, 12–50 h; zero, more than 50 h.

Using the method presented in Table 5.2, the contents are released into the workroom, which is assigned a numerical group to the bath based on the tank capacity.

The combination of pollutant emission rate and hazard potential classification is applied to select the appropriate capture design velocity for a specific activity and tank geometry (Table 5.3). The exhaust airflow is calculated based on the required design rate, the hood type, and the tank width-to-length ratio (Table 5.4), where this ratio of width to length is the distance that the hood must suck the air.

The ANSI approach described in the previous section uses a 16-step tank classification scale ranging from A1 (highest capture velocity) to D4 (lowest capture velocity). Although this procedure is a step forward, it requires information on open-surface tank processes, and therefore, many designers apply the simpler former criteria of airflow per tank area unit.

In 1953, the New York State Department of Industrial Hygiene published a set of tables supporting the use of the ANSI method. These tables provide information on typical plating processes including the component released from the bath, the physical form of the pollutant, the rate of gassing, and the operating temperature of the tank. The designer can obtain this data and use Tables 5.1 and 5.2 to make a hazard-emission classification for each open-air tank separately and select an appropriate airflow from this classification.

TABLE 5.3
Minimum Capture Velocity for Tank

Group	Velocity at the Opening of the Enclosure Hood [fpm, (m/s)]		Side Exhaust Capture Velocity [fpm, (m/s)]
	One side open	Two sides open	
A-1 and A-2	100 (0.5)	150 (0.75)	150 (0.75)
A-3, B-1, B-2, and C-1	75 (0.38)	100 (0.5)	100 (0.5)
B-3, C-2, and D-1[a]	65 (0.33)	90 (0.45)	75 (0.38)
A-4, C-3, and D-2[a]	50 (0.25)	75 (0.38)	50 (0.25)
B-4, C-4, D-3[a], and D-4	General ventilation required for the room		

[a] Where complete hot water control is desired, the design is based on the next highest group.

TABLE 5.4
Minimum Exhaust Airflow per Unit Tank Area for Side Exhaust Hoods

Minimum Required Capture Velocity [fpm, (m/s)] (from Table 5.3)	Cubic Feet per Minute per Square Foot (Cubic Meters per Second per Square Meter) to Maintain the Minimum Required Capture Velocity in the Ratios of Tank Width (W)/Tank Length (L)				
	0–0.09	0.1–0.24	0.25–0.49	0.5–0.99	2–1
Hood along one or both parallel sides of the tank when the hood is against a wall or baffle; also, for a manifold along the centerline of the tank					
50 (0.25)	50 (0.25)	60 (0.3)	75 (0.38)	90 (0.45)	100 (0.5)
75 (0.38)	75 (0.38)	90 (0.45)	110 (0.55)	130 (0.65)	150 (0.75)
100 (0.5)	100 (0.5)	125 (0.63)	150 (0.75)	175 (0.88)	200 (1)
150 (0.75)[a]	150 (0.75)	190 (0.95)	225 (1.10)	250 (1.27)	250 (1.27)
Hood along one or both parallel sides of the self-standing tank when the hood is not against a wall or baffle					
50 (0.25)	75 (0.38)	90 (0.45)	100 (0.5)	110 (0.55)	125 (0.63)
75 (0.38)	110 (0.55)	130 (0.65)	150 (0.75)	170 (0.85)	190 (0.95)
100 (0.5)	150 (0.75)	175 (0.88)	200 (1)	225 (20/1)	150 (0.75)
150 (0.75)[a]	225 (1.10)	250 (1.27)	250 (1.27)	250 (27/1)	250 (1.27)

[a] If a minimum exhaust airflow of 250 (1.27) is specified, the capture velocity of 150 (0.75) will not be achieved, but it is considered adequate for control.

Once the appropriate hazard-emission classification is obtained for a given tank process, the designer determines the critical design velocity. For enclosure hoods (Figures 5.1a and 5.1b), the design velocity is the speed at the hood opening. For side-exhaust hoods (Figures 5.2 and 5.3), the design velocity is the capture velocity derived from Table 5.3. The capture velocity is also considered in the design velocity. It is strongly recommended that canopy hoods not be used to control ventilation of open-surface tanks. Canopy hoods should not be applied for tanks containing highly toxic materials in places where there are air drafts, or in a position that the worker must bend over the tank or process.

Once the critical design velocity is determined, Table 5.4 is used to select the appropriate minimum exhaust airflow. The minimum exhaust rate depends on the hood type selected for the process and the width to length ratio of the tank. A review of Table 5.4 shows that using the small tank, keeping the low width to length ratio, and using the hoods with integral baffles or installing baffles or guards to prevent drafts can reduce the airflow rate. The supply-exhaust concept is an effective approach to suck air from large surface tanks that are difficult to operate with conventional side-mounted hoods. In a supply-exhaust system, a plenum chamber and a pressurized air supply placed at the edge of a one-sided tank blows a uniform air jet into the tank and transfers the pollution to a hood located on the other side of the tank. Although this design can be effective, some studies have reported that some poor installations enable exposure to pollutants. Fortunately, since the early 1990s, the design principles of supply-exhaust systems have been carefully examined, and the results of laboratory studies combined with mathematical modeling of the flow pattern have provided insight into optimal design.

5.2.2 Spray Painting

Spray painting by air spray exposes workers to vapors of volatile compounds, and the paint mists with adhesives, fillers, and pigments. Engineering control is typically achieved with ventilated spray booths, which are classified as simple partial enclosures. These booths are frequently used to control other industrial processes, including chemical weighing and mixing, solvent cleaning, and coating operations. Because of their wide range, the designer should know the types of spray paint hoods and their applications. The discussion in this section is limited to ventilation control in liquid paint spray painting.

5.2.2.1 Hood Design

Many companies in the United States manufacture spray paint hoods, ranging from small benchtop units for the jewelry industry to room-sized booths for automotive and locomotive painting. Each company has a standard product line of hoods with numerous design options. Hoods are designed to provide a side flow of clean air passing through the operator's breathing area with a uniform velocity distribution along the hood surface. So, it reduces exposure of workers to toxic air pollutants. Downdraft hoods (downflow) are used for painting objects with large sizes and irregular shapes, such as pieces in foundries, automobiles, trucks, and locomotives.

A typical spray booth is a heavy sheet metal enclosure with an open opening (Figure 5.4). The side and roof panels may have openings for pieces to enter the spray section using a conveyor belt. This feature is especially important in automated painting operations. The hood size depends on the size of the pieces to be painted. It is important that the area of the hood opening is large enough so that the pieces do not block the airflow through the hood section, which may result in poor capture efficiency. These pieces must require a holding device or a round table for full spray coverage.

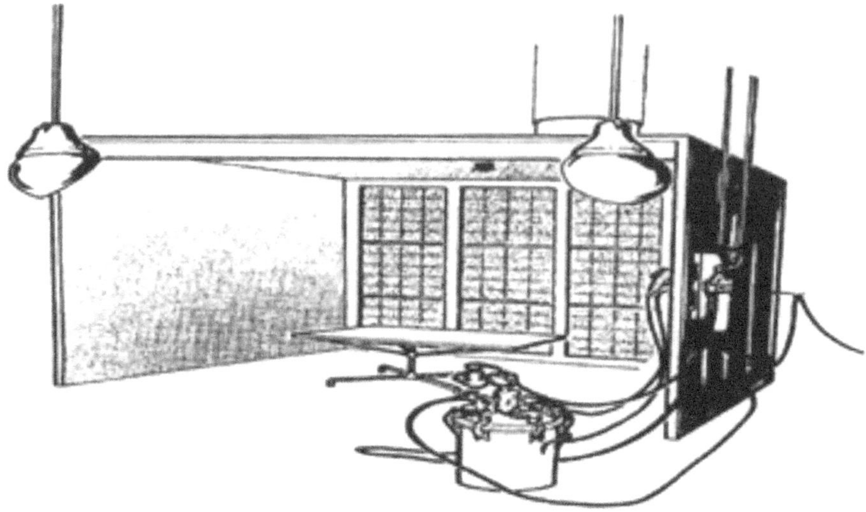

FIGURE 5.4 Main features of a typical spray booth.

The depth of the booth is critical to good performance, as it helps to ensure a uniform velocity profile at the hood opening and it results in effective capture of airborne contaminants. On the return (face) from flat surfaces sprayed in a shallow booth, pollutants may not be captured—a deep booth helps to improve such pollution (Figure 5.5).

Suspended sheet metal baffles behind the hood are the simplest method of improving air distribution (Figure 5.6). Baffles are easy to clean and install, and those have minimum resistance to airflow. The required numbers and spacing of baffles can be determined as a function of the booth cross-sectional area (Table 5.5). In addition to the provision of a uniform velocity profile in the hood, the baffles capture a small portion of the coarse paint spray mist.

Several disposable particulate filters are installed at the rear of the hood, which forms a uniform chamber at this part and helps to create a uniform velocity across the hood surface. The filter is usually a coarse fiber media with a collection efficiency of less than 50% against paint overspray. As these filters become coated with paint, their resistance increases, and airflow is reduced. Performance should be monitored continuously, and filters should be replaced periodically. Filter life may be extended by installation of baffles in front of the filters to minimize the collection of coarse paint mist.

The most effective method for cleaning paint mist in air is a water wash booth, while it also provides a good velocity profile in the hood (Figure 5.7). These booth designs vary but typically include a water wash curtain, countercurrent sprays, and impingement plates to remove the water mist. Studies have shown that these systems are relatively effective against paint mist; however, simple water wash booths are not effective for removing organic vapors.

The efficiency of a co-current wet scrubber with short-throttle used in a downdraft paint spray booth for alkaline coating operations with high solids and paint mist produced by air-spray guns and rotary diffusers was evaluated in a study. The alkaline

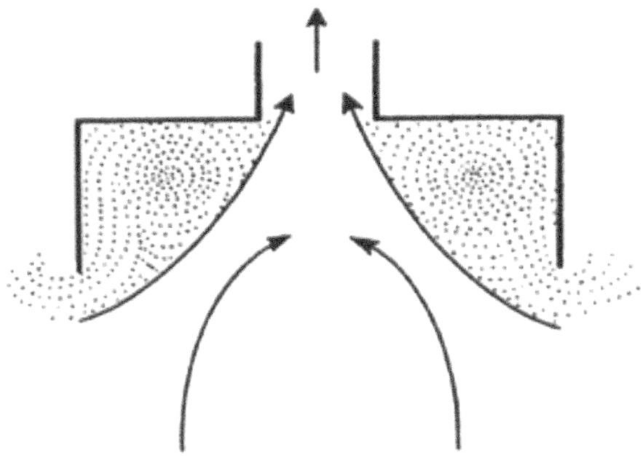

FIGURE 5.5 Hood depth is an important design parameter. As shown in this view, vortices form with leakage in a shallow booth.

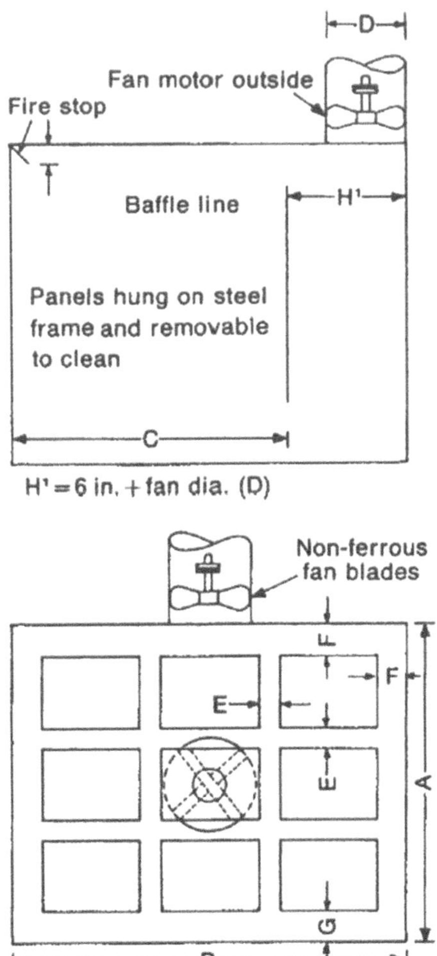

FIGURE 5.6 Hood baffles for air distribution.

TABLE 5.5
Spray Booth Baffle Design

Chamber Dimensions, [ft (m)] A × B[a]	Number of Panels, D	Spacing [inches (cm)] Between Panels, E	Top and Side, F	Bottom, G
3 × 3 (0.9 × 0.9)	1			
$4 \times 6\frac{1}{2}$ (1.2 × 2)	3	4 (10)	6 (15)	12 (30)
$6 \times 6\frac{1}{2}$ (1.8 × 2)	4	4 (10)	6 (15)	12 (30)
$8 \times 6\frac{1}{2}$ (2.4 × 2)	6	4 (10)	10 (25)	12 (30)
$10 \times 6\frac{1}{2}$ (3 × 2)	9	6 (15)	10 (25)	18 (46)
12 × 9 (2.7 × 3.7)	9	8 (20)	12 (30)	18 (46)
14 × 12 (4.3 × 3.7)	12	8 (20)	18 (46)	24 (60)

[a] For cases that chamber dimensions have not been provided, use the closest size in the table for the number and spacing of baffles.

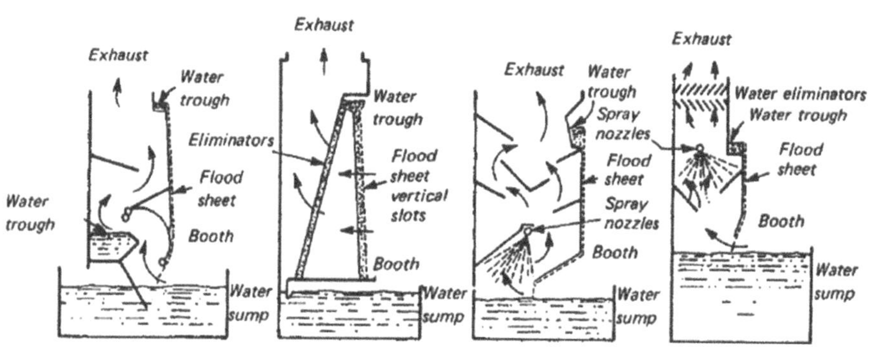

FIGURE 5.7 Typical spray booth designs with water wash.

coating paint mist produced by air spraying had a mass median aerodynamic diameter (MMAD) ranging from 4.7 to 6.6 μm, which depends on the gun pressure. The MMAD of the coating paint produced by the rotary sprayer operating at different rotational speeds ranged from 20 to 35 μm. The collection efficiency of this single-stage wet scrubber was reported to be greater than 90% for particles with an MMAD greater than 2 μm. The researchers emphasized the importance of optimizing spray operating conditions to minimize the production of mist with diameter smaller than 2 μm.

5.2.2.2 Airflow

The selection of spray booth opening speeds and airflows varies based on the used reference. A comparison between the ventilation manual and ANSI/OSHA

TABLE 5.6
Ventilation Requirements for Spray Booths in cfm/ft² (m³/s/m²)

	ACGIH		ANSI OSHA	
Booth Type	**Air Spray**	**Airless**	**Air Spray**	**Airless/ Electrostatic**
Desktop Spray Booth	150[a]–200[b] (0.76–1.02)	100[a]–125[b] (0.51–0.64)	150[c]–200[d] (0.76–1.02)	100 (0.51)
Large Spray Booth	100[e] (0.51)	60 (0.31)	100[c]–150[d] (0.51–0.75)	-
Spacious	100–150 (0.51–0.75)	60–100 (0.31–0.51)	100[c]–150[d] (0.51–0.75)	-
Operator Outside Booth	100 (0.51)	60 (0.31)	-	-

[a] Chamber area less than 24 ft² (0.37 m²).
[b] Chamber area greater than 24 ft² (0.37 m²).
[c] Cross-flow draft up to 50 fpm (0.26 m/s).
[d] Cross-flow draft up to 100 fpm (0.51 m/s).
[e] 75 fpm (0.38 m/s) for very large and deep chambers (in these conditions, the operator may need a respirator).

requirements is represented in Table 5.6. These criteria are not sufficient to provide protection against paint containing heavy metals, such as lead and cadmium or catalytic systems used in epoxy or urethane paints. When the proper opening speed is selected, the open area consisting of the chamber opening and any openings for conveyor access must be calculated. Proper installation and operation are essential to achieve acceptable performance for a spray booth. The following exposure control measures are important:

Minimize the distance between the spray booth nozzle and the piece.
Consider spray conditions to produce particles larger than 12 μm in diameter to minimize overspray.
Use the pull arms on the spray gun to access the empty spaces of the workpiece.
Use robotic methods for operation to eliminate the presence of workers in the booth, if possible.
Stand next to the sprayed piece.

The replacement air should be supplied to the spray booth in sufficient quantities to fulfill the suction requirements. The air supply should usually be provided by end-point filtration to minimize surface defects in the paint. The place and type of diffuser for the replacement air are important considerations in booth operation and product quality control. Although it is generally a good practice to replace the air at some distance from the hood, this is not the case with paint spray booths. The distance increases the potential for particle pickup and contamination of the painted surface. Use of the hood in corners often results in asymmetric flow and poor velocity profiles at the opening. Spray booths should not be placed close to transport doors where there may be disruptive drafts.

The correct selection of the fan is important. The fan is part of the paint spray booth assembly. Such fans are usually selected based on a short duct without elbows or joints, and direct discharge to the outside. In practice, there are rarely such conditions that the booth-fan system is compatible, and proper airflow is achieved at the time of the installation completion. A fan delivering air with flow rate of 5000 cfm (2.36 m^3/s) at a pressure loss of 1/8 in H_2O (31.14 Pa), if there is additional resistance, may only provide airflow rate of 3500 cfm (1.65 m^3/s) at a pressure loss of 1 in H_2O (249.1 Pa). It is best to purchase and size the fan based on the total system loss calculations.

Although spray hood design is important, improper work practices can affect the performance of any hood, which results in worker exposure to airborne pollutants. For example, when spraying small pieces in a shop, the operator typically places the freshly sprayed pieces on a rack behind him. Air moving into the hood first passes the rack where highly volatile solvents are evaporating from the pieces. This air with solvents moves into the operator's breathing zone, which leads to significant exposure to pollutants. Also, simple changes in the position of the worker in front of the spray hood can have a significant impact on performance of the hood and potential worker exposure. In a study with a tracer gas in a large hood, researchers reported that moving the operators from "back stream" to "side stream" significantly reduced worker exposure.

5.2.3 Granular Materials Processing and Handling

Mining the mineral materials and producing and distributing the granular chemicals are major potential sources of airborne toxic emissions in the workplace from many handling and conveying operations. The primary feed and storage operations of Portland cement (Figure 5.8) illustrate the complexity of the problem, with dozens of potential dust emission points at each stage of the process. The discussion in this chapter will focus on an alternative procedure for calculating exhaust flow rates at material handling points.

The Ventilation Manual, in a series of design pages, suggests exhaust flow rates for bucket elevators, conveyors, bins, hoppers, and loading stations based on the hood opening area, belt width and speed, bin volume, and plate area. Based on the experience of researchers, the recommended airflows based on these design pages may be inadequate for effective dust control. The concept of induced air provides a principle for understanding the mechanisms of dust generation and emission in material handling activities, and it introduces a rational principle for the design of local exhaust for such facilities.

Early attempts for dust control at granular material handling points were based on the use of enclosure hoods with a selected velocity at the opening for the entire open surface area. Implementation of this approach requires the New York State Occupational Health Department's initial criteria, 150–200 cfm/ft^2 (0.76–1.02 $m^3/s/m^2$), at all hood openings. In many installations, these recommendations do not provide adequate control. Packaging system manufacturers have suggested 400–500 cfm/ft^2 (2.54–2.03 $m^3/s/m^2$) at all hood openings.

Some studies report that systems installed based on the approach of velocity at opening do not provide effective control because this design approach does not

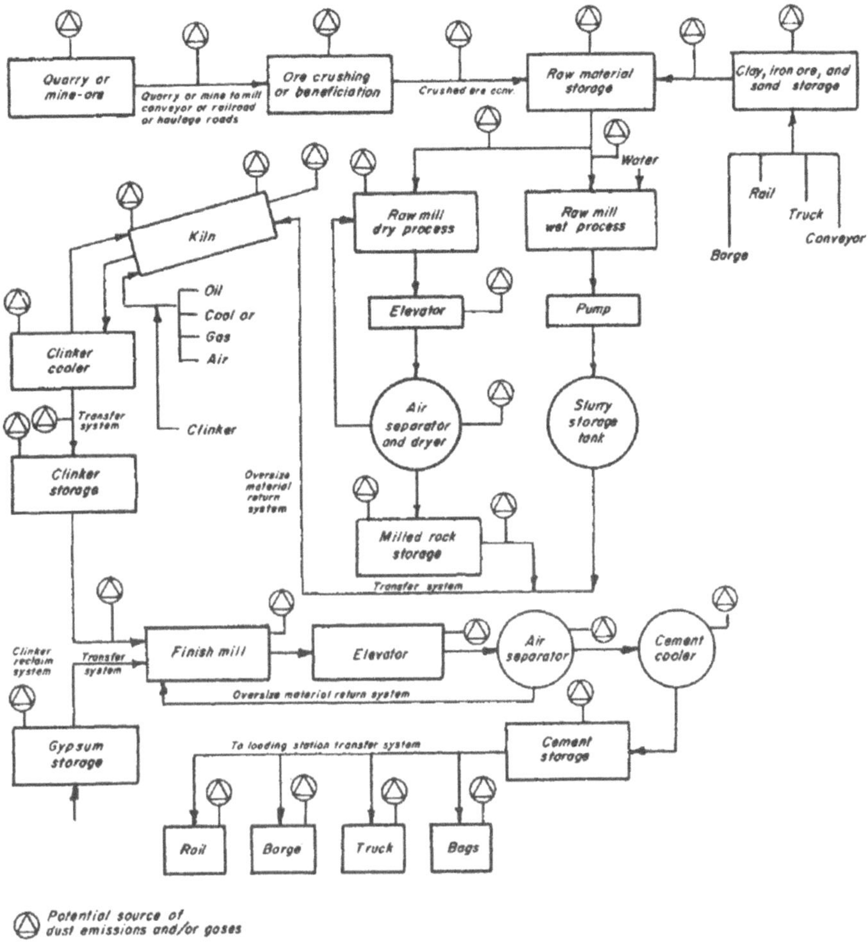

FIGURE 5.8 Multiple granule transfer points in primary storage and feeding operations for Portland cement production.

predict the amount of air induced into the system. Pring, as one of the first researchers, suggested that falling granular materials would pull air from the top of the system enclosure and expel it from the bottom that ultimately carry the dust with them. If the system components are considered small, such as a bucket, the induced air reverses its path, and the dust is carried back through an opening placed in upstream of the flow, as shown in Figure 5.9. Hemeon also investigated this problem and found that the maximum volume of induced air to such system is proportional to:

The cube of the mass flow rate of solids in the flow, the cube of the specific gravity of solids, the cube of the particle diameter, the two-thirds power of the cross-sectional area related to solids, and the two-thirds power of the fall distance.

Although this information is valuable in identification of the system parameters affecting the amount of induced air, it does not provide a useful design approach.

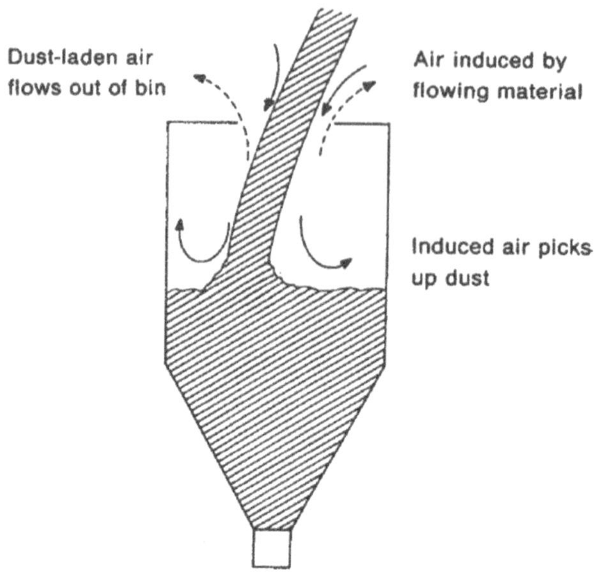

FIGURE 5.9 Induced air illustrates how induced air is transported from moving granular material to the bucket, causing workplace pollution.

Laboratory investigations performed by Dennis on the induced airflow caused by falling granular materials in a material handling operation led to Anderson's empirical equation for induced airflow:

$$Q_i = 10A_u\left(\frac{RS^2}{D}\right)^{1/3} \tag{5.1}$$

where Q_i is induced airflow rate (cfm), A_u is an open surface area of the chamber at upstream of the discharge point (ft), R is material flow rate (tons/h), S is material fall height (ft), and D is average particle diameter (ft) ($D > 1/8\,\text{ft}$).

The use of this design equation for a set of material transfer points is shown in Figures 5.10–5.14. Anderson provided examples of the differences between the exhaust volume calculated by the conventional orifice velocity and this induced air method. The total system airflow obtained by the induced air method may not significantly differ from the approach given in the ventilation manual. However, as an advantage of the induced air approach, it can help to identify places where required control and exhaust are assured to minimize dust emissions. Therefore, it is prudent to calculate the exhaust required for the granule transfer points using both the ventilation manual and the induced air method and select the larger value for design.

A common practice in dry granular material handling is placing the final product in paper bags of 50–100 lb. (45.4–22.7 kg). This operation is one of the dustiest in the industry, and it is difficult to control because decreased exposure of workers depends not only on the installation of ideal local exhaust ventilation but also on the operator's

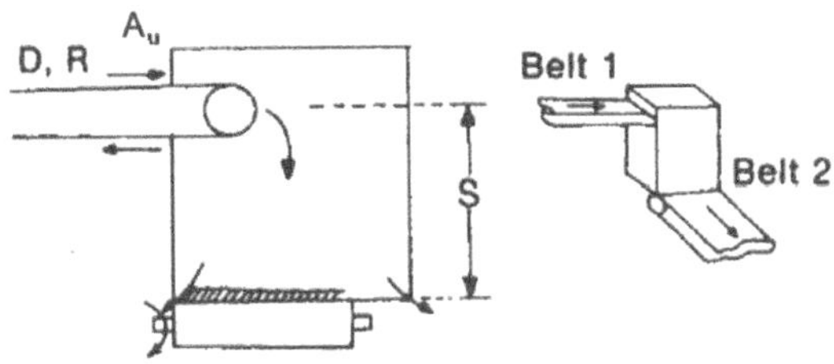

FIGURE 5.10 Induced air, conveyor to conveyor, chute to chute.

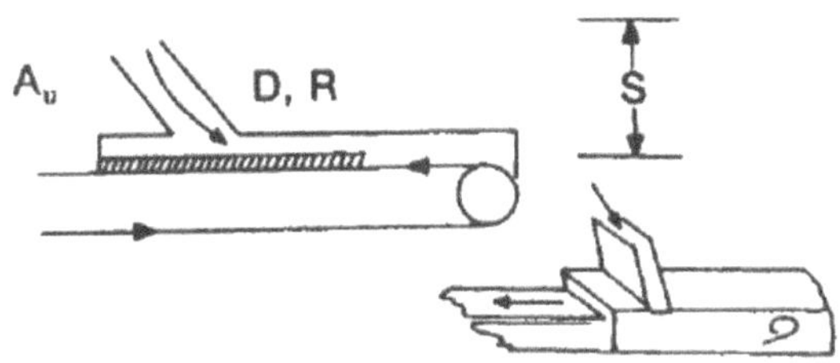

FIGURE 5.11 Induced air, chute to conveyor.

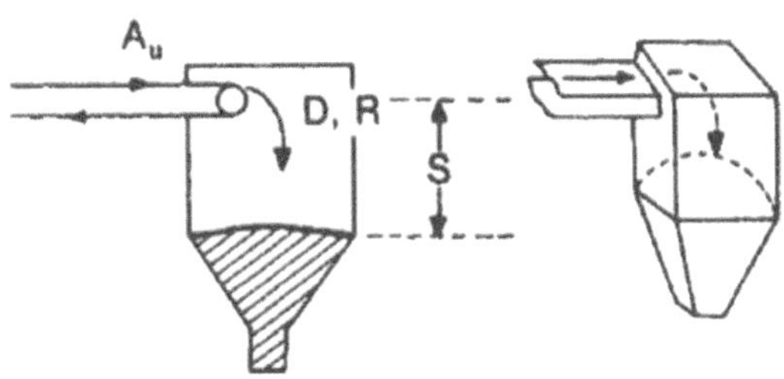

FIGURE 5.12 Induced air, conveyor to bucket.

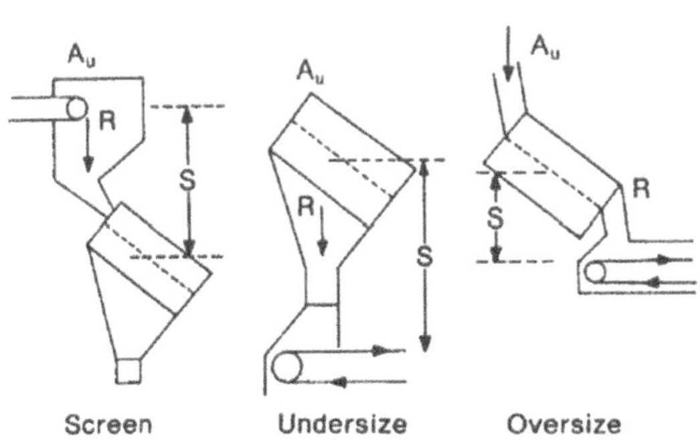

FIGURE 5.13 Induced air, sieves.

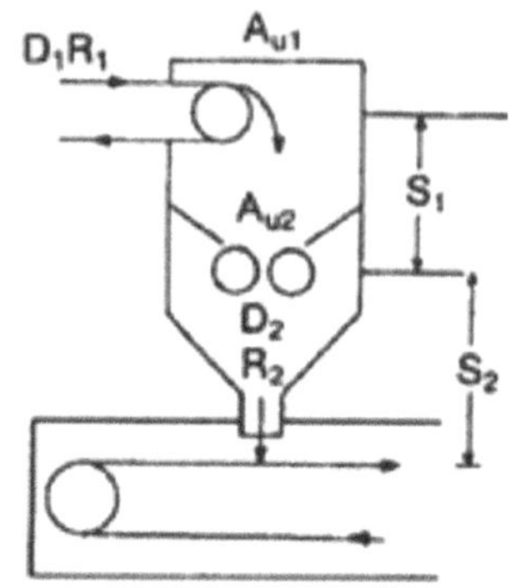

FIGURE 5.14 Induction air, shredder.

adherence to strict work procedures. The bagging operation studied at the NIOSH Pittsburgh Research Laboratory led to many innovations in ventilation control. Dust exposure during these operations varied from 1 to 5 mg/m^3 with standard ventilation controls. In one study, exposure was decreased by 80–90% using a dual-bag nozzle system to reduce product blowback caused by pressure buildup in the bag. Researchers also investigated the use of a bag clamp to decrease product blowback during bag filling.

A new approach by the NIOSH team to reduce dust concentrations in a wooden bag loading floor is use of a clean air scrubber with a recirculation airflow rate of 6000–10000 cfm (4.72–2.83 m^3/s). In two pilot programs presented with this scrubber, a reduction of 98% and 86% in operator exposure was reported. A forced-exhaust ventilation system at a wooden bag loading floor workstation decreased the air concentration from 0.82 to 0.2 mg/m^3. The air blower nozzle provided air at a flow rate of 120 cfm (0.06 m^3/s) with an exhaust velocity of 1200 fpm (6.10 m/s) with a capture hood operating at a flow rate of 2500 cfm (1.2 m^3/s). Another ventilation innovation investigated by the NIOSH group with positive results was a "bottom-up" general dilution ventilation system.

5.2.4 Welding, Soldering, and Brazing

The ventilation controls required for welding operations depend greatly on the welding process, the base metal composition, the base metal coating, and the confinement degree at the weld surface. Before World War II, most welding was done by shielded metal arc welding (SMAW) (a type of arc welding without pressure in which an arc is produced between a shielded electrode as a filler metal and the weld pool, and the smoke and slag due to the coating decomposition prevent the molten and hot metal from mixing with the oxygen and nitrogen in the air) (Figure 5.15). Using this method, the welder is exposed to respirable metal particles in the form of iron oxides and other alloying metals present in the welding smoke column, concentrations of various pollutants released from the rod and its coating, and trace amounts of ozone and nitrogen dioxide fixed by the arc. Low-hydrogen electrodes are used in selected alloys. The coatings on these electrodes contain soluble fluorides that release hydrogen fluoride and particles of soluble fluoride salts. These substances are respiratory irritants and cause systemic effects.

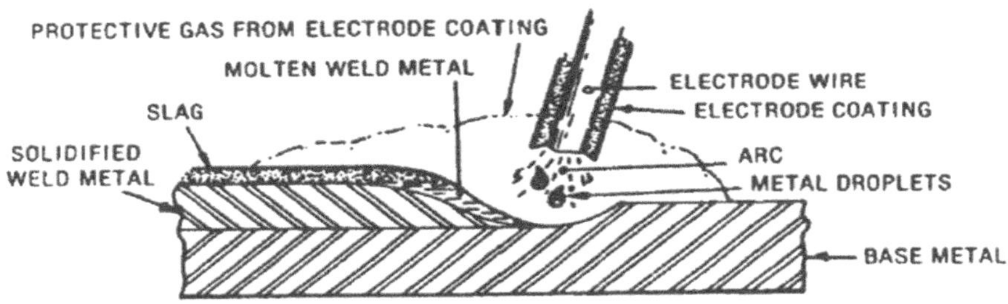

FIGURE 5.15 Shielded metal arc welding (SMA). The consumable electrode used in this process is coated with flux, which has several functions including the formation of a gas to protect the molten weld metal. The arc is performed manually by the welder, who feeds the electrode on the work surface. This process, which is usually used on mild (low-carbon) steels, may release large amounts of metal fumes from the electrode and base metal, as well as gases from the welding wire coating.

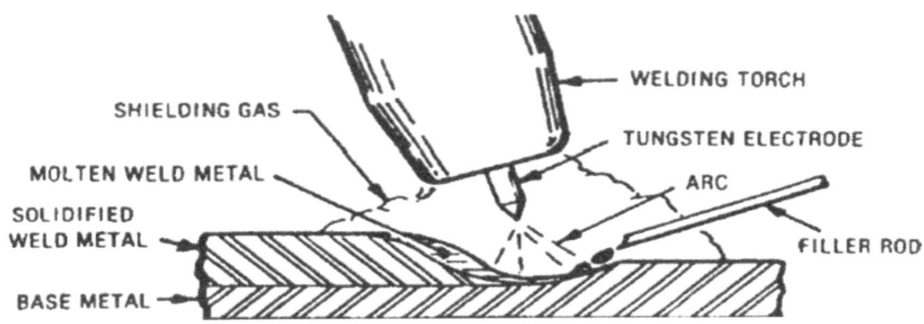

FIGURE 5.16 Tungsten-inert gas arc welding (TIG). In this process, an arc is created between the workpiece and a tungsten electrode. A welding rod is separately fed to the weld by the welder. A shielding gas, such as argon or helium, coats the weld surface. This process is used on certain alloys and in very critical applications of lighter parts than SMA. Airborne pollutants include relatively low concentrations of metal fumes from a variety of alloys, steels, and nonferrous materials welded by this method. Significant concentrations of ozone may be emitted in poorly ventilated workstations.

Technical innovations in welding engineering in the 1940s and 1950s led to the development of methods for covering the arc with inert gas. Two methods of tungsten inert-gas (TIG) welding (Figure 5.16) and metal inert-gas (MIG) welding (Figure 5.17) expose workers to metal fumes and toxic gases, such as ozone, nitrogen dioxide, and carbon monoxide. In the fourth method, submerged arc welding (SAW) (a type of arc welding without pressure in which one or more arcs are created between the filler metal and the weld pool, side by side or in succession so that, in this process, the arc and molten metal are protected by a layer of granular powder (Figure 5.18)), the arc is protected by a granular flux that produces minimal air pollution and does not normally require local exhaust ventilation. Plasma welding (Figure 5.19) operates at higher current densities than TIG and MIG processes, resulting in a concomitant increase in the production of toxic air pollutants. Arc-related processes such as cutting, gouging, and flux-cored welding (a type of arc welding in which an arc is

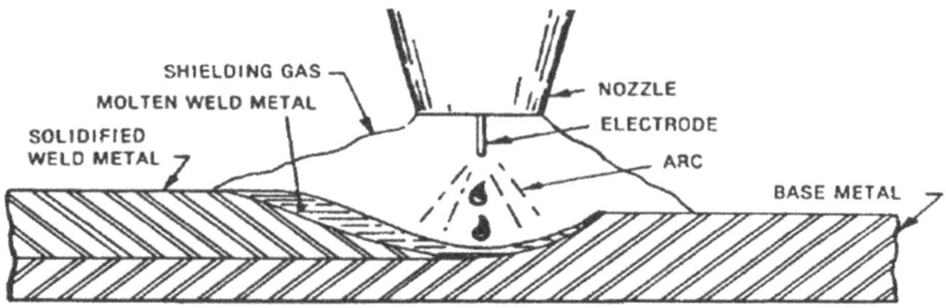

FIGURE 5.17 Metal-inert gas arc welding (MIG). The arc is created with a consumable electrode, and the inert shielding gas flowing in an annular space in the torch covers the welded pieces. This process is often used at high current densities for a variety of metals, including steel and aluminum. High concentrations of ozone are produced, especially when welding aluminum with argon as the shielding gas.

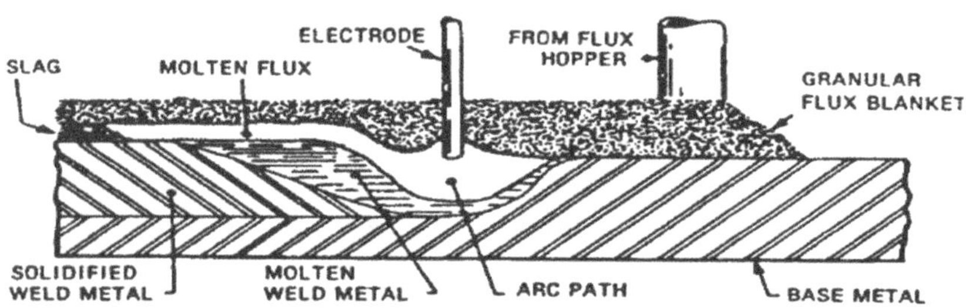

FIGURE 5.18 Submerged arc welding (SAW). This process does not typically require local exhaust ventilation. SAW is used for welding heavy sheet steel pieces such as shipbuilding. The consumable electrode produces an arc under the granular flux to provide protection. The flux is destroyed, and minimal concentrations of dissolved fluoride particles and hydrogen fluoride gas may be emitted.

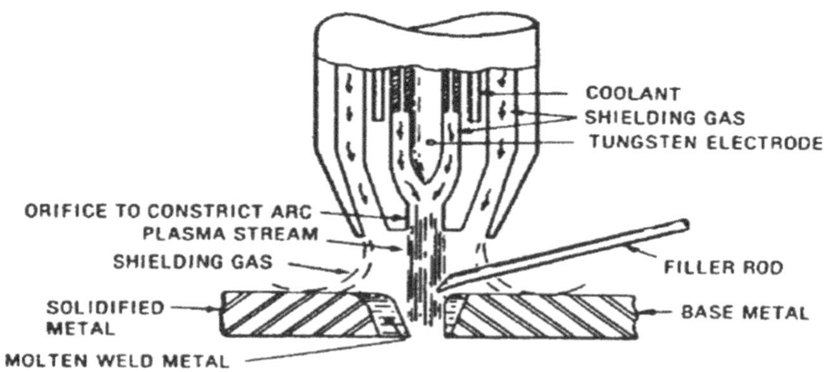

FIGURE 5.19 Plasma arc welding (PAW). In this welding equipment, the torch has both an inert stream of argon and a separate orifice stream of argon. Orifice gas is subjected to high voltage, resulting in a highly ionized gas stream or plasma with arc temperatures exceeding 60,000°F. This process can be used for a variety of metals, including titanium, stainless steels, aluminum, and refractory metals. Because of exposure to ozone, nitrogen dioxide, and metal fumes, this process requires local exhaust ventilation.

formed between a tubular electrode as filler metal and the weld pool, and the fumes and slag due to the powder decomposition, with or without a shielding gas, prevent the molten metal from mixing with oxygen) also require greater attention to control of welding exposures.

Similar welding processes such as brazing may require air ventilation to control pollutants such as volatile products from filler rod or solder, thermal flux degradation products, and carbon monoxide and nitrogen oxides produced by gas torch operations.

Information on each welding process is essential to design the most effective ventilation system. An overview of these processes and the major airborne pollutants is described in Table 5.7.

In a comprehensive review of ventilation control in welding operations, pollutants produced during welding operations were divided into three groups based on the toxicity (Table 5.8). The suitability of various types of ventilation methods and the minimum exhaust ventilation flow rates for each method were determined based on the pollutant group, as presented in Table 5.9. General exhaust ventilation was not considered as a control method option in this study. The use of general exhaust ventilation for welding can be somewhat problematic. Because design criteria are not validated, installations rarely provide adequate control and require frequent updating. In addition, the airflow required for effective general exhaust designs is associated with very high operating costs. One local exhaust ventilation method is listed in Table 5.9, which is usually feasible for control.

Of the four local exhaust ventilation methods listed in Table 5.9, the common method is the free-hanging open hood (Figure 5.20). In this system, an external hood connected to a flexible duct is provided with a positioning system that allows the welder to move the hood closer to the weld. As shown in Figure 5.20, the required airflow is reduced if the hood is placed within an interval of 6 inches (15 cm) from the welded pieces. Obviously, this approach needs the welder's cooperation in positioning the hood and achieving good control. At multiple workstations, either a central system or separate systems can be used.

Since 1975, several manufacturers have offered systems consisting of a hood, flexible duct, positioner, fan, and air cleaner. The air cleaner is usually a high-efficiency particulate filter or an electrostatic precipitator. If aluminum is welded, an electrostatic precipitator may not be suitable because the electrodes may be coated with an insulating coating of aluminum oxide. Depending on the application, these batch systems may be used in recirculation mode. The free-hanging open hood approach is commonly applied in repair and maintenance shops where large equipment may be welded onto a portion of the shop floor.

Welding operations in production, repairment, and maintenance of small pieces can be performed in a small workstation with rear-slot ventilation (Figure 5.21), which is an attractive approach since the welder does not need to use a hood for effective control. The hood should have side baffles to minimize the effects of cross-draft air currents and to improve pollutant collection. Such stations arc often used for welding long pipes, strapping, and hooks, and welders often complain regarding interference of the side baffles. This problem can be overcome by slotted baffles, hinged side baffles, or heavy fabric side curtains. A welding station with rear-slot

TABLE 5.7
Potential Welding Hazards[a]

Hazard	SMA[b]	Low Hydrogen	GTA[c]	GMA[d]	Submerged	Plasma	Gas
Metal fumes	M–H	M–H	L–M	M–H	L	H	L–M
Fluorides	L	H	L	L	M	L	L
Ozone	L	L	M	H	L	H	L
Nitrogen dioxide	L	L	L	M	L	H	H
Carbon monoxide	L	L	L	H and L if CO_2	L	L	M–H
Chlorinated hydrocarbons' decomposition	L	L	M	M–H	L	H	L
Radiant energy	M	M	M–H	M–H	L	H	L
Noise	L	L	L	L	L	H	L

[a] Hazard codes: L: low; M: medium; H: high.
[b] Shielded metal arc welding.
[c] Gas shielded tungsten arc.
[d] Gas shielded metal arc.

TABLE 5.8
Classification of Pollutants for Welding

Pollutant Group	Classification Criteria	Sample Materials
I	Dusts, fumes, and mists with exposure limits of 5 mg/m^3 and higher; gases and vapors with exposure limits of 100 ppm and higher	Iron oxide
II	Dust, fumes, and mists with exposure limits of 0.1 mg/m^3 and higher (up to 5 mg/m^3); gases and vapors with exposure limits of 1 ppm and higher (up to 100 ppm)	Fluorides, carbon monoxide, and nitrogen dioxide
III	Dust, fumes, and mists with exposure limits lower than 0.1 mg/m^3; gases and vapors with exposure limits lower than 1 ppm	Beryllium, mercury, ozone, and cadmium

ventilation is also an effective design for brazing operations. A third local exhaust method is the low-volume/high-velocity hood, which is applicable to both TIG and MIG welding (shown in Figure 5.22). This method provides effective capture with minimal airflow, using a suction ring around the top of the welding area. The suction airflow varies depending on the manufacturer and the specific operation, but it is typically in the range of 50–100 cfm (0.5–0.25 m^3/s) and approximately 10% of the free-hanging open-top hood. There is reluctance among some welders to use this method because the suction may remove shielding gas from the welding point unless the

TABLE 5.9
Ventilation Requirements for Various Welding Methods and Pollutant Groups

Activities	Free-Hanging Open Hood	Workbench with Slotted Hood at the Back	Low Volume/ High Velocity Hood Equipped with Gun	Enclosure Hood
Gas welding	Group I and II - capture velocity of 100 fpm (0.51 m/s) Group III - not recommended	Group I and II - capture velocity of 100 fpm (0.51 m/s) Group III - not recommended	Not applicable	Group I and II - capture velocity at opening of 100 fpm (0.51 m/s) Group III - capture velocity at opening of 150 fpm (0.76 m/s)
Brazing and soldering	Group I and II - capture velocity of 100 fpm (0.51 m/s) Group III - not recommended	Group I and II - capture velocity of 100 fpm (0.51 m/s) Group III - not recommended	Not applicable	Group I and II - capture velocity at opening of 100 fpm (0.51 m/s) Group III - abduction velocity at opening of 150 fpm (0.76 m/s)
Arc welding with a coated electrode	Group I and II - capture velocity of 100 fpm (0.51 m/s) Group III - not recommended	Group I and II - capture velocity of 100 fpm (0.51 m/s) Group III - not recommended	Not applicable	Group I and II - capture velocity at opening of 100 fpm (0.51 m/s) Group III - capture velocity at opening of 150 fpm (0.76 m/s)
Plasma arc welding and submerged arc welding	Group I and II - capture velocity of 100 fpm (0.51 m/s) Group III - not recommended	Group I and II - capture velocity of 100 fpm (0.51 m/s) Group III - not recommended	Not applicable	Group I and II - capture velocity at opening of 100 fpm (0.51 m/s) Group III - capture velocity at opening of 150 fpm (0.76 m/s)
Inert gas arc welding and flux cored welding	Group I and II - capture velocity of 100 fpm (0.51 m/s) Group III - not recommended	Group I and II - capture velocity of 100 fpm (0.51 m/s) Group III - not recommended	Group I and II - exhaust flow rate determined by manufacturer or user Group III - not recommended	Group I and II - capture velocity at opening fpm 100 (0.51 m/s) Group III – abduction speed at the opening of 150 fpm (0.76 m/s)

(*Continued*)

TABLE 5.9 (Continued)

Activities	Free-Hanging Open Hood	Workbench with Slotted Hood at the Back	Low Volume/ High Velocity Hood Equipped with Gun	Enclosure Hood
Oxygen cutting, arc cutting and surfacing	Not recommended	Not recommended	Not applicable	Group I and II – capture velocity at the opening of 100 fpm (0.51 m/s) Group III – capture velocity at the opening of 150 fpm (0.76 m/s)
Thermal spraying	Group I - 200 fpm (1.02 m/s) at the opening with activity in the hood - Group II and III - not recommended	Not recommended	Not applicable	Group I – capture velocity at the opening of 125 fpm (0.64 m/s) Group II and III – capture velocity at the opening of 200 fpm (1.02 m/s)

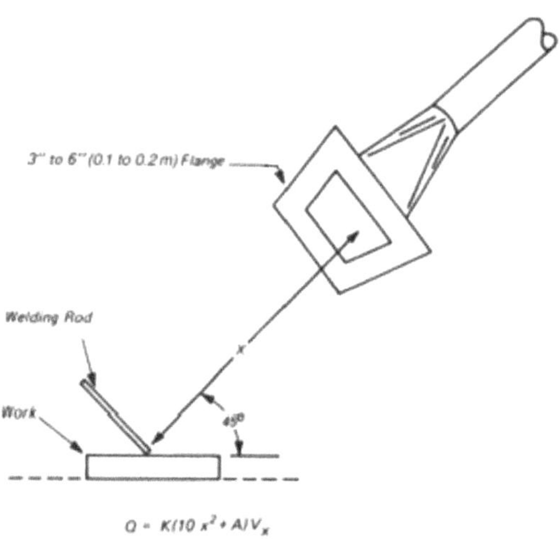

Q is the exhaust volume (cfm or m^3/s), X is the distance from the center of the hood opening to the farthest point of the pollutant emission (ft or m), A is the area of the hood opening without flange (ft^2 or m^2), V_x is the minimum capture velocity (fpm or m/s), K is 1 for unflagged hoods, 0.75 for flanged hoods, entry loss = entry loss coefficient for conical hood × duct velocity pressure; Duct velocity = minimum 2000 fpm (10.2 m/s).

FIGURE 5.20 Movable hood with adjustable position by welder.

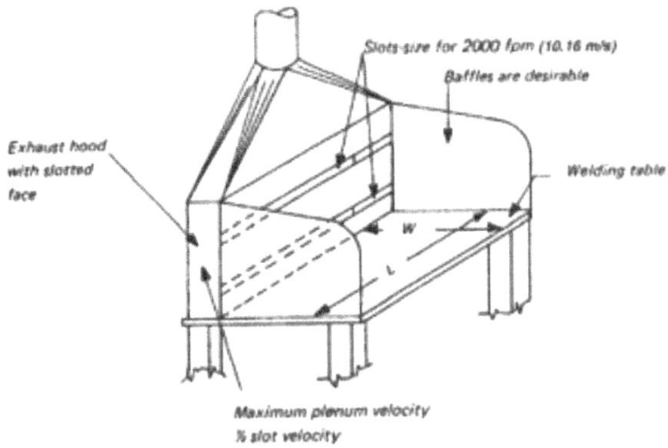

Q is exhaust volume (350 cfm/ft (0.54 m^3/s/m) hood length), W is table width (ft) (maximum 24 inches (600 mm)), L is table length, entry loss = 1.78 slot velocity pressure + entry loss coefficient, and duct velocity = minimum 2000 fpm (10.16 m/s).

FIGURE 5.21 Slot-back exhaust hood for welding and brazing of small pieces.

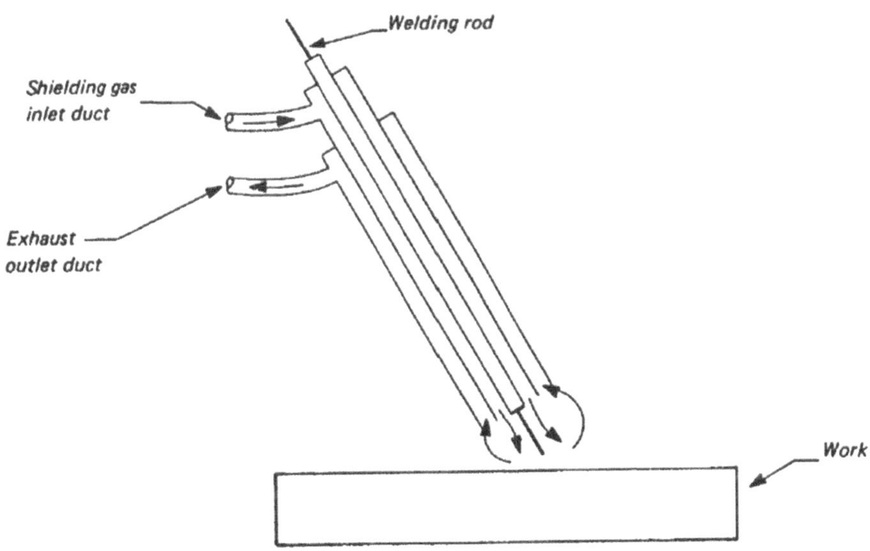

FIGURE 5.22 Low volume/high velocity exhaust for inert gas shielded process. Exhaust flow requirements should be met for each welding operation and welding torch configuration by laboratory tests with air pollutant sampling and analysis.

exhaust rate is regulated. If highly toxic metals are processed, the best practice is use of an enclosure hood. In practice, it may resemble a modified welding table in a spray booth. The downdraft table (Figure 5.23) is a local exhaust hood designed primarily for gas and arc cutting processes. The metal slab is placed on the table, and the metal fumes are sucked down. Large droplets of molten metal flow into the plenum chamber, and the fumes are sucked out of the plenum chamber box. Use of a canopy hood for a welding station is a common mistake in industry. These hoods are not acceptable for control of toxic air pollutants.

5.2.5 Chemical Processing Industries

Although there is not the wide range of pollutant generation points and variety of air pollutants in other industries except chemical processing industries, standard ventilation designs are available for only a limited number of activities in these industries. This section reviews the common activities in a chemical processing industry and possible ventilation control approaches.

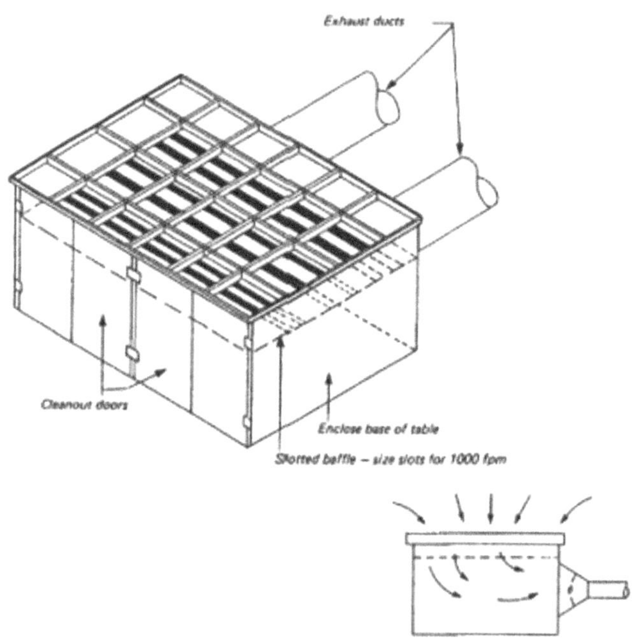

Q is suction volume (150 ft^2/cfm (0.75 $m^3/s/m^2$) net table area), entry loss is 1.78 slot velocity pressure + 0.25 duct velocity pressure, and duct velocity is minimum 2000 fpm (20 m/s).

FIGURE 5.23 Cutting table with downward airflow in gas or plasma cutting operations.

5.2.6 Chemical Processing Industry Activities

A chemical processing industry may produce dozens of chemicals at various scales, from tens to hundreds and thousands of kilograms per year (Figure 5.24). Raw materials include granular chemicals, liquid chemicals such as solvents, and reactive gases. Granular materials used in small quantities are usually distributed in drums and may require weighing before use. These materials can usually be produced as prill or flake to minimize dust generation. Regardless of the material form, a local exhaust hood is usually required at the weighing station to control dust (Figure 5.25). Granular chemicals are transported to the plant by truck, bulk rail car, or conveyor, and may be directly transferred by pneumatic conveyor to a mixing or reaction tank or storage bins for subsequent transition to the tank. This bulk transfer is usually done in enclosed systems with minimal dust emissions to the environment.

Liquid reagents may also be transferred to a storage tank at the plant or, if used in small quantities, by drums stored in special storage areas. Bulk solvents are often received at a loading dock equipped with a multi-way piping system with flexible hoses from the truck or train to the storage tank. Significant exposure to volatile liquids may occur when the compounds are discharged from the transfer lines. Even this work, which is usually done in outdoor places, requires ventilation control.

The first major process equipment in a plant that requires ventilation control is the reaction vessel. For example, consider the non-continuous process for the synthesis of a dye, as shown in Figure 5.24. Various chemicals are added to the reactor in a strictly planned manner. Liquid is added by closing all ports and valves and placing the reactor under vacuum. The liquid is then transferred from a storage tank to the reactor in an enclosed piping system with minimal worker exposure.

Granular materials are often added to the vessel. In large manufacturing operations, granular chemicals may be pneumatically transferred from a storage bin into the vessel. In the non-continuous process, as shown in Figure 5.24, these materials are added manually through a vessel feed port or manway. In this case, exposure can occur. Pouring granular materials into a reactor that has already been fed with liquid solvent can induce airflow into the vessel, resulting in the movement of particles and solvent-laden air into the work area. That is a significant exposure point and should be controlled by local exhaust ventilation.

A common control approach is an external hood placed at the feed port where geometry and airflow permit effective capture of solvent vapors and respirable dust (Figure 5.26). The collection efficiency of well-designed hoods is greater than 95%, as measured by tracer gas studies.

Small amounts of solvent may be added to the reactor from a barrel using a pump that is discharged through an open access port into the reactor. Exposure to chemical vapors can occur during the preparatory mounting of the pump in the barrel and the removal of the wetted wand from the barrel at the end of the pumping operation. Two approaches exist to control ventilation during such operations. Installation of a bung hole hood on the barrel (Figure 5.27) or use of a partial enclosure for the barrel can be useful—the second approach is usually more effective. Periodically, samples should be gathered from the reactor to check the status of the reaction or the quality of the final product. This is often performed by an access port at the top of the reactor and a dipstick with a sampling cup.

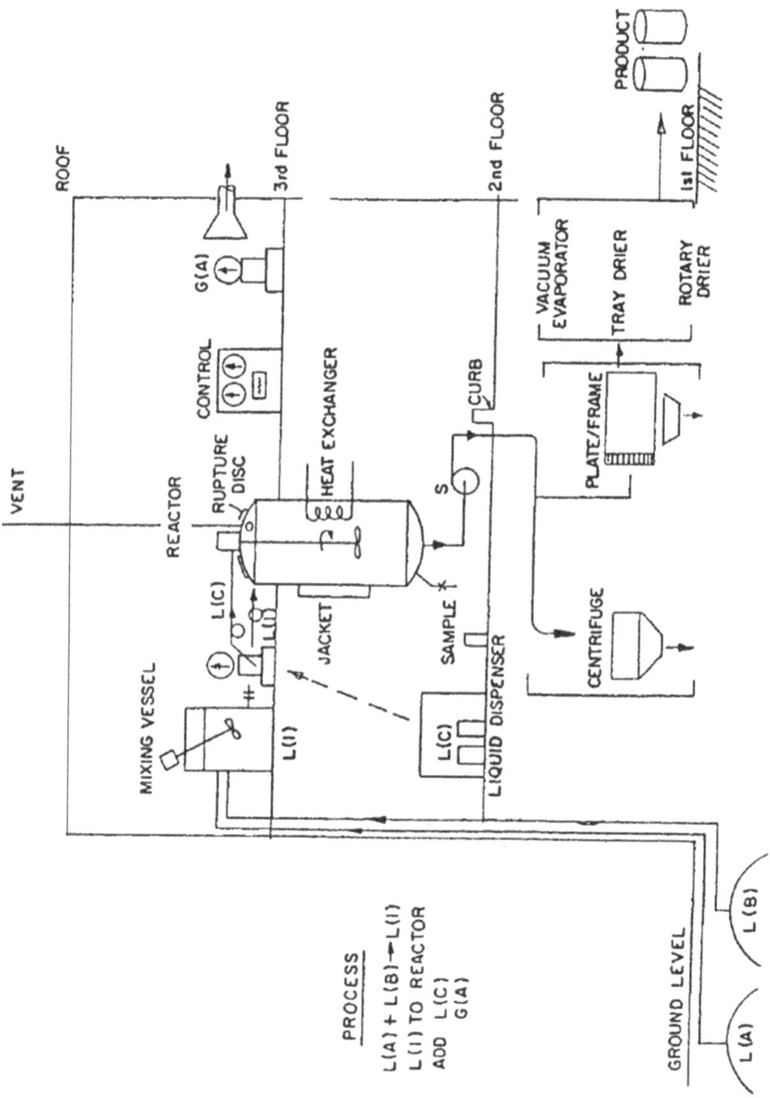

FIGURE 5.24 Chemical process industry for non-continuous processing (L: liquid; G: granular; S: suspension).

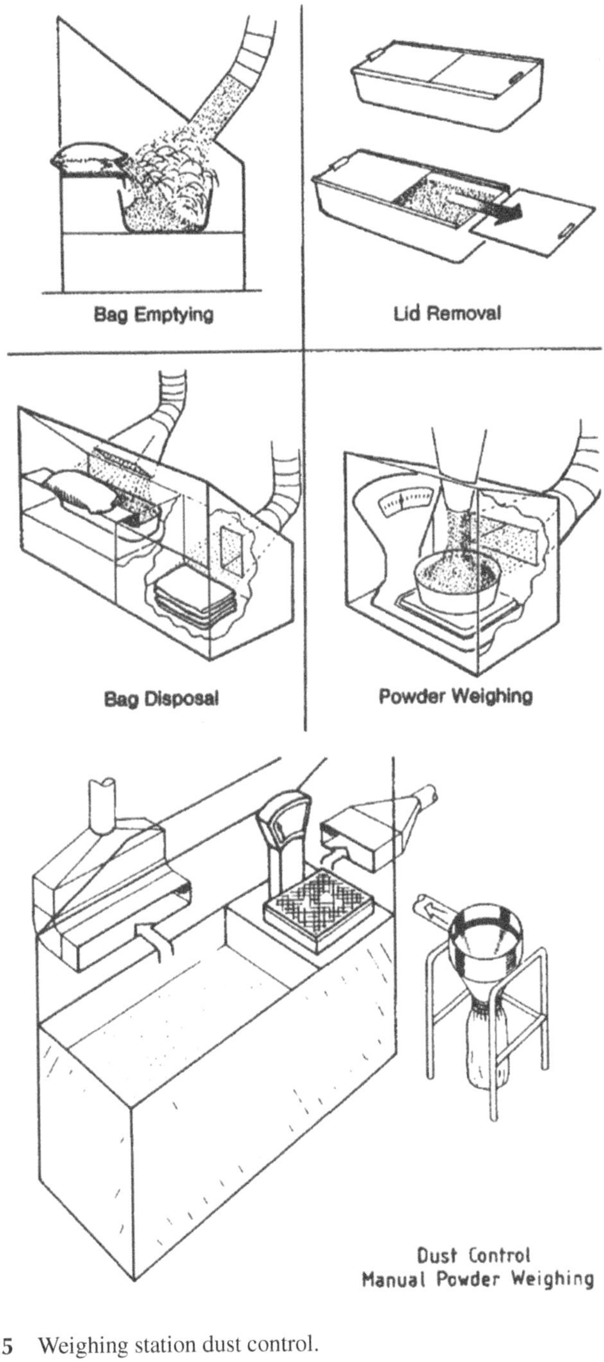

FIGURE 5.25 Weighing station dust control.

FIGURE 5.26 Reactor feed port ventilation.

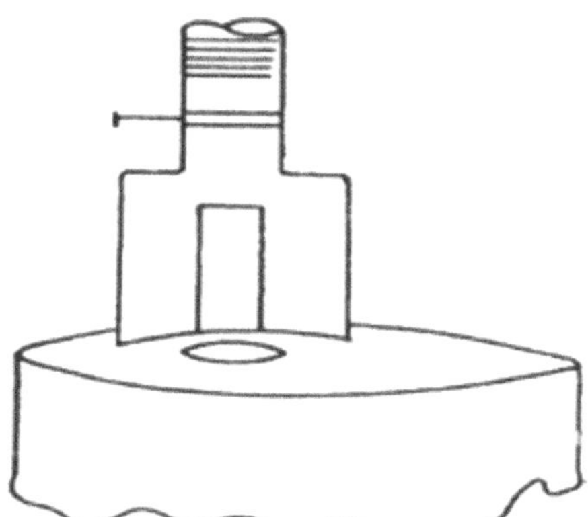

FIGURE 5.27 Barrel valve suction (Q = 150 cfm (0.07 m^3/s), C_e = 0.5).

Despite the potential exposure to reactor headspace pollutants and sample vapor, exposure is usually negligible when the reactor is equipped with feed port ventilation. Exposure in the reactor headspace can also be avoided by placing the reactor in a vacuum with an internal airflow through the feed port. The second sampling method involves the use of a special sampling line near the bottom of the reactor. This sampling place should be equipped with a small suction chamber (Figure 5.28).

The final product, in this example, is a suspended particle for recovery from the reaction fluid. As shown in Figure 5.29, three common methods for preformation of this operation include the Nutsche filter, the centrifuge, and the plate and frame filter. Each operation results in potential exposure to residual solvents. The latter two cases require special attention.

If a centrifuge is used, the suspension in the reactor is pumped to the centrifuge, where it flows into a rotating basket with a filter pad (Figure 5.30). The particles are

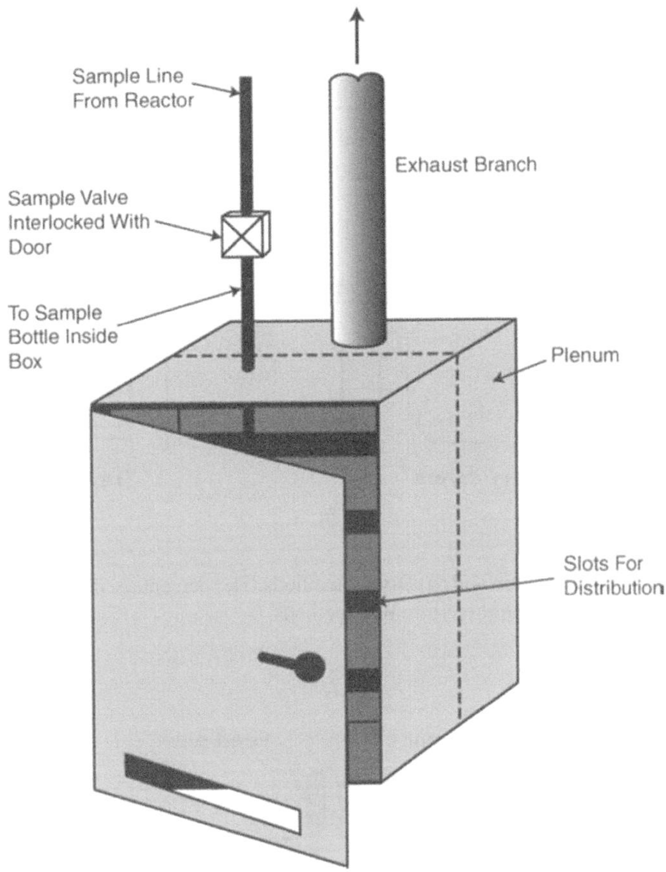

FIGURE 5.28 Sampling hood.

trapped on the filter as the solvent passes through the filter and is pumped to a storage tank. The cake on the inner surface of the basket is repeatedly washed by spraying solvent through a spray header. If the solvent poses a fire or explosion hazard, the centrifuge cavity is flushed with carbon dioxide or nitrogen. When the cake is completely flushed, a scraper scrapes the basket surface and the recovered product flows into a container. Worker exposure to solvent occurs when access doors are opened to manually scrape the basket or inspect the condition of the filter basket. Secondary exposure happens to the product collection barrel or bucket.

It is difficult to apply the control at the access port on top of the centrifuge. One option is exhaust of the return air from the centrifuge to a storage tank using a transfer line. If this is not possible, an external hood should be installed at the access port, as shown in Figure 5.31. Solvent vapor control in the product collection barrel can be accomplished by a barrel suction station.

The second commonly used method for product recovery is the plate and frame filter. This common system, available in a variety of designs, uses several filter pads

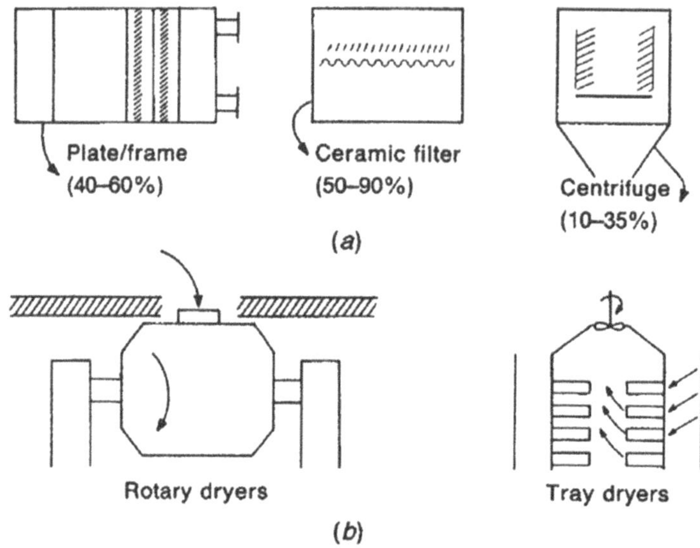

FIGURE 5.29 (a) Filtration and (b) Drying methods. The percentage of solvent remaining in the wet cake is shown for all three filtration methods.

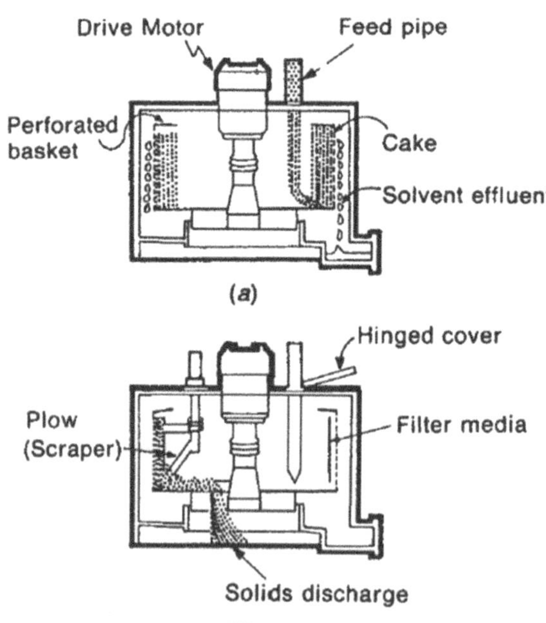

FIGURE 5.30 Details of the centrifuge basket in two main operational stages: (a) active filtration and (b) product harvesting. This product is harvested manually through a hinged cover with a small shovel.

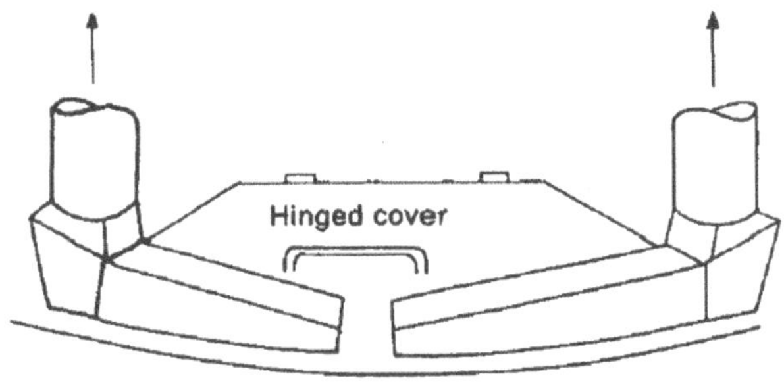

FIGURE 5.31 Extraction hood for centrifuge with hinged cover in access door (Q = 250 cfm (0.12 m^3/s) per leg, C_e = 0.6).

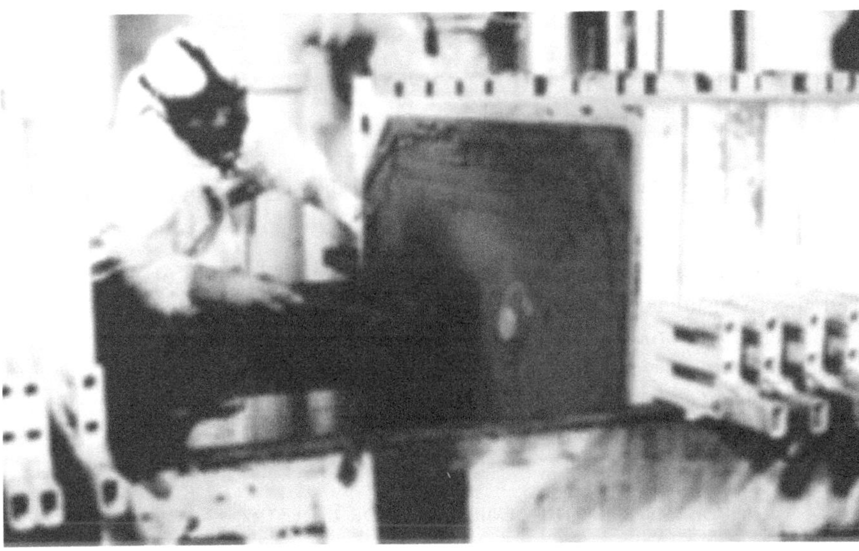

FIGURE 5.32 Product release from plate and frame filter.

arranged in parallel to capture particles. The slurry in the reactor is pumped to a plate and frame filter, the particles are collected on the filter, and the removed solvent is directed to a storage tank. The plate and frame framework also allow the collected product to be washed with clean solvent or water. When the washing is complete, the pressure due to the plate closure is released and the plates are separated. The product cake wetted with solvent is manually scraped from the plates and dropped into a handwheel (Figure 5.32). During this process, solvents evaporated from the wet cake result in significant concentrations in the workplace air that require ventilation control.

Effective control of solvent vapors from plate and frame filters is extremely difficult. One approach is to use a top-exhaust plenum enclosure with a plastic perimeter curtain or horizontal sliding sash, which must be positioned by the operator for effective control (Figure 5.33). If overhead space for a plenum enclosure is not available, two fishtail hoods can be placed at the ends of the press (Figure 5.34).

FIGURE 5.33 Ventilation of a plate and frame filter housing with rigid sliding panels or plastic curtains.

The control effectiveness depends on the proper positioning of the curtain by operators to ensure adequate airflow across the worker breathing zone.

Drying of the product cake in a non-continuous operation is typically accomplished by a tray dryer, cone dryer, or fluid bed dryer. Loading in each operation involves solvent exposure. Unloading may lead to exposure to respirable dust during packaging and transportation.

5.2.7 Gas Chambers in the Semiconductor Industry

A variety of toxic gases, including arsine and phosphine, is used in semiconductor manufacturing operations. The gases are typically vented from pressurized cylinders through a pressure regulating manifold. An exhaust hood is the primary control to minimize worker exposure to these gases in the event of a leak during storage and distribution activities. These chambers, known in the semiconductor industry as gas chambers, are equivalent to pollutant hoods, and those should provide a high level of control with minimal airflow. It should be noted that gas chambers are designed to control the emission of released gases, not the intense emission of the contents in the entire gas cylinder. Manufacturers of standard gas chambers design these chambers to accommodate a maximum of four gas cylinders. The gas cylinder is placed in the chamber and connected to a manifold mounted on the panel, which allows for accurate measurement of gas flow at various manufacturing operations.

The main design variables in exhaust hood design are the air inlet and outlet configurations. In the gas chamber design, air is reciprocated through an opening at the bottom of the door and is sucked through a direct duct connection at the top of the

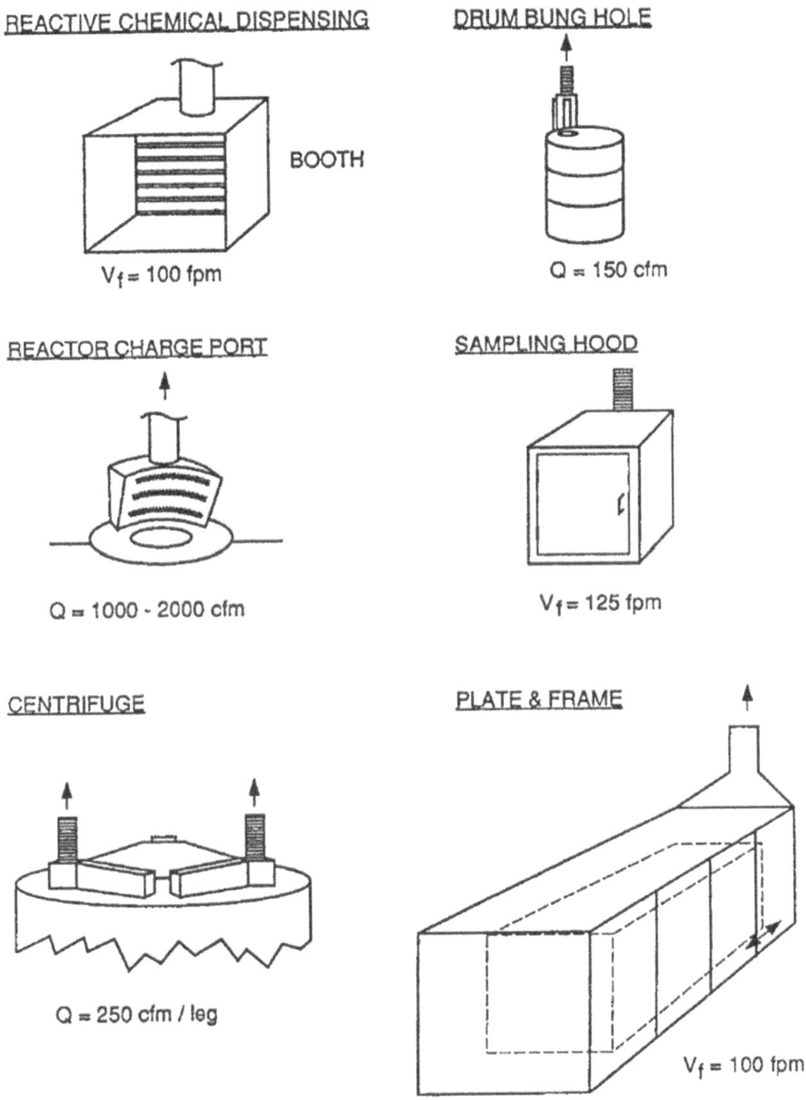

FIGURE 5.34 Summary of exhaust ventilation for chemical processes.

inlet chamber. The entry is usually a series of slots in the door, and a particle filter may also be installed, which acts as a diffuser. Although these hoods are widely used in industry, little information is available on their performance. In one study, the optimal geometry of the gas chamber and their exhaust airflow were evaluated using tracer gas.

A two-cylinder gas chamber equipped with various inlet and outlet configurations and places for a wide range of cylinder sizes was used in one study (Figure 5.35). The

FIGURE 5.35 Two-cylinder gas test chamber showing the entrance door with access port, diffuser entry at the door base, shallow cone exhaust, and transparent side panels for flow visualization studies.

chamber was equipped with a door in the front section that was used for cylinder replacement. Pressure and other items were adjusted through the access port to minimize the potential for personnel exposure to toxic gas.

The test chamber was equipped with three interchangeable air entry configurations: (1) a simple slotted entry in the door; (2) a perforated plate entry in the chamber base; and (3) a diffuser entry placed in the door. The perforated plate inlet was placed at the chamber bottom to test the "piston" airflow from the plate to a channel interface at the top of the chamber. It provided a free space for any gas leakage.

The gas chamber for semiconductor gases has a flat exhaust channel interface. Geometry may induce angular vortex flows and possibly increase purge time. To minimize these effects, conical suctions are incorporated into the chamber design. Shallow and deep exhaust cones, as the most common types of aerodynamic designs, were tested with a flat exhaust duct interface (Figure 5.36).

5.2.7.1 Entry Loss

The entry coefficients (C_e) were calculated at different flow rates for each of the different inlet and outlet configurations, as presented in Table 5.10. The air inlet geometry had the greatest effect on the entry coefficient, while differences in outlet

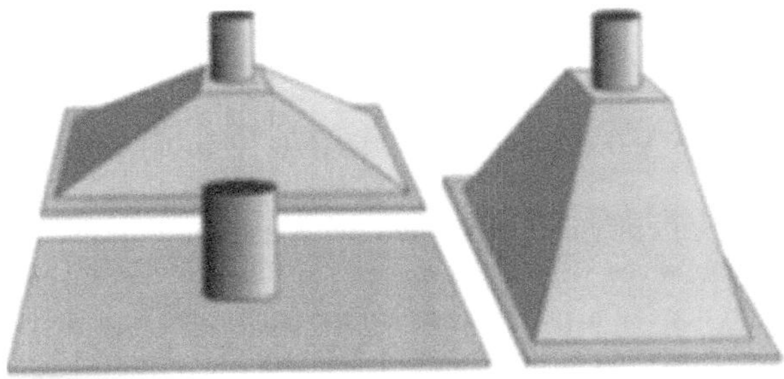

FIGURE 5.36 Three air suction outlets for installation at the top of the test chamber. Front left: flat cone; rear left: shallow cone; and right: deep cone.

TABLE 5.10
Entry Coefficient C_e for Gas Chamber Inlet and Outlet Configurations

Entry type	Output	C_e
	Flat	0.44
Slotted	Shallow Cone	0.47
	Deep Cone	0.47
	Flat	0.7
Diffuser	Shallow Cone	0.75
	Deep Cone	0.72
	Flat	0.68
Perforated base	Shallow Cone	0.72
	Deep Cone	0.72

geometry had little effect on the entry loss. C_e for the slotted entry ranged from 0.44 to 0.47, while the best values were 0.68 to 0.75 for the diffusers and perforated base entries.

5.3 DESIRABLE EXHAUST FLOW RATE

In a series of tests for a study to determine the minimum exhaust airflow rate for control, a tracer gas was released at a critical leakage place at a rate of 1.9 lpm (liters per minute). Inside the chamber, the downstream concentration of the tracer gas was measured continuously at a sampling point in the exhaust duct. When the system and the downstream concentration were stabilized, the leakage was stopped, and the time was recorded and plotted. The concentration loss in downstream was then used to determine system performance. For these tests, 5% of the original concentration was applied to evaluate performance.

The purge time data for the various test configurations, without the cylinder in place, are summarized in Figure 5.37. These data indicate several points:

- Cleaning of the leakage from the slotted entry was consistently longer than that from the other two entries
- The cleaning time for all three entries decreased significantly as the airflow increased from 150 to 250 cfm (0.07 to 0.12 m^3/s) for the two-cylinder chamber
- For the diffuser entry and perforated plate, a slight improvement in performance was achieved with an increase in airflow to 350 cfm (0.17 m^3/s)
- Little difference in the performance of the perforated base and diffuser front section was observed at flow rates of 250 and 350 cfm (0.12 and 0.17 m^3/s)
- The type of exhaust configuration showed weak relationship with cleaning time

The cleaning time for the gas chamber was evaluated. The results of this test with flat plate and shallow cone exhausts are presented in Figures 5.38 and 5.39. The deep cone was not evaluated because it had not shown a significant advantage in previous experiments compared to the shallow cone. As a result of this investigation, a diffuser entry at the bottom of the chamber door with an exhaust flow rate of 250 cfm (0.12 m^3/s) is recommended for the two-cylinder chamber.

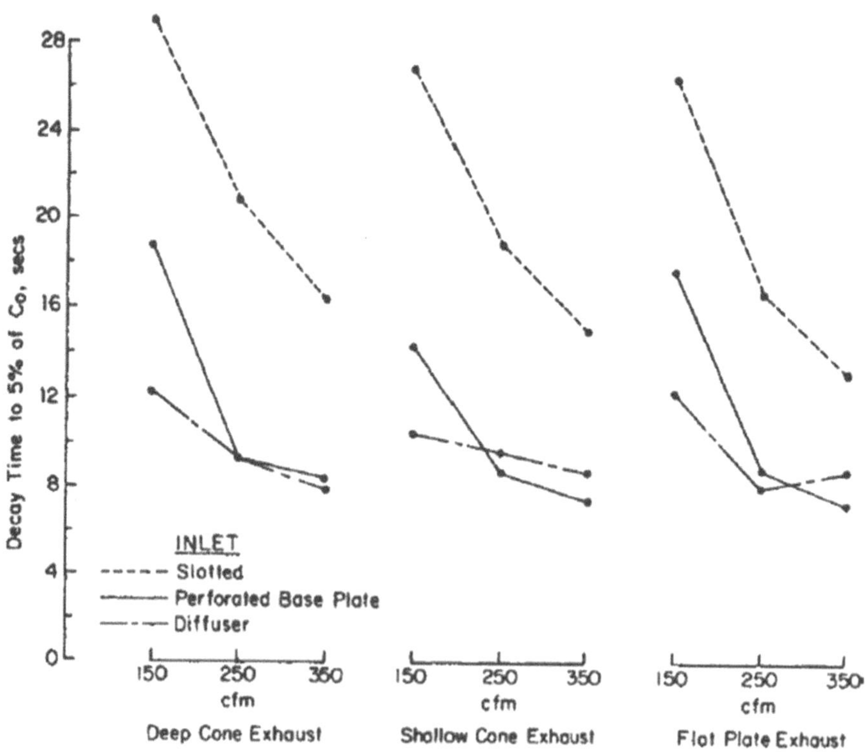

FIGURE 5.37 Cleaning times of three entries for each test configuration without cylinder (150 cfm = 0.07 m^3/s; 250 cfm = 0.12 m^3/s; 350 cfm = 0.17 m^3/s).

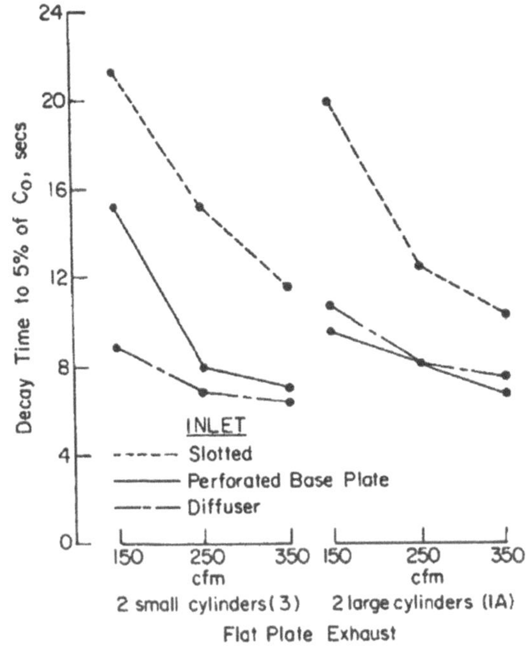

FIGURE 5.38 Cleaning times of three entries for flat plate exhaust with small and large cylinders (150 cfm = 0.07 m^3/s; 250 cfm = 0.12 m^3/s; 350 cfm = 0.17 m^3/s).

5.4 LOW VOLUME/HIGH VELOCITY SYSTEMS FOR PORTABLE TOOLS

In a low volume/high velocity capture system, an industrial ventilation extractor hood (kitchen hood) is used as an integral part of the tool or in a position very close to the operating point of the cutting tool. The hood is used empirically for high capture velocities, usually greater than 10,000 fpm (50 m/s) at the point of pollutant emission. These velocities are usually applied at airflows less than 50 cfm (0.02 m^3/s) using a small hood opening area.

This approach was implemented in the United States in the 1950s to control highly toxic materials such as beryllium released from machining operations. Although these tools are applied, the general adoption of low-volume/high-velocity systems was due to the use of portable tools in the UK and Scandinavia since 1965. Currently, on-site grinding and sanding activities are widely utilized in industrial operations and pneumatic drilling in mining.

This ventilation-controlled method has several advantages. The collection system is an integral part of the tool and does not rely on the operator's position to the hood's position near the work area for proper control. Furthermore, the required exhaust airflow to achieve high collection efficiency is often 10% of that for conventional outdoor hoods. Because of the high cost related to conditioned replacement air in northern latitudes, this is a very cost-effective approach in an industry that involves many workstations. Ultimately, it may be the only practical method for control of highly toxic air pollutants in operations where local exhaust ventilation workstations cannot be installed (Figure 5.40).

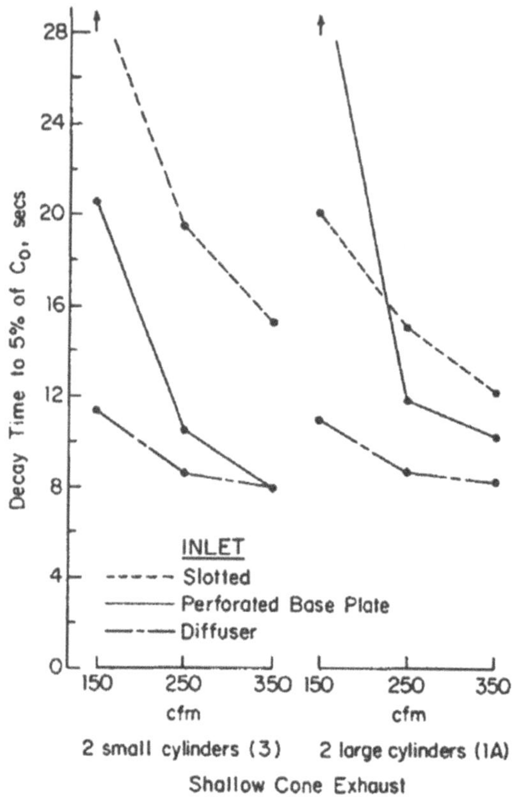

FIGURE 5.39 Cleaning times of three-entries for shallow cone exhaust with small and large cylinders (150 cfm = 0.07 m^3/s; 250 cfm = 0.12 m^3/s; 350 cfm = 0.17 m^3/s).

At first glance, it seems like an advantage in the system since this design results in reduced air mover operating costs. However, it is later found to be quite the opposite, because the energy cost of an air mover is proportional to the airflow and the total resistance. In low volume/high velocity systems, the airflow, as mentioned in the section above, is often reduced by a factor of 10, but the system resistance may increase tenfold; hence, the air horsepower will be about the same as in a conventional system, and the air mover operating cost will be about the same. The main disadvantages of this system include the high initial cost of the high static pressure air mover, special suction nozzles and associated equipment, and the noise generated by the air moving through the restricted nozzle surface at high velocity. When used on portable equipment, operators complain about the extra weight and bulky suction hose.

The design of a low volume/high velocity system is based on experience and a general understanding of the air pollution emission mechanisms. The hood is connected to the exhaust system with a long, flexible, high-strength duct, which allows its use on portable equipment and its lateral movement for activities performed on the moving surfaces of stationary equipment. The flexible duct is connected to the main duct, which is a heavy-walled tube with a smooth, uninterrupted internal surface. Since the tube wears out quickly in particle collection activities due to the abrasive action of the

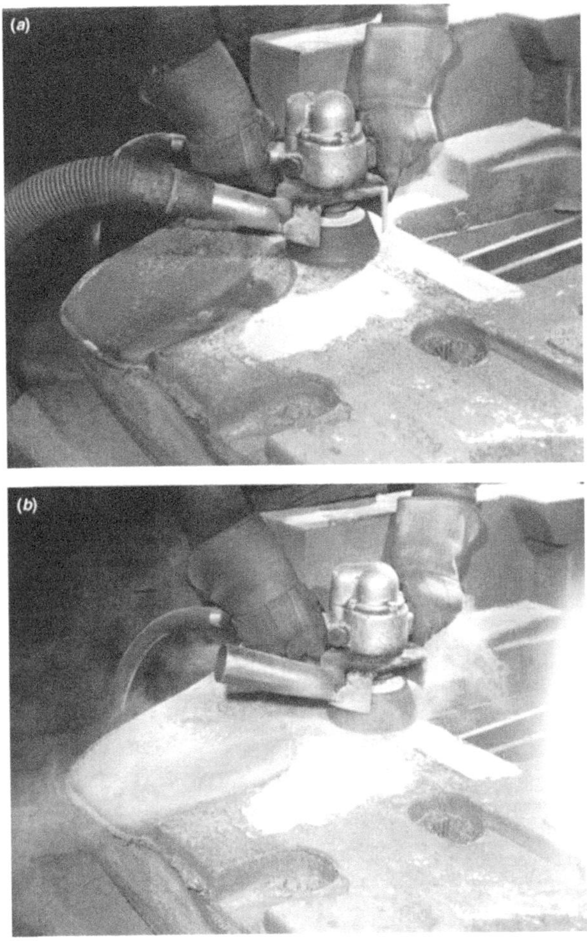

FIGURE 5.40 (a) The effectiveness of a low volume/high velocity hood on a cup grinder in a casting operation is shown; (b) Dust released from the operation is visible when the exhaust hose is removed from the cup grinder.

debris carried by the high velocity airflow, the main duct should be equipped with large reciprocating elbows and low-angle entries. Also, because of the static charge accumulation, the flexible hose in the tool should be conductive and antistatic.

In the UK, extractor hoods are designed to provide a static pressure of 5 inches of mercury (16,950 pascals) for the required airflow. Typical system losses at downstream of the hood due to the flexible hose, smooth duct, and air cleaner are 5 to 7 inches of mercury (23,730–16,950 pascals) and for a static pressure system, 10–12 inches of mercury (40,680–33,900 pascals). A single tool can be used with an exhaust cleaner. A multi-hood system requires more power than a multi-stage turbine fan or a positive displacement blower.

The design of common low volume/high velocity systems in the US is empirical; the minimum recommended volumes for the various tools in Table 5.11 are

TABLE 5.11
Minimum Hose Volumes and Sizes

Hand Tools	CFM	Inner Diameter of Plastic Hose (in)
Sander disc		
1000 rpm–3 in	125	1
5000 rpm–7 in	200	2
2500 rpm–9 in	200	2
Vibrating pad sander - 4 × 9	150	$1\frac{1}{2}$
Wood router		
50000 rpm - $\frac{1}{8}$ in	125	$1\frac{1}{2}$
30000 rpm - $\frac{1}{4}$ in	150	$1\frac{1}{2}$
20000 rpm–1 in cutting tool	150	$1\frac{1}{2}$
Belt sander 4000 rpm–3 in	125	$1\frac{1}{2}$
Pneumatic chisel	125	$1\frac{1}{2}$
Radial wheel grinder	150	$1\frac{1}{2}$
Dead surface grinder $\frac{1}{4}$ in	125	$1\frac{1}{2}$
Conical wheel grinder 2 in	150	$1\frac{1}{2}$
Abrasive stone grinder 4 in	200	2
Abrasive type brush 6 in	250	$1\frac{1}{2}$
Radial wire brush 6 in	175	$1\frac{1}{2}$
Hand wire brush - 3 × 7	125	$1\frac{1}{2}$
Safety saw blade	175	$1\frac{1}{2}$
Plaster molding cutter	150	$1\frac{1}{2}$
Electric saw saber saw	150	$1\frac{1}{2}$
Rotating frame grinding machine - 2 × 18	600	3
Sanding saw 3 in	150	$1\frac{1}{2}$

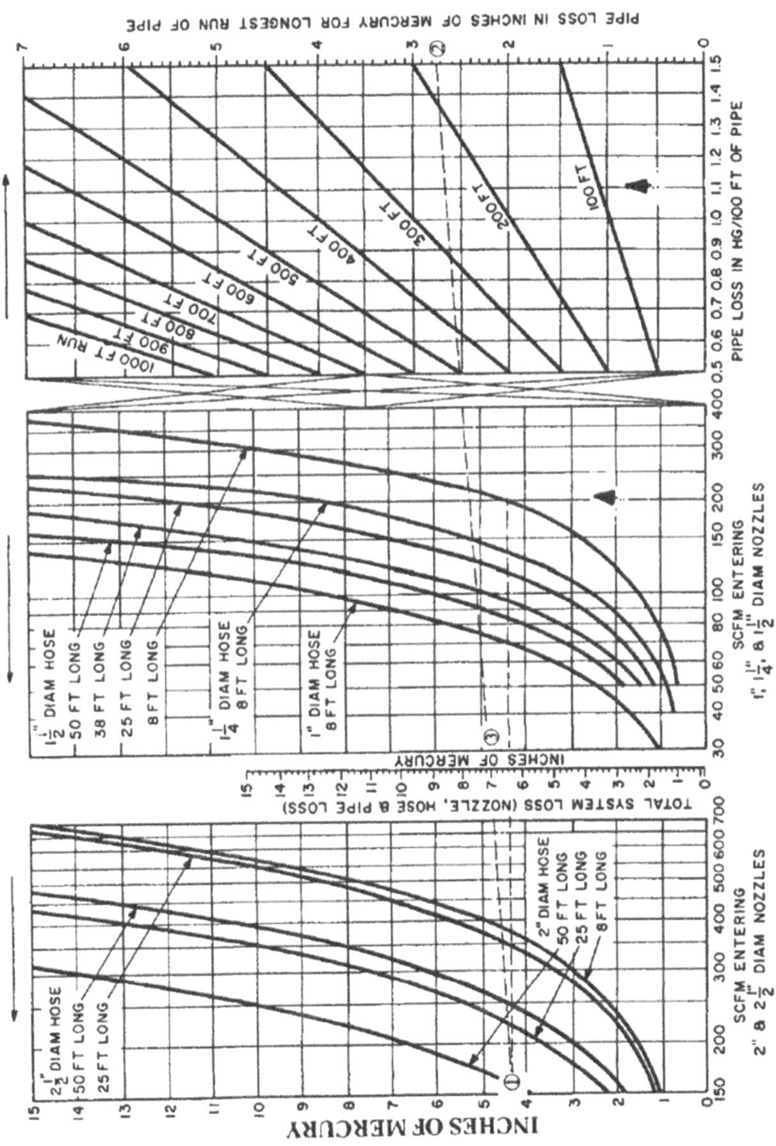

FIGURE 5.41 Low volume/high speed system design procedure.

approximately three to four times the recommended values in the UK. The static pressure of the extractor hood is 6 inches of mercury (20,340 Pascals) and American manufacturers and ventilation manuals recommend pressures in the range of 7–14 inches of mercury (23,730–47,460 Pascals). The design procedure for these systems is given in Figure 5.41. The nozzle entry loss is extracted from the diagram and is calculated along with the losses due to the flexible hose, air cleaner, and piping system leading to the multistage turbine. Since the system operates at approximately half-atmospheric conditions, the air volumes must be corrected to standard conditions.

The performance of low volume/high velocity systems has been investigated by American and British researchers.

These studies demonstrate high collection efficiencies for a variety of particulate pollutants using air volumes in the same range as those listed in ventilation manuals. The use of LVHV systems in portable sanding tools has rapidly increased, and many studies have investigated their effectiveness in dust control on automotive bodywork, aircraft composites, and woodworking. However, little attention has been paid to their application in stationary machine tools, where they have great potential. In addition to their use in particle control for portable tools, these systems have also been used for other air pollutants such as soldering and welding fumes. Since metal fume pollutants are released at low velocities, adequate control can be achieved with very low airflows. In one study, a soldering operation was investigated in which the pyrolysis products of solder fluxes were captured with high efficiency at flow rates of 1–2 cfm.

Recent efforts have been made to design machine tools based on the concept of low volume/high speed. In one study, a lathe was designed with a hollow blade to suck air with a flow rate of 50 cfm and by a hood static pressure of 7.5 in of mercury. In this system, the blade was mounted on a hollow rotating rod connected to a vacuum source and filter. This system is currently used in the United States for "resistant" milling boards made of Styrofoam (a white, moisture-, sound-, and temperature-insulating polymer obtained from petrochemical processes) and structural plastic machining materials.

6 Designing Laboratory Hoods

A. H. Khoshakhlagh, H. M. Moda, S. Yazdanirad, and A. Gruszecka-Kosowska

6.1 OBJECTIVES OF THE LESSON

Upon completing this lesson, readers should be able to:

- Identify the types of laboratory hoods
- Describe the characteristics of a laboratory exhaust system
- Describe the configuration of a laboratory hood system
- Explain the types of ventilation speeds
- Describe the factors affecting the performance of a laboratory hood
- Describe the role of layout in design

6.2 DESIGNING LABORATORY HOODS

Early studies on the suction of harmful air pollutants from laboratory activities were initially crude and rudimentary. Researchers conducted their experiments near a window or door to disperse air pollutants. In later years, a storage cupboard was converted into a work area and equipped with an exhaust chimney for natural ventilation. Some improvements in control were made using lights in the hood to indicate the movement of air convection. These mechanically ventilated workstations were first introduced in the late 19th century, and the first dedicated laboratory hoods were produced. Major design innovations to improve the capture of pollutants began in the 1950s.

This chapter will examine the benchtop hood used in chemical laboratories, also known as a "laboratory hood", "chemical laboratory hood", or "laboratory chemical hood". Equivalent terms for the laboratory hood, such as "fume hood" and "laboratory fume cupboard", are widely used in the European Union and, to a lesser extent, in the United States. This inconsistency with occupational health terminology was resolved in the British Standards Institution (BSI, 1994) by redefining fume as a gas, vapor, and particulate pollutant.

Since then, the laboratory hood has become the primary ventilation control device in education, materials control, and research. In this chapter, the focus of the chemistry laboratory hood is on the integration of the hood into the overall ventilation design and the effective use of the hood by laboratory personnel. Chemical storage hoods, auxiliary supply hoods, and ductless hoods are also discussed in detail.

DOI: 10.1201/9781003501091-6

6.2.1 Chemistry Laboratory Hood Design

In this chapter, the design of an advanced "benchtop" laboratory hood will be examined. The designer should understand why the main features of a standard laboratory hood should be considered to properly and effectively design a laboratory ventilation system.

6.2.2 Vertical Sliding Window Hoods

The first laboratory hoods were a simple fume cupboard, as shown in Figure 6.1. These hoods are classified as partial enclosures. The size and geometry of these laboratory hoods vary greatly. In most cases, the exhaust duct interface is placed at the top of the hood. Because of this geometry, the profile along the hood opening is poor, and the exhaust efficiency of the emitted pollutants in the hood is highly dependent on the emission point.

The first significant development in the design of this simple hood was the use of a vertical sliding hatch (Figure 6.2). The hatches are made of laminated or enameled safety glass. When access is not required, the hatch is designed to be partially or completely closed to improve pollution control efficiency, to dilute flammable airborne pollutants in the hood below the explosion limit, and to protect the operator from chemical splashes or projectiles from minor explosions or exothermic reactions.

With the hatch in the up position, the velocity can be highly variable across the hood opening, and the hood pollutant efficiency may be low. A simple modification to flatten the velocity profile is the use of a baffle with slots at the back (Figure 6.3), which in early designs had a slot at the work surface. Although this results in a flat horizontal velocity profile at the work surface, it does not improve the vertical velocity profile. Improvements in the velocity profile are achieved by adding a secondary adjustable slot at the top of the hood bonnet.

The uniformity of the velocity profile using this slot geometry allows the velocity at the opening to be maintained relatively constant at all places with the fully

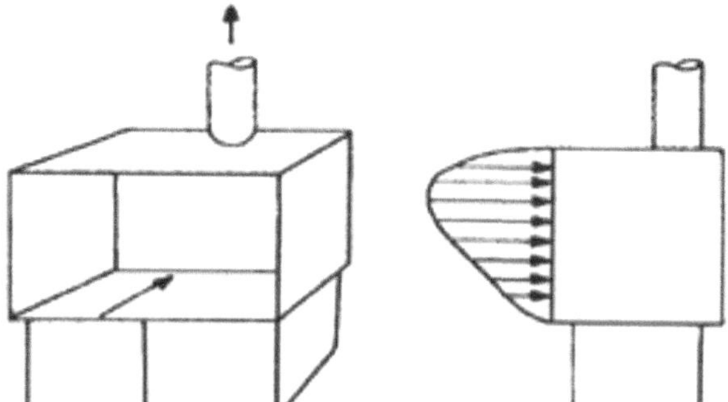

FIGURE 6.1 Simple exhaust hood or enclosure with a "duct connector" at the top. This laboratory hood provides a dedicated workspace but with poor ventilation control. Since most of the air enters through the top half of the hood, the velocity profile is not uniform. A duct connector placed at the back improves the velocity profile, but this geometry requires more space and is rarely used in laboratory hoods.

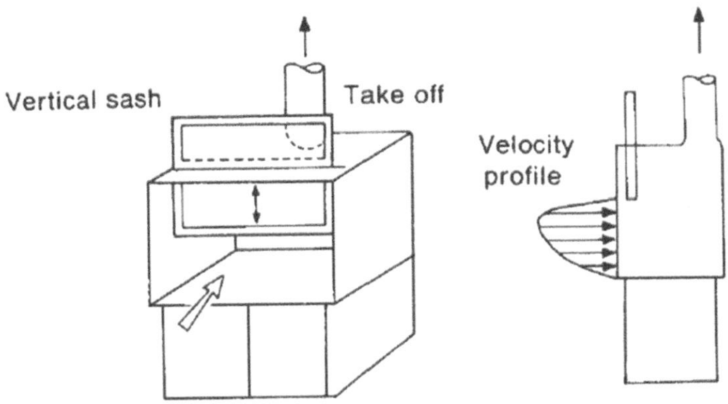

FIGURE 6.2 Use of a vertical sliding hatch allows the velocity and opening area to be varied. In early designs, the hatch was placed through a slot at the top of the hood, which could result in leakage under high pollutant production rates. In more recent designs, the hatch opens into the air bag or hood bonnet, eliminating the leakage path.

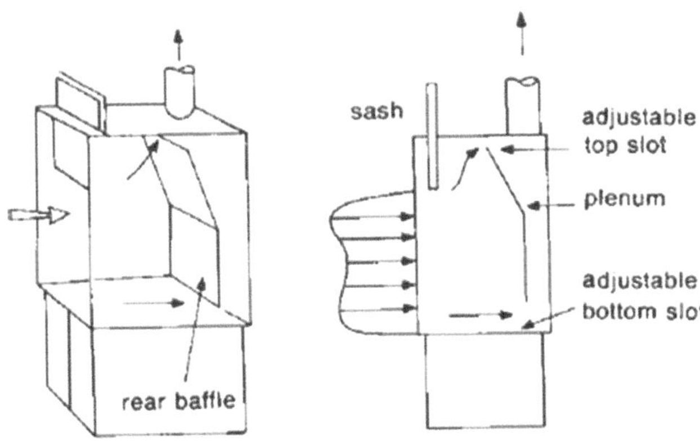

FIGURE 6.3 Plenum chamber slot arrangement introduced in the 1930s; a uniform velocity profile is formed across the hood opening. The slot size for velocities of 1000–2000 fpm (5.08–10.16 m/s) produces a drag factor that results in a uniform distribution of airflow along the slot length. Typically, the slots placed at the top and bottom are adjustable, and the middle slots are fixed. The slots should be adjusted to the optimum position during installation and maintenance unless conditions change. A perforated plate associated with the plenum chamber is an alternative to slots, although manufacturers are hesitant about this option because it is clogged by dust and requires periodic cleaning. With proper slot dimensions, a velocity profile at the hood opening is obtained with an average velocity variation less than 10%. Such a flat velocity profile ensures that the collection efficiency will be uniform along the hood opening.

open hatch. The number of slots required to obtain a reasonable velocity profile depends on the hood geometry and slot location (Figure 6.4). The effect of additional slots on hood performance has not been adequately studied. Poor design of the intermediate slots may contribute to the "circulation" of pollutants, resulting in leakage at the opening.

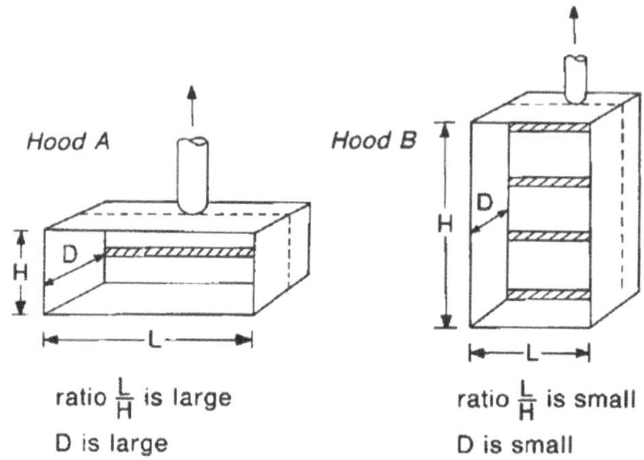

FIGURE 6.4 Hood (a) is a deep hood with minimal height and may require only one of two slots to properly distribute the airflow across the hood surface. A large shallow hood, such as hood (b), requires several slots to create a uniform velocity across the opening. Turbulence losses at the leading edge of the work surface or floor are reduced by an airfoil section (Figure 6.7). A properly designed hood with all these design features is shown in Figure 6.8. Studies have indicated that opening window handles may be associated with a significant impact on pollutant escape, and hood designers added an airfoil to the opening window to improve performance.

In general, the opening window and baffle at the back lead to improved hood performance. However, when the opening window is in the lower position, the air inside the hood may have a backward characteristic towards the hood opening, which can be indicated by smoke (Figure 6.5). In a poorly designed hood, pollutants may be expelled from the hood opening by this mechanism. Using the vertical opening window, another problem was added to the hood design. As the hatch is pulled down, the opening area decreases and the velocity in the hatch increases. This increase in velocity may be detrimental to open flames, sensitive balances, filter aids, and certain chemical activities. A screen should be fitted over the gaps to prevent material to be swept from the work surface into the duct.

To solve this problem, the bypass hood (Figure 6.6) was introduced in the 1950s. In this design, when the hatch is in its upper position, the opening covers the bypass, and all suction is through the hatch at the design velocity. When the pop-up window is pulled down, at the point where the pop-up window is approximately one-third closed, the bypass network is uncovered, and a fraction of the suction occurs through the bypass—and the fraction increases as the pop-up window approaches the closed position. This design ensures that the velocity at the opening does not increase significantly and the total hood suction volume from the bypass remains constant at all opening window positions, thereby simplifying the design of the laboratory supply system. It is important to design the bypass network and its enclosure to ensure that when the louver is closed, the minimum airflow through the hood is typically 50 cfm per foot of hood width (0.08 m^3/s per meter of hood width) or 20% of the airflow from the fully open louver.

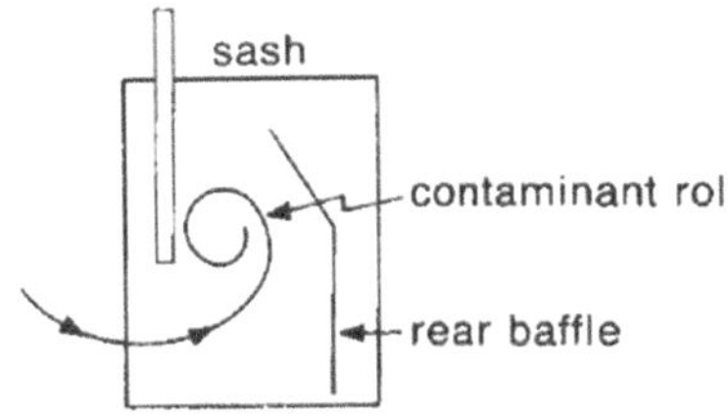

FIGURE 6.5 Air circulation characteristic behind an opening window in a hood that may lead to the escape of pollutants into the work environment.

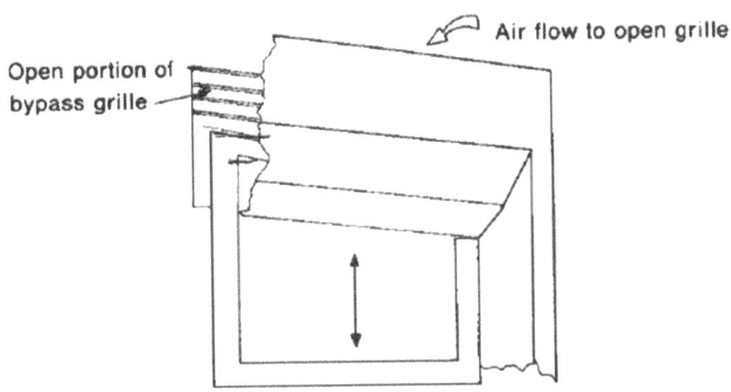

FIGURE 6.6 In a simple bleeder hood, the critical velocity at the opening does not exceed 300 fpm (1.56 m/s) when the window is down. In this position, there is also an increase in velocity at the opening until the upper edge of the window reveals the secondary path network, which allows air to enter the hood. The velocity at the hood opening then remains relatively constant until the window is closed. Manufacturers offer a variety of designs to provide this bleeder feature. In all cases, the bleeder mesh in the hood must be protected from debris generated by explosions.

Entry modification to improve the airflow pattern at the hood opening is a valuable step in the hood evolution. Based on the performed studies, the results show that the square vertical ports and the standard front edge of the hood's work surface make a slight bottleneck in the hood opening. The turbulent eddy currents cause flow changes and impair the hood's capture efficiency. To minimize this problem, the side legs are "adjusted" by an airfoil section (flat or curved blades like a wing), as shown in Figure 6.7. Although this change has little effect on the entry coefficient, it reduces the loss due to the pollutants by the boundary eddy currents. Moreover, a properly designed hood with all these design features is shown in Figure 6.8.

6.2.3 Horizontal Sliding Sash Fume Hoods

In the design of the laboratory hood, described in the above section, the researcher can create a suitable work environment by adjusting the vertical position of the sliding sash in the hood opening. It is also possible to achieve this goal with horizontal sliding plates located in the paths of the hood opening (Figure 6.9).

FIGURE 6.7 The turbulent eddy currents at the side posts of the hood can be minimized by modifying the inlet geometry of the hood. An ideal side stand features a bell-shaped opening. In practice, it is usually made with a simple angular shape that is cheap to build and install—but effectively minimizes inlet eddy currents and improves contaminant control at the hood inlet. Although the same simple geometry used in the side struts can be used to improve the inlet geometry at the front edge of the work surface, a more effective approach is to use an airfoil design. This design breaks up turbulent eddies, creates a well-defined airflow path for speeding through the hood's work surface when the window is down, and may also be used to determine the minimum work distance from the hood surface. The space between the work surface and the airfoil should be large enough for the operator to pass hose and electrical cables.

FIGURE 6.8 Modified fixed volume/bypass vertical sliding sash fume hood that includes various design features including an airfoil, bypass feature, and multiple slots in the back.

Although horizontal sliding sash fume hoods have been available for decades, they have only recently become widely used. In fact, they are an effective means of minimizing the opening surface, reducing airflow and operating costs. In addition to minimizing the opening surface, a claimed advantage for horizontal sliding sash fume hoods is the availability of a screen as a shield to protect against splashes and small explosions.

FIGURE 6.9 Horizontal sliding sash laboratory hood.

Critics of horizontal sliding sash fume hoods claim that most researchers reject this hood design because it requires different measures than the conventional vertical sash fume hood. If not properly secured, the window may be removed by laboratory personnel, causing the hood to become unsafe as a result. A combined vertical/horizontal sash fume hood can be the solution to this problem.

The experience of a large university, in the installation and commissioning of more than 1200 chemistry laboratory hoods, reported a successful result in achieving acceptance of horizontal sash fume hoods by a group of chemists. Large changes in the occupancy density of a new research building led to an unplanned increase in the number of hoods, which ultimately led to increased initial costs. It was decided to install 150 vertical-horizontal combination hoods which, due to their low suction requirements, could be serviced at the available capacity. Three test hoods were installed before final design and installation decisions were made. As a result of the successful installation, the experts were encouraged to use these hoods and share their opinions on the design. The "initial evaluation" was favorable, and 150 hoods were installed with excellent results. The designers concluded that the pilot of three test hoods played a significant role in the adoption of horizontal sliding sash fume hoods.

A vertical sash fume hood that has a sharp edge at the bottom of the window creates swirling eddy currents behind the window. In a horizontal sliding sash fume hood, air may pass through two or more places at the vertical sharp edge inlet. In a vertical sash fume hood, edge-induced turbulence may occur away from the pollutant generation point, which may minimize the vortex effect. In a horizontal sliding sash fume window system, vertical turbulence is created along the full height of the window and thus passes through the area where the pollutant is generated, which may

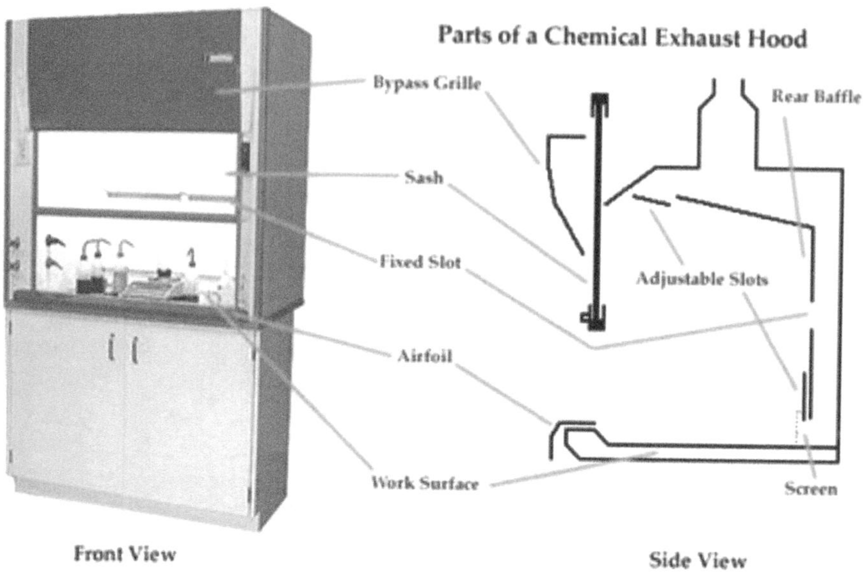

FIGURE 6.10 Front and side view of a laboratory hood.

result in significant concentrations of pollutants escaping from the hood. Eddy currents are formed at the edges of the vertical and horizontal sliding sashes, which can be minimized using an airfoil similar to the one shown in Figure 6.7.

Horizontal sash designs are different; the main differences are the width, the number of plates, and the number of paths. A minimum screen width of 18 inches (0.46 m) is recommended to minimize edge effects. However, the American National Standards Institute (ANSI) states that a sheet used as a safety guard should not be more than 14 inches (0.36 m) wide. In all cases, an airfoil must be installed in the front window.

Figure 6.10 shows the front and side view of a laboratory hood.

6.2.4 Auxiliary Air Supply Hoods

An auxiliary air supply hood, which is also called a compensating hood or supply air, is designed to minimize the amount of alternative air to provide the laboratory space in the cool season. The main use of this hood is in laboratories with high demand for air that is not supplied by the existing system inside or in places where there is a strong thermal load. The system presented in Figure 6.11a can be used as a complete hood system or as a canopy system to augment conventional hoods, which are typically designed for use with vertical sliding window openings. However, it can also be applied to horizontal sliding window systems.

When used with a horizontal sliding opening window hood, complete survival of energy is not achieved and may cause additional turbulence around the hood opening. In this design, instead of receiving 100% of the total suction of the hood from the room, at least 35–40% of the laboratory space is taken, and a maximum of 60–65% of the suction is from the outside air, which is directly transferred to the opening of

FIGURE 6.11 (a) An auxiliary air supply hood with a suitable design with a side path. Air from outside the canopy enters the system to create a laminar flow pattern that covers the opening of the hood and mixes with room air when it enters the hood. The hood is designed to supply air at room temperature during the heating season. Unconditioned air is supplied during cooling periods, thereby saving air conditioning costs for supply air. When a person stands in front of the laboratory hood, a small negative pressure is created between the body and the hood, causing turbulence and the escape of pollutants from the hood. An auxiliary air source fills this space, eliminating turbulence and improving braking efficiency; (b) Egg crate grille window.

the hood. The issue of economy is also raised in the summer season, since unconditioned outside air or semi-conditioned air may be used for most of the suction. In winter, this section should be heated to room temperature to ensure the comfort of the operator and to minimize condensation problems in the hood.

If possible, the canopy source can be used as secondary air, i.e., air that is conditioned and typically must be extracted to meet general ventilation requirements. In any case, the air must be conveyed to the outside opening of the hood by an appropriately designed canopy source. A canopy is typically used with an egg crate grille window (Figure 6.11b) to approximate the laminar flow along the hood opening. Designs that deliver air directly into the hood cavity should be avoided because they are disruptive, pressurizing the hood and allowing pollutants to escape into the room.

An auxiliary air supply hood with a suitable design reduces air conditioning costs and improves hood capture efficiency. Disadvantages of this hood include higher initial and maintenance costs due to the excess air replacement system, the potential entry of airborne dust into the laboratory, and unfavorable conditions for operators during winter unless the weather is mild. Additionally, poor canopy designs can disturb the suction velocity profile at the hood opening. In high-humidity environments, condensation inside the hood poses a potential hazard if the ventilator fails while the auxiliary ventilator continues to operate in a poorly designed system.

6.3 VELOCITY AT THE OPENING OF LABORATORY HOODS

Discussing face velocity of a laboratory hood may seem unnecessary, since extensive discussions are available about face velocity of industrial hoods. However, experience has shown that special attention should be paid to face velocity criteria for laboratory hoods due to the unique application, the range of toxic chemicals used, and the location of the hood in the laboratory.

By definition, a lab hood is an enclosure with an adjustable panel accessible from the front. Face velocity is an important design parameter for a laboratory hood. Capture velocity is the velocity generated outside the physical boundary of the hood to entrain the contaminated air, which is often incorrectly used to determine face velocity. "Opening surface" indicates the working span, where the design velocity at the opening should be considered. Other terms, such as "slotted window opening" or "door surface", are also used. The design airflow is based on the average velocity at the opening and the design surface of the hood at the opening with a vertical or sliding opening window configuration. As mentioned earlier in this chapter, hood airflow must meet minimum fire protection standards when the window is closed. If the hood is the only source of general ventilation in the laboratory, the minimum airflow should be provided under closed window conditions.

The standard opening height for a desktop vertical sliding casement window is 30 inches (0.76 m), though it typically ranges from 26 to 36 inches (0.66 to 0.91 m).

Laboratories leading in energy efficiency have found that opening heights of up to 20 inches (0.51 m) may be acceptable for routine laboratory work. In this configuration, the standard height of the total opening is 30 inches (0.76 m).

Research into face velocity for laboratory hoods provides three suggested approaches. The first approach is a logical format for calculating face velocity for a specific chemical used in a laboratory environment. This calculation requires the user to consider a wide range of important variables that affect the containment of pollution by the hood. In the second approach, the graded groups or the quality of the hoods are examined and the minimum speed at the opening is determined for each group. Classifications may be based on: (1) application realm with inferred risk level, (2) toxicity of chemicals used under the hood, or (3) a classification scheme that includes consideration of chemical toxicity and laboratory geometry. A third approach is to use a recommended face velocity for all hoods based on the principle that a minimum face velocity is required to overcome background air movement in the laboratory to control any chemical contaminant in the hood.

Although the responses to these approaches—presented over several decades—are diverse, general trends have emerged in this field. Chemical-by-chemical procedure provides extensive insight into hood performance design criteria. However, it has not been widely adopted, probably because of its complexity. Many laboratories use broad classification schemes, but critics of this approach state that it is difficult to predict the types of chemicals and methods of use under laboratory hoods, even in the short term. This procedure also limits the use of the hood to selected chemicals, thus complicating laboratory performance and monitoring. The most common method for determining the minimum face velocity is the third approach, which uses a face velocity that allows the hood to be effective across a wide range of chemicals, activities, and laboratory environments.

The suggested velocities at the face are very different from these approaches. The minimum velocity required to overcome the emission of gases and vapors from the hood is 20 fpm (0.1 m/s). During the 1950s, when stratification designs were popular, a wide range of velocities from 50 to 150 fpm (0.26 to 0.76 m/s) were widely used. The velocity of 100 fpm (0.51 m/s) was proposed during the 1960s under the single-value approach and was widely supported until the 1980s. In fact, at 100 fpm (0.51 m/s) most chemicals in industrial laboratories are controlled using logical design procedure. This speed is also suitable for overcoming the turbulent airflow created by a chemist working under the hood. For these reasons, a velocity of 100 fpm (0.51 m/s) is considered appropriate for good design.

Energy constraints introduced in the mid-1970s created a wave of research into methods of reducing hood operating costs by decreasing the total amount of suction air. A common approach is to reduce face velocity; however, applying a design face velocity of 100 fpm (0.51 m/s) has been widely criticized. The results of the studies showed that face velocity up to 60 fpm (0.3 m/s) is sufficient if an aerodynamic hood is used in an almost ideal environment and also if the operator follows precise working methods. Critics agree with this logical control approach under ideal conditions with low face velocities, but replicating these operating conditions in the laboratory, or in the "real world," is challenging.

This procedure is clearly stated in the new laboratory design guidelines, which recommend a face velocity of 80 to 100 fpm (0.41 to 0.51 m/s), depending on the direction of the supply airflow and the uniformity of hood face velocity. The working distance behind the opening window should be maximized so that the air replaced by perforated ceiling panels, grids, or diffusers results in a "throw" hood face velocity of less than 30–50 fpm (0.26–0.02 m/s). In addition, air movement caused by doors, windows, and traffic, as well as the number of devices used under the hood, should be minimized.

Although, there is still disagreement about the minimum required face velocity for laboratory hoods, most researchers agree that increasing face velocity above 120 fpm (0.61 m/s) does not improve the capture efficiency, and at velocities above 150 fpm (0.76 m/s) the performance of the hood may decrease because of the turbulence in the hood opening. Expected operating characteristics at hood face velocities from 60 fpm to greater than 150 fpm (0.76–0.31 m/s) are described in ANSI standards.

For effective restraint, a uniform velocity should be applied along the opening of the laboratory hood with an aerodynamic design. Acceptable variations in face velocity under actual laboratory conditions include a maximum opening of the hood window not exceeding 10% (ACGIH), or a face velocity less than 90% or greater than 120% of the acceptable value (ANSI).

Although face velocity is an important predictor of performance, it is not the only influential factor. Other factors discussed later in this chapter that may have a significant effect on containment include hood design, facility layout, cross drafts, air distribution, and traffic.

6.4 SPECIAL LABORATORY HOODS

The laboratory hoods, described in the above section, are valuable control devices. These hoods are a physical barrier to prevent leaks, splashes, and minor accidents caused by exccssive pressure, as well as the relatively low face velocity that will inhibit the contaminants released from chemical activities. Although laboratory hoods have impressive features, they should not be the only control solution in laboratories. Typically, many processes such as evaporation, distillation, and hydrolysis are performed in laboratories, and a wide range of local exhaust hoods can be selected for these activities. The selection of these special-purpose hoods—which have the advantages of low initial cost, low airflow rate, and good capture efficiency—is explained below.

Sample hydrolysis using strong oxidizing agents, such as perchloric acid, is frequently performed in different environments of biological and metallurgical laboratories. This process requires excellent containment and air cleaning to ensure that no violent reactions take place. Organic materials cannot be used in the construction of the hood or duct, and spray nozzles must be installed to wash the hood's ductwork network—as well as the aerator must be used to prevent the accumulation of unstable perchlorates. The hood used for the perchloric acid hydrolysis process should be exclusively dedicated to this activity. In one study, a scrubber was designed specifically for the hydrolysis of perchloric acid, which was located directly under a standard laboratory hood. After cleaning, the scrubber's output stream was entered into the laboratory hood's output stream.

A common activity in chemistry laboratories is paper chromatography. In this process, a sheet of paper with a liquid sample is placed in a larger glass container, where the environment is saturated with a solvent to create a chromatogram. Changing samples leads to the release of solvent vapors into the air. This can be done under a standard lab hood but can be more effectively controlled with a portable external hood placed directly behind the glassware during transfers. This design provides excellent control using moderate airflow rates of 250–300 cfm (0.12–0.14 m^3/s). The reduction of installation and operating costs in the use of this special hood compared to the usual laboratory hood is significant.

A common procedure in material control laboratories is to evaporate a large sample of material. Instead of performing this procedure under the hood of the laboratory, it can be controlled more effectively using point local exhaust ventilation. One of the more challenging examples to control in the laboratory is distillation or reaction equipment mounted on a large rack, which has the potential to release air contaminants from multiple points. A "walk-in hood" can be used for this, but access restrictions and large suction volume are disadvantages. An alternative approach is to install the equipment outdoors and install one or more flexible drops that can be installed at critical emission points. Typically, a 3-inch (7.62 cm) diameter flexible duct with a suction volume of 200 cfm (0.09 m^3/s) is suitable for controlling most accidental emissions. Finally, for maximum control of contaminants, a fully enclosed hood is ideal.

6.5 LABORATORY SUCTION SYSTEM CHARACTERISTICS

6.5.1 System Configuration

Whether for use in fans separately for each hood or for multiple hoods fed by one fan in a system, one of the first decisions that must be made is the design of the laboratory ventilation system. Each installation should be considered individually. A third option is to use fans for hoods with discharge to a homogenizing chamber powered by a high-volume diluting-boosting fan.

6.5.1.1 Single Hood System

At first glance, the "single hood/single fan" system may seem desirable. This system provides maximum flexibility for every operator and laboratory manager. In fact, the expert has access to a single system, with the ability to turn it on or off entirely at their discretion. In addition, the hood is suitable for use in off-shift periods. If a classification scheme is used to select hood face velocity, the hood can be upgraded to a higher classification by changing the fan velocity or installing a new fan to achieve a higher face velocity. If air cleaning is necessary throughout the hood's suction flow, the additional resistance to airflow caused by the air cleaner may be managed by adjusting the size of the sheave or air mover. If a new special hood is needed, such as a special hood for the perchloric acid hydrolysis process, a single hood system can be considered for this purpose.

The design of such systems is simple, and modular design can be used in the entire center. If repairs and maintenance are necessary, the system must be turned off, and as a result, the operator's activities are disrupted. Disadvantages of the single hood/single fan system include the higher initial cost and the additional space required for several ducts and fans. If there are multiple hoods in a lab and one hood is turned off, the absence of replacement air may result in flow back into the lab through the inactive hood duct. At best, it may cause extreme temperatures in the vicinity of the hood. In the worst case, if the duct leads to an area with polluted air, it can introduce toxic air contaminants into the laboratory. Installing backflow dampers are not effective in preventing this, because these devices have low performance in the closed position. Finally, due to the hood utilization factor, it is difficult to design a suitable supply system for the "single hood/single fan" approach.

6.5.1.2 Central System

In a central system, several hoods are fed by a fan, which can be used as long as the air contaminants released from the different hoods meet suitable conditions. Examples of fires, explosions, or health hazards caused by the mixing of hood exhaust streams have been reported in laboratories. In general, the initial cost of a centralized system may be lower than that of single-hood systems, although the balanced design of such a system requires more engineering skill and maintenance. Since all hoods in a central system operate continuously, the air exchange system can be a constant flow design to meet all system needs.

6.5.1.3 Progressive and Collection System

A combination system that is becoming increasingly popular due to its ability to deliver high flow rates of progressive air and ensure excellent dispersion of fan exhaust flow from a point source is the progressive and collector system. In this system, where separate fans are used to feed the hood, the air is not discharged directly to the atmosphere, but to a large equalizing chamber or a multi-way network, where a high flow and low static pressure fan is also installed. A long single vent with high velocity discharge increases contaminant dispersion compared to multiple hood exhaust.

If the single unit design approach (single hood/single fan) is chosen, the hoods should not be used with an integral fan mounted on the hood, since this configuration will pressurize the duct path. Data and leakage of contaminated air into occupied spaces may occur. In addition, integrated fans used with laboratory hoods are usually direct-drive, low static pressure fans, which cannot control the resistance to airflow provided by the long duct paths.

In any system, the fan must be installed in such a way that the duct path inside the building is under negative pressure, which means that a fan is installed in the roof for a multi-story building or, for a single-story building, the fan is installed in the ground or roof. If a short part of the duct route inside the building is under positive pressure, the duct must have welded seams, flanged connections with gaskets, or sealed connections with an impervious material.

Fans for large laboratories using single-hood systems are typically installed in rooftop chambers to facilitate maintenance. This placement protects the fans from weather exposure, ultimately aiding in their upkeep and repair. However, this practice may pose a risk to maintenance personnel in case of duct failure downstream of the fan inside these chambers. The chamber must be equipped with progressive ventilation and all the connections in the duct paths under positive pressure must be sealed inside the chamber.

A wide range of structural materials have been introduced for these special environments. Although stainless steel is a common hood sheathing material, currently, mild steel plates coated with polyvinyl chloride, polyethylene, or powder coating paints are used. Solid plastic construction materials are rigid polyvinyl chloride, polypropylene, and composites based on amino and epoxy resins. Powder coating with epoxy colors is also widely used. The use of asbestos-based materials is obsolete, although exposure to asbestos fibers may still occur when old hoods are renovated and removed.

The primary path of the duct in the hood should be installed with an elbow close to the hood to prevent material from falling into the hood. If the purpose of using the hood is to control the condensing steam, the first few meters of the duct should lead to a drainage point and then trap the exhaust. All the longitudinal seams in the horizontal ducts must be positioned on top of the duct during installation.

The size of the duct should be designed according to the speed of 1000–2000 fpm (5.08–10.16 m/s). If the particle load in laboratory systems is high, higher transfer speeds in the duct are required.

The choice of fan will depend on the type of system used. Axial flow fans are often used in single-hood systems due to their low cost, straightforward geometry, and ease of installation. This choice ignores many of the advantages of a centrifugal fan, including the flexibility to change fan speed. Large systems that use multiple hoods are typically equipped with centrifugal blowers for durability and high performance. Fan blades can be made of polyvinyl chloride, fiberglass-reinforced polyester, or steel with an impermeable plastic coating. If flammable gases or vapors or explosive dusts are present, a fiberglass-reinforced polyester, aluminum, or bronze impeller may be required.

The re-entry of odorous or toxic air contaminants from suction fans is a common issue in many industrial and academic laboratories. Design data helps minimize this concern, but ultimately, only proper air and gas cleaning can solve this problem. The conventional industrial approach to air cleaning is to direct the entire exhaust flow of the hood to an air cleaning device. This approach will not be effective for a general purpose laboratory hood since the outlet stream has very low concentrations of contaminants in a relatively high airflow. Such systems do not have the ability to clean the air.

The most effective solution to this problem is to direct the low-volume emissions from the reaction occurring under the hood to an air cleaning unit in the hood before they reach the main suction system. Most of the general chemistry laboratories can control the values of 0.1–1 M. Pollutant emission rates, assuming a 20-minute reaction time, are in the range of 0.1–1 lpm. Acidic and alkaline air contaminants can be washed in simple packed columns.

Air purification technology related to respiratory protection has grown rapidly since the early 1950s. A wide range of adsorbents and surface adsorbents, available in cartridges, chin canisters, and chest canisters may be effective against certain exhaust gas flows. Unfortunately, it is difficult to convince the user to clean the air or "wash" the contaminant stream before it reaches the hood suction. The recent trend towards "green chemistry" by the community of teachers and researchers in the field of chemical agents may support this approach in the future.

6.5.2 Factors Affecting Hood Performance

Multiple factors that affect the field performance of the laboratory hood and should be considered by the designer are discussed below. A series of factors related to the physical location of the hood and other components in the laboratory may disrupt the airflow in the hood opening. In addition, the performance of the hood also depends on its proper use by the user.

6.5.3 Laboratory Arrangement

6.5.3.1 Corridors

The passage door installed near the opening of the laboratory hood may seriously disrupt the performance of the hood. A normally opened door acts like an air mover. If a 30-inch (0.76 m) × 80-inch (2.03 m) door is rotated through a 90-degree arc in 2 seconds, its edge velocity is approximately 200 fpm (1.02 m/s), and it moves the room air at a flow rate of 1200 cfm (0.57 m^3/s). This effective pumping rate is approximately equal to the airflow through a 5-foot hood with an opening window height of 30 inches (0.76 m) operating at an orifice velocity of 100 fpm (0.51 m/s). The results of a study showed that the speed of induced air caused by the door opening adjacent to a hood varies between 175–450 fpm (0.9–2.29 m/s).

This door, located near the hood, can create airflow along the hood opening temporarily equivalent to hood suction, which disrupts the airflow pattern and propels contaminants out of the hood at high speeds. Therefore, it is important that the hoods are located away from the laboratory access doors. If it is necessary to install the door near the hood, using a door governor can ensure low opening and closing speeds, minimizing its disruptive effects. Another solution is to limit access to the laboratory through a Crittall door. Crittall-style doors allow for intelligent division of open spaces, creating distinct areas without causing space problems. Since the hood may be located near a fire, explosion, or toxic chemical spill, it is essential that the exit route does not pass through the hood.

6.5.3.2 Pedestrian Traffic

Traffic around the hood can impact its performance, and this should be considered in the laboratory layout. While wearing a button-down lab coat, you move back and forth across a cross-sectional area of approximately 10–20 square feet. At a slow walking speed of 2.3 mph (mi/h) or 200 fpm (1.02 m/s), 2000–4000 cfm (0.95–1.89 m^3/s) of air is moved. This is equivalent to the suction of an 8-foot-long laboratory hood with a 30-inch (0.76 m) window opening, which operates at 100 fpm (0.51 m/s) at the opening. If a person walks in front of the laboratory hood, the induced air movement at 260–450 fpm (1.34–2.29 m/s) can cause the contaminants to escape from the hood (Figure 6.12). To minimize this problem, a clear zone should be established 1 m from the hood opening, accessible only to the user.

6.5.3.3 The Windows

Most laboratories are equipped with air conditioning, and even if there are windows, the opening windows are permanently closed. However, if a window is opened in a laboratory, an outdoor wind gust of 3–5 mph (300–450 fpm or 1.52–2.29 m/s) can blow across the hood opening, which can be disruptive and cause pollutants to escape from the hood.

FIGURE 6.12 The impact of traffic on hood performance.

TABLE 6.1
Key Factors to Minimize Airflow Disturbances at the Hood Opening

	Minimum Distance from Hood Opening	
Points to Consider	**ft**	**m**
The typical passing through	3.3	1
The front desk used by the user	4.9	1.5
The hood used by the user	9.8	3
Opposite wall	6	2
Adjacent wall	1	0.3
Corridor	3.3–4.9	1–1.5

6.5.3.4 Equipment

Optimal performance of a laboratory hood occurs when the air movement in front of the hood approaches a laminar flow perpendicular to the hood opening. Placement of laboratory tables, hoods, and partitions can interfere with this airflow. To minimize this interference, the minimum clearances specified in Table 6.1 should be maintained.

Another important laboratory feature that affects hood performance is the introduction of replacement air. However, due to the compact arrangement of hoods and the variety of hood uses, it is difficult to achieve an adequate air supply without interfering with hood performance due to drafts of air in the hood opening. In many cases, nets and diffusers cannot be used without creating strong air drafts. The only solution is to use perforated walls or ceiling supply plates.

A low velocity airflow from the power supply to the hood opening is rarely achieved in a laboratory. The installation of the supply/suction system is equally difficult due to the pressure differences between the laboratory spaces and the side offices. A general rule of thumb is that the air should enter as far from the hood as possible, with reciprocating movement in the laboratory, to achieve maximum effective ventilation.

Another principle is to keep the discharge speed of the air supply networks below 400 fpm (2.03 m/s) and ensure that the air velocity from the network or diffuser, ranging from 30 to 50 fpm (0.15–0.26 m/s), does not exceed the opening of the hood. As mentioned in the above section, if the laboratory is small, replacement air can be introduced into the system through a roof-mounted perforated uniformizer chamber at a speed of 70–100 fpm (0.36–0.51 m/s). Such a system provides excellent mixing in the laboratory space and minimizes disturbance effects at the hood opening.

6.6 WORK PRACTICES

Even if the initial design of the hood is perfect and it is placed in a "sheltered" location within the laboratory, contaminants may still escape. Poor work practices can negatively affect hood performance; for example, a common practice is using the fume hood as a storage area for chemicals (Figure 6.13). Typically, multiple chemical

FIGURE 6.13 An example of a cluttered hood that could disturb its performance.

bottles are placed deep in the hood, blocking the bottom suction slot. Blocking this gap disrupts the airflow balance between the lower and upper openings, reducing the hood's performance.

Chemical storage areas should be located elsewhere, in accordance with safety guidelines. A flammable liquid storage rack requires a suction flow of only 50 cfm (0.02 m^3/s). If it is necessary to store a lot of bottles inside the hood, a shelf can be installed above the bottom slot. However, shelves should not be placed on the side plates of the hood.

Another source of disruption is storing different boxes of materials or equipment inside the hood. If a large item, such as an oven, mixer, or hot plate, must be placed in the hood, it should be positioned carefully to minimize disruption to the hood's velocity profile. Equipment should be placed at the back of the hood on stands or blocks, raising the height by 1–2 inches (25–50 mm) to avoid blocking the bottom opening of the hood. The equipment should also be positioned to create the smallest possible cross-section in the natural direction of airflow.

The depth of the working surface is an important variable in the performance of the hood. For example, in industrial ventilation systems, a spray booth with a larger or deeper spray area in the hood requires a lower face velocity—the same guidelines apply to laboratory hoods. If the work is done directly in the opening, the contaminant may escape from the hood and not be captured. If this operation is carried out deep in the hood, the ability to contain the contamination in the hood is improved. Therefore, if work can be done at the back of the hood, the airflow speed at the hood opening can be reduced. However, users should be cautioned not to lean on the hood or share it with others.

The appropriate working depth can be determined by marking the desired position on the working surface of the laboratory hood with adhesive tape, placing a physical marker on the surface, or using the extended part of the airfoil. This process will physically specify the minimum working depth. Work practices have an important influence on the hood's ability to contain pollutants, as outlined in the comprehensive standards.

These standards recommend that: (1) all equipment be placed inside the hood compartment before starting work, and (2) equipment be positioned at least 6 inches (150 mm) behind the opening window and at least 2 inches (50 mm) from the back gap. These actions limit the work surface and are often ignored by the user as a result.

The height of the convective airflow generated by the heat-producing device in the hood can strain the suction system and disrupt its performance. In one study, the role of four Bunsen burners (a standard source used to generate heat in the laboratory) placed in a 4-foot hood was investigated. These lights resulted in an air temperature of 38°C in the upper part of the hood, which is higher than the room temperature. The results showed that the convection currents created were disruptive to the effective collection and containment of contaminants. Additionally, the tracer smoke released in the hood during this temperature rise entered the room through the openings via a bypass mechanism.

In another study, three types of laboratory lights (Figure 6.14) were used to generate a range of heat loads. The standard containment test was performed using sulfur hexafluoride tracer gas to evaluate the effect of heat load on hood performance. Each increase of 10,000 Btu/h (British thermal unit per hour; 2931 W) in heat load over the test range of 69,342 Btu/h (20,324 W) resulted in a 4 ppm increase in the tracer air concentration in the breathing zone of the test dummy.

The three lamps used in this study include an economic lamp with a thermal load of 5000 Btu/h (1465 W), a Maker burner (a laboratory gas lamp distinct from a typical Bunsen burner, featuring a compression in the tube and a grid on top; the flame consists of several short blue inner cones and a single large outer cone, making it hotter than a Bunsen flame) with a heat load of 10,000 Btu/h (2931 W), and a blast burner with a heat load of 40,000 Btu/h (11,724 W). Additional studies using hotplates, ovens, water baths, and other heat-producing laboratory devices are needed to determine operating parameters for the safe use of the hood.

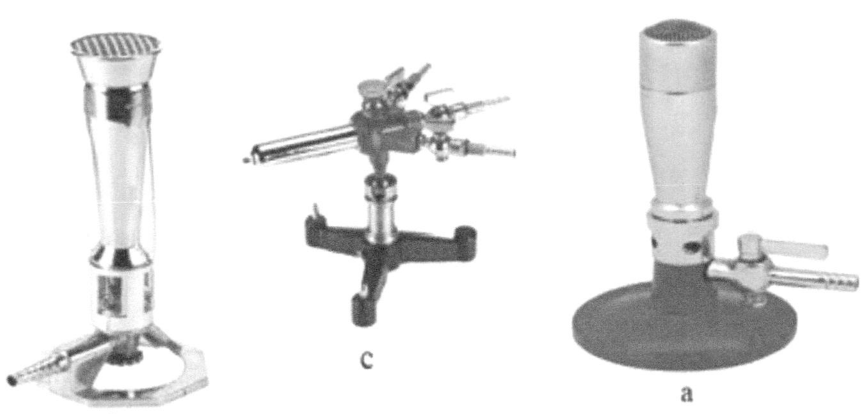

FIGURE 6.14 Visualization of the three types of heaters: (a) Economic lamp; (b) Maker burner; (c) Blower lamp.

In a study, it was reported that the use of electric heaters should be limited to 100 W/ft of hood length to minimize the thermal effect. If the heat load in the hood is high, the top gap in the hood should be adjusted to draw a larger portion of the airflow. As previously mentioned, the chemist should place all equipment and supplies under the hood before starting a process, because changes during the chemical procedure may lead to the escape of contaminants. Also, the proper action is to close the opening window as much as possible; as a result, a complete enclosure is formed and the escape of pollutants is minimized. A study found that, when training chemists on proper work methods, emphasizing the lowering of the window opening as the key factor in reducing hood leakage is essential. Closing the window opening serves as a barrier, providing protection against sprays, splashes, and minor explosions.

6.7 PERFORMANCE OF LABORATORY HOODS

The quantitative measurement of hood performance is currently done using gas detectors. The tracer gas is released from several points or a spreader located inside the hood. Leakage of the tracer gas from the laboratory hood is measured using a multiple-point sampling grid or a probe installed on the manikin's head in front of the hood opening. In the ASHRAE 110 standard, sulfur hexafluoride is used as the tracer gas, and the concentration of the tracer gas outside the hood is measured with an electron capture detector. This detector operates based on the reduction in the flow of electrons, which are absorbed by electron-affinity components. The application of this standard is suitable for hood performance testing under the following conditions: (1) an ideal test room without draft airflow or cross currents, (2) an empty laboratory space with an operational HVAC system, and (3) a real work environment in the laboratory.

7 Design of Single Hood Systems

A. H. Khoshakhlagh, H. M. Moda, S. Yazdanirad, and A. Gruszecka-Kosowska

7.1 OBJECTIVES OF THE LESSON

Upon completing this lesson, readers should be able to:

- Describe the general design principles of single-hood systems
- Outline the design approach for single-hood systems
- Identify the key factors that affect the design of single-hood systems
- Design a hood system with slits

7.2 DESIGN OF SINGLE HOOD SYSTEMS

In this chapter, local exhaust ventilation systems for single-hood configurations are described using the pressure-velocity design method. This method has been widely used for over 50 years, primarily due to its simplicity in initial calculations and the ease of redesigning system components when necessary. The concept that all losses from turbulence and friction are directly proportional to pressure-velocity provides a rationale that is easy for novice designers to understood.

7.2.1 Design Approach

The standard design procedure for a hood system follows a five-step process:

Step 1: Select the geometry of the hood.

Step 2: Calculate the required airflow. The airflow specified in the design document is a critical element of any hood design.

Step 3: Specify the minimum duct velocity. Information regarding the required velocity is included in both the duct specifications and the design document. For hoods used to handle gases and vapors, the typical velocity in the duct ranges from 1,500–2,000 fpm (8–10 m/s). For particle transfer, the minimum velocity is typically between 3,500–4,500 fpm (18–23 m/s).

DOI: 10.1201/9781003501091-7

Step 4: Select the duct size. The size of a round duct is chosen to achieve the minimum required duct velocity for a given airflow. The maximum cross-sectional area of the duct is calculated by dividing the airflow by the minimum velocity. Selecting a standard duct diameter with a smaller surface area ensures that the minimum velocity is met. The actual velocity in the duct is then calculated based on the duct's cross-sectional area and the airflow.

Step 5: Design the duct route. Proper placement is crucial to minimize energy losses caused by direct duct paths and elbows, particularly in cases where space limitations arise due to architectural or equipment constraints. Multiple route options should be considered, and it is advisable to consult with plant engineers and process managers during this process. An important consideration is the location of the exhaust stack (chimney). Two common issues must be addressed. The first is preventing the re-entry of extracted contaminants, and the second concerns potential infiltration through the roof or walls of the building. The locations of such intrusions should be carefully examined and coordinated with building management.

Step 6: Calculate the energy losses. Once the duct placement is finalized, losses due to friction and turbulence caused by the airflow through the system can be determined using equations, graphs, or tables. This static pressure drop can be calculated directly or as a multiple of the velocity pressure, which is then converted to static pressure.

7.2.2 Designing a Simple Single-Hood System (Banbury Mixer Hood)

The first example of system design is the projectile mixing hood (Figure 7.1), commonly used in the rubber and plastics industries. This hood is employed to mix elastomeric materials with accelerators, antioxidants, antiozonants (organic compounds that prevent or delay the damage caused by ozone), pigments, plasticizers (polymer additives that increase softness and flexibility by reducing viscosity or elastic modulus), and curing agents (vulcanizing agents that create cross-links in rubber). Feeding or charging these granular materials into a projectile mixer is a dusty operation, necessitating effective local exhaust to control airborne particulates. The standard method involves using a partial chamber on the port of the feeding (Figure 7.2). In addition, two smaller hoods are typically installed at the ends of the trunnion to control the dust released from the dust rings. In this example, only the feeding port is covered with ventilation. The ventilation system to be designed consists of a hood, fan, and network of connecting ducts. The primary goals of the design process are to determine the minimum required airflow (cfm) and total airflow resistance (in. H_2O) for the system. This design data will then guide the selection of an appropriate fan.

Step 1: Select the geometry of the hood. The design data presented in Figure 7.2 can be used to develop the system design for the hood on a Banbury mixer.

FIGURE 7.1 A Banbury mixer is used to mix rubber and plastic materials with various additives. This figure shows the charging platform at the top level, where raw materials (stock) and additives are fed to the mixer. Once feeding is complete, the charging door is closed, a ram seals the chamber, and the materials are mixed. When mixing is complete, the bottom dump door opens, and the mixed material flows into a mill located below the Banbury mixer on the lowest level.

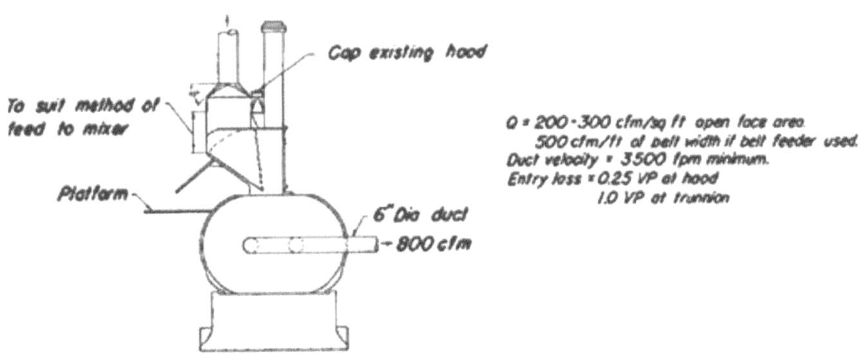

FIGURE 7.2 This design sheet for a projectile mixer outlines four key design components for the engineer: (1) hood design and placement, (2) minimum airflow required for effective control, (3) minimum velocity in the duct, and (4) hood inlet losses.

Step 2: Calculate the required airflow. In the example above, inspection revealed that the Banbury mixer was manually fed, without the use of a belt. Additionally, there were no strong air drafts from doors, windows,

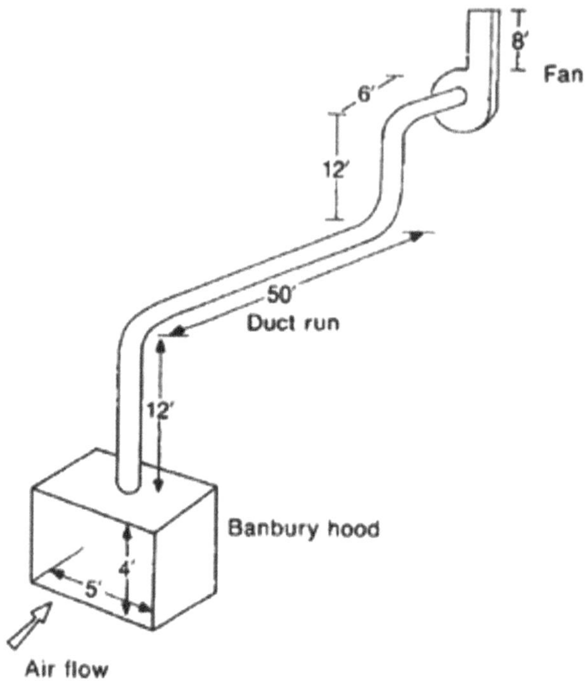

FIGURE 7.3 Layout of the Banbury, duct, and fan. The hood opening measures 5 feet in width and 4 feet in height, with the lengths of the ducts indicated. The sources of static pressure drop in this system include: hood inlet losses, inertia losses, direct duct frictional losses, and losses caused by elbows.

or traffic. Based on this information, a minimum suction of 200 cfm/ft^2 at the opening surface (equivalent to a face velocity of 200 fpm) was selected. If the projectile mixer was installed near a receiver door with strong air drafts, or if the air drafts were created by pedestal-type cooling fans, a higher velocity value would be required. In this case, a face velocity of 300 fpm would be chosen, and a belt feeder with a flow rate of 500 cfm/ft^2 would be necessary. As shown in Figure 7.3, the hood is 5 feet wide and 4 feet high, so the airflow can be calculated using the following equation:

$$Q = VA = 200\frac{\text{ft}}{\text{min}}\left(5\,\text{ft} \times 4\,\text{ft}\right) = 4000\,\text{ft}^2\,/\,\text{min} \tag{7.1}$$

Where Q is flow rate (ft^3/min), V is velocity (ft/min), and A is surface (ft^2).

Step 3: Specify the minimum duct velocity. The minimum recommended duct velocity is 3500 fpm.

Step 4: Calculate the duct size. The maximum permissible dust cross-sectional area is determined by dividing the airflow by the minimum duct velocity:

$$A = \frac{Q}{V} = \frac{4000\,\text{ft}^3/\text{min}}{3500\,\text{ft}/\text{min}} = 1.14\,\text{ft}^2 \quad (7.2)$$

For any duct with an area of less than 1.14 feet^2 and a flow rate of 4000 cfm, the minimum required duct velocity is 3500 fpm. Using Table 7.1, a round duct with a cross-sectional area slightly smaller than 1.14 feet^2 corresponds to a 14-inch duct, which has a cross-sectional area of 1.069 feet^2. The actual velocity in the duct and the corresponding velocity pressure must be calculated, as they are necessary for the design process. As a result, the corresponding velocity will be:

$$V = \frac{Q}{A} = \frac{4000\,\text{ft}^3/\text{min}}{1{\cdot}069\,\text{ft}^2} = 3740\,\text{ft}/\text{min} \quad (7.3)$$

The corresponding velocity pressure will be:

$$p_v = \left(\frac{V}{4000}\right)^2 = \left(\frac{3740}{4000}\right)^2 = 0.87\,\text{in.}\,H_2O \quad (7.4)$$

Step 5: Design the duct route. The placement of the selected duct is shown in Figure 7.3. The duct path is as follows: it runs upward for 12 feet beneath the factory roof; it then makes a 90° turn and runs 50 feet along the roof; after another 90° turn, the duct continues for 12 feet to reach the roof; a third 90° turn brings the duct 6 feet to the fan located in the ceiling. Finally, an 8-foot duct leads from the fan to the chimney.

TABLE 7.1
Choosing the Correct Diameter of the Duct

	Surface		Perimeter	
Diameter (in)	in^2	ft^2	in	ft
12	113.1	0.7854	37.70	3.142
13	132.7	0.9218	40.84	3.403
14	153.9	1.069	43.98	3.665
15	176.7	1.227	47.12	3.927
16	201	1.396	50.26	4.189
17	226.9	1.576	53.41	4.451
18	254.4	1.767	56.55	4.712
19	283.5	1.969	59.69	4.974
20	314.1	2.182	62.83	5.236
21	346.3	2.405	65.97	5.498

Step 6: Calculate the energy losses. Energy losses occur in simple systems due to both turbulence and friction. Turbulence losses are caused by changes in the direction and speed of the airflow, while friction losses result from the movement of air through the duct. Both types of losses are directly proportional to the velocity pressure in the system. For this reason, each loss can be expressed as the product of a loss coefficient and the velocity pressure in the duct. For the projectile mixer hood design, these losses, measured in inches of water, are calculated using this method.

Hood static pressure in the duct. As explained earlier, the measured hood static pressure in the duct, just downstream of the hood, includes both the hood inlet losses and the acceleration coefficient (velocity pressure):

$$p_{s,h} = p_v + h_e \tag{7.5}$$

Where, $p_{s,h}$ = hood static pressure (in. H_2O); p_v = pressure velocity (in. H_2O); h_e = hood inlet drop (in. H_2O) (Table 7.2).

The hood inlet losses are usually expressed as the product of the hood inlet loss coefficient (F_e) and the velocity pressure. The formula for calculating the hood inlet loss is:

$$h_e = F_e \times p_v$$

Therefore:

$$p_{s,h} = p_v\left(1 + F_e\right) \tag{7.6}$$

The inlet loss factor of F_e for the projectile mixer hood is equal to 0.25, as shown in Figure 7.2. The static pressure of the hood is obtained using the following equation:

TABLE 7.2
Pressure Losses Table

Pressure Velocity p_v (in. H_2O)	Velocity V (fpm)
0.82	3622
0.83	3644
0.84	3666
0.85	3688
0.86	3709
0.87	3731
0.88	3752
0.89	3774
0.90	3795
0.91	3816

$$p_{s,h} = p_v\left(1 + F_e\right) = p_v\left(1 + 0.25\right) = 1.25\,p_v$$

The static pressure of the hood must be generated within the system to ensure the airflow rate of 4000 cfm is achieved.

7.2.2.1 Duct Friction

The frictional loss caused by airflow through a duct with a flow rate of 4000 cfm and a diameter of 14 inches can be expressed as the product of the loss coefficient and the velocity pressure. As shown in Figure 7.3, the total straight duct length upstream of the vent is 80 feet (6 + 12 + 50 + 12), with an additional 8 feet downstream of the fan. Since airflow through the duct experiences the same frictional losses regardless of whether the duct is upstream or downstream of the fan, the entire 88 feet duct length can be considered as a unit. The loss caused by direct duct friction, expressed as velocity pressure, is calculated from Figure 7.4. By locating the intersection of the 4000 cfm flow rate curve and the 14-inch duct diameter curve in Figure 7.4, the corresponding frictional losses per foot of duct are found to be 0.0155 p_v. Therefore, for the projectile mixer system with an 88-foot straight duct, the total frictional losses will be:

$$88\,\text{ft} \times 0.0155\frac{p_v}{\text{ft}}\ \text{or}\ 1.36\,p_v \tag{7.7a}$$

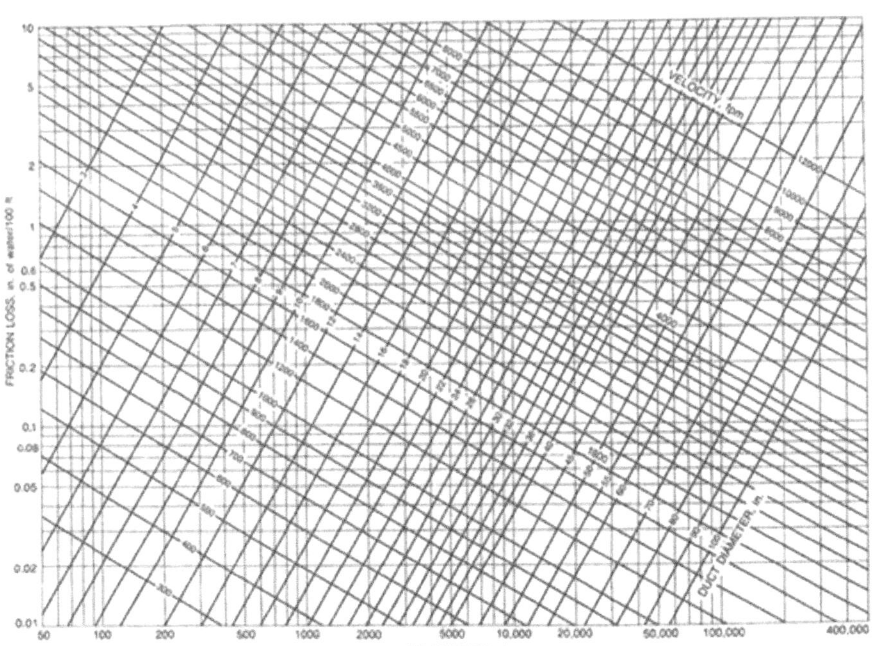

FIGURE 7.4 Friction loss due to straight duct, using the velocity pressure method.

The losses per foot can be calculated by the following equation:

$$p_{s,f} = 0.0307\,\mathrm{p_v}\frac{V^{0\bullet533}}{Q^{0\bullet612}} \tag{7.7b}$$

Where D is the diameter of the duct in inches. In this case we will have:

$$p_{s,f} = 0.0307\,\mathrm{p_v}\frac{3740^{0\bullet533}}{4000^{0\bullet612}} = 0.0155\,\mathrm{p_v}$$

7.2.2.2 Elbow Losses

The losses caused by turbulence in the elbows are directly proportional to the velocity pressure. Typical loss coefficients for elbows are shown in Figure 7.5.

Elbows are typically constructed from sections of straight or stamped smooth duct, as shown in Figure 7.5. The larger the radius of curvature (R) and the more sections the elbow contains, the lower turbulence loss. Pressed elbows are generally more expensive than modular (pieced) elbows. In addition, large sweep elbows, although effective, are costly to manufacture and occupy more space. As a result, most local suction ventilation systems use 4- or 5-piece elbows with a curvature

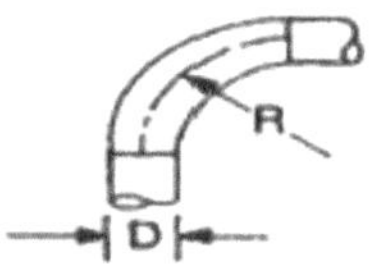

Round Elbow

R/D	Pressure drop, p_v fraction
2.50	0.22
2.25	0.26
2	0.27
1.75	0.32
1.50	0.39
1.25	0.55

FIGURE 7.5 Geometry of a 90° elbow showing duct diameter (D), radius of curvature (R), and typical loss coefficients for specific R/D ratios.

radius of 1.5 to 2 times the diameter of the duct (D). For the projectile mixer and in all sample designs, elbows with R/D ratio of 1.5 and sections will be used. Therefore, a 14-inch diameter elbow will have a 21-inch radius, resulting in a turbulence loss of 0.27 velocity pressure (p_v) (Figure 7.5). Given that there are three 90° elbows, the total elbow losses due to the turbulence will amount to 0.81 p_v.

The total frictional and turbulence losses of the hood and duct components in the suction system of a projectile mixer are as follows: direct duct frictional losses ($p_{s,f}$): 1.36 p_v; acceleration (inertia) losses (p_v): 1 p_v; hood inlet losses (h_e): 0.25 p_v; elbow losses: 0.81 p_v; total static pressure drop (ps): 3.42 p_v.

The previously calculated velocity pressure is 0.87 inches of water. Based on this, the actual static pressure required to maintain an airflow of 4000 cfm through this system would be:

$$p_s = (3.42)(0.87\,\text{in.}\,H_2O) = 2.98\,\text{in.}\,H_2O$$

In the design of a complex local exhaust system, it is important to implement a consistent and clear method for identifying the hoods and various duct sections (Figure 7.9). Each hood is assigned a unique number, while each duct connection or interface is labeled with a letter. Duct sections are designated by a junction-to-junction notation, typically a two-part combination (i.e., 1A, AB), which helps label the specific sections under design. Each column in the calculation sheet corresponds to a particular section of the ductwork (Figure 7.6).

7.2.3 Design of a Slotted Hood System for a Degreaser Tank

Due to the toxicity of chlorinated hydrocarbon solvents commonly used in degreasers, the standard approach is to install a side-slotted hood on the degreaser tanks (Figure 7.7). The inlet losses for these hoods are more complex than those of simple

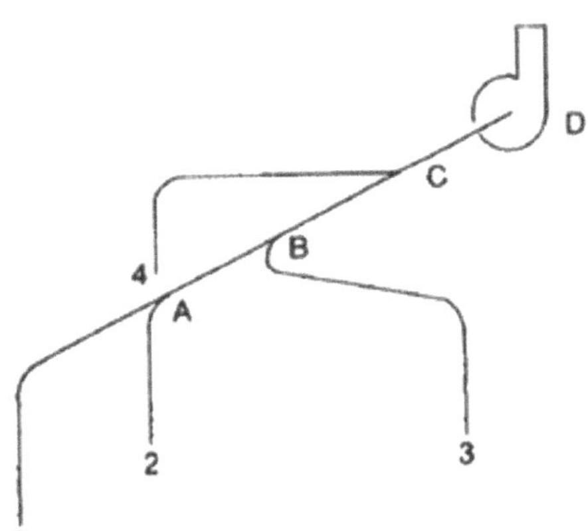

FIGURE 7.6 The main components of the duct in a local exhaust ventilation system are named as follows: hoods are assigned a number, and connections are designated by a letter.

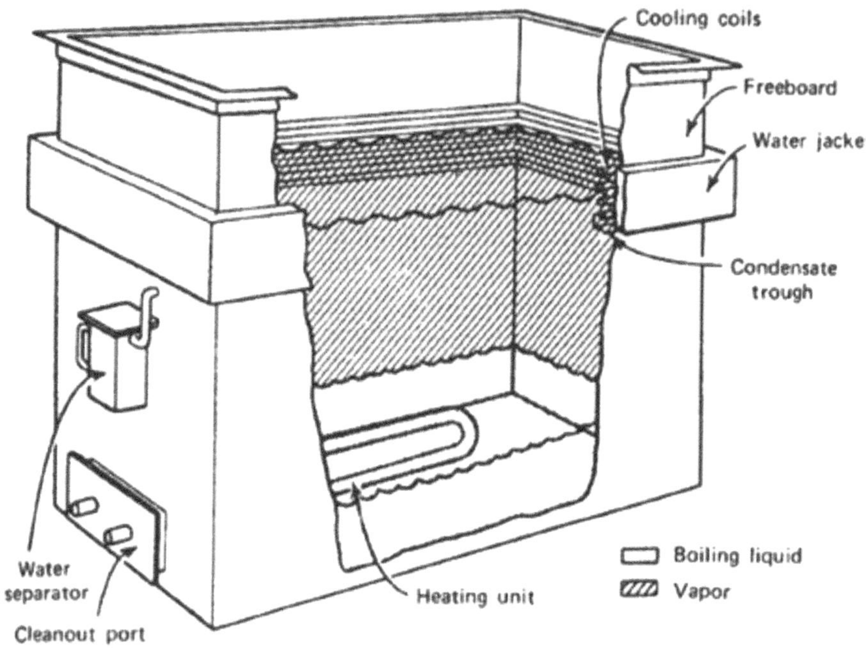

FIGURE 7.7 The solvent in the vapor phase of the degreaser is heated using either steam or electricity, producing a dense vapor that fills the tank. The tank is equipped with a cold water condenser, where the hot steam condenses in the covered part of the tank and is then recirculated. As the parts move from the vapor cloud, the hot solvent vapor condenses on the colder surfaces of the parts, dislodging the oily residue, which then flows to the bottom of the tank. To control emissions, tank losses are typically controlled with side slotted exhaust hoods.

hoods, as they involve additional components such as the gap, equalizing chamber, and duct interface.

7.2.4 Loss Components in a Complex Hood

The inlet losses in a slotted hood are influenced by the relationship between the slot dimensions, the velocity in the duct, and the geometry of the equalizing chamber. A general view of the slot, equalizer chamber, and duct interface is shown in Figure 7.8. The inlet losses for a slotted hood can be divided into two components: the inlet losses from the slot itself ($h_{e,s}$) and the inlet losses at the interface between the equalizing chamber and the duct ($h_{e,d}$). The equation for these losses will be:

$$h_e = h_{e,s} + h_{e,d} \tag{7.8}$$

As air flows through a slotted hood, both the air velocity and velocity pressure change as they pass through different sections: the slot $p_{v,s}$, the equalizing chamber $p_{v,p}$, and finally the duct $p_{v,d}$. In typical cases where the velocity in the gap is high, a residual

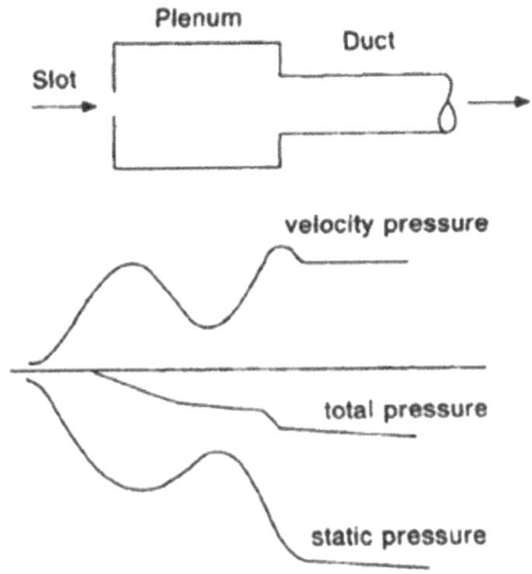

FIGURE 7.8 The pressure in slotted hoods consists of the velocity pressure in the slot $p_{v,s}$, the velocity pressure in the equalizing chamber $p_{v,p}$, and the velocity pressure in the duct $p_{v,d}$.

velocity remains in the equalizing chamber $p_{v,p}$ (Figure 7.8), which then accelerates as it enters the duct $\boldsymbol{p_{v,d}}$. The total required velocity pressure can be expressed as:

$$p_v = p_{v,s} + \left(p_{v,d} - p_{v,p}\right) \tag{7.9}$$

The hood static pressure will be:

$$p_{s,h} = p_v + h_e \tag{7.10}$$

By combining Equations 7.9 and 7.10, we obtain the following:

$$p_{s,h} = p_{v,s} + \left(p_{v,d} - p_{v,p}\right) + h_{e,s} + h_{e,d} \tag{7.11}$$

If the velocity in the duct exceeds that in the slot ($p_{v,s} < p_{v,d}$),

- For a shallow equalizing chamber ($p_{v,p}$ is approximately equal to $p_{v,s}$):

$$p_{s,h} = p_{v,d} + h_{e,s} + h_{e,d} \tag{7.12}$$

- For a deep uniformizer chamber ($p_{v,p}$ is almost zero):

$$p_{s,h} = p_{v,s} + p_{v,d} + h_{e,s} + h_{e,d} \tag{7.13}$$

If the velocity in the duct is lower than the velocity in the slot ($p_{v,s} > p_{v,d}$),

- For a shallow uniformizer chamber ($p_{v,p}$ is almost equal to $p_{v,s}$) and both $p_{v,p}$ and $p_{v,s}$ are larger than $p_{v,d}$:

$$p_{s,h} = p_{v,s} + h_{e,s} + h_{e,d} \tag{7.14}$$

(It cannot be used for static recovery, $p_{v,d} - p_{v,s}$)

- For a deep uniformizer chamber ($p_{v,p}$ is almost equal to zero):

$$p_{s,h} = p_{v,s} + p_{v,d} + h_{e,s} + h_{e,d} \tag{7.15}$$

The most common scenario involves a hood with a shallow equalizing chamber, where the velocity in the duct exceeds the velocity in the slot. In this case, the static pressure of the hood can be determined as follows:

$$p_{s,h} = p_{v,d} + h_{e,s} + h_{e,d} \tag{7.16}$$

If the slot is a sharp-edged orifice, the input losses will be the slot:

$$h_{e,s} = 1.78\, p_{v,s} \tag{7.17}$$

The entrance losses of the equalizer chamber to the duct will be:

$$h_{e,d} = F_{e,d} \times p_{v,d} \tag{7.18}$$

Where, $F_{e,d}$ is the coefficient of loss of the equalizing chamber to the duct, which depends on the cone angle of the interface. For example, if the taper angle available for a round duct is 120°:

$$h_{e,d} = 0.25\, p_{v,d} \tag{7.19}$$

By replacing Equations 7.17 and 7.19 in Equation 7.9, the hood inlet losses will be obtained:

$$h_e = 1.78\, p_{v,s} + 0.25\, p_{v,d} \tag{7.20}$$

And therefore, the static pressure of the hood from Equation 7.10 will be:

$$p_{s,h} = 1.78\, p_{v,s} + 0.25\, p_{v,d} \tag{7.21}$$

7.2.5 Degreaser Hood Design Using the Pressure-Velocity Calculation Screen

The system design for the degreaser hood, using pressure-velocity calculation sheets, is shown in Figure 7.9. The duct design for the degreaser tank follows a similar

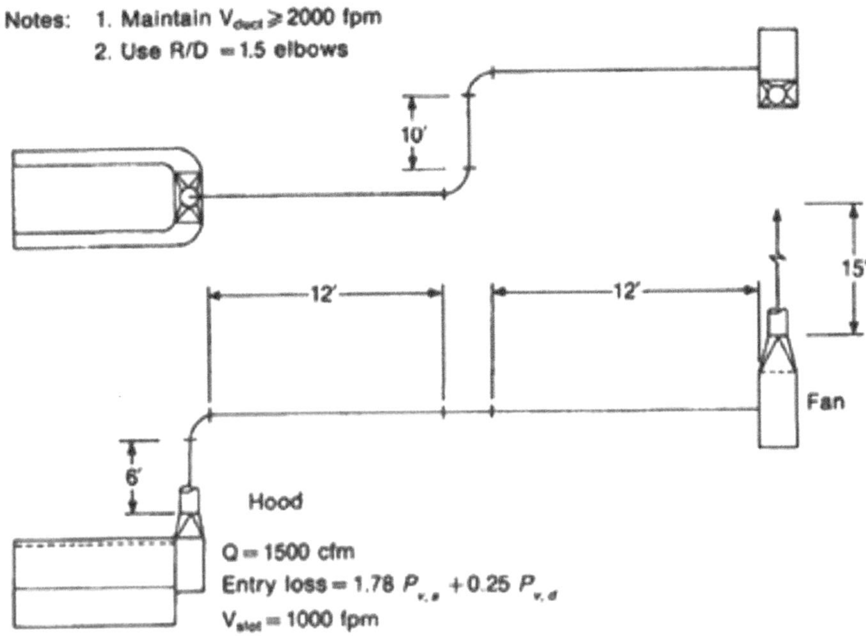

FIGURE 7.9 Design page for a vapor phase degreaser.

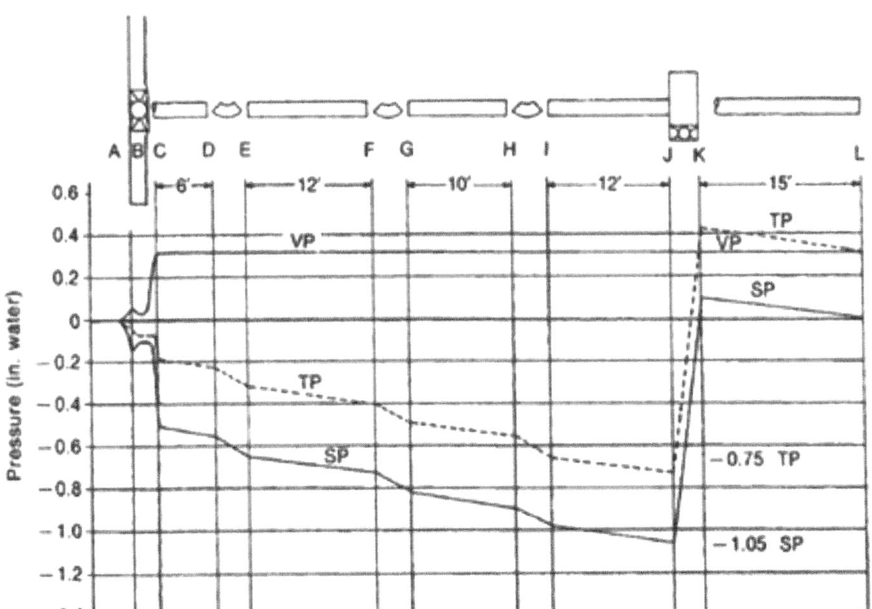

FIGURE 7.10 Duct pressure diagram for vapor phase degreaser system using 11-inch duct.

approach to that of the projectile mixer hood. The losses associated with the slot are calculated using lines 15–21 on the standard velocity pressure calculation page.

The design process shows that this system will achieve an airflow rate of 1500 cfm, with static pressure losses of 1.15 inches of water. By selecting an 11-inch diameter duct, the minimum duct velocity of 2000 fpm, as shown in Figure 7.9, is maintained. In the particle control systems, it is important to maintain this minimum velocity in the duct to prevent dust settling. For systems handling gases or vapors, the duct velocity is less critical, and a lower speed can be chosen if space permits a larger duct. In such cases, the energy loss at a lower velocity can be easily calculated. Opting for a larger duct (12 inches) reduces the duct velocity from 2270 fpm to 1910 fpm. This decrease in velocity results in a corresponding reduction in pressure from 1.15 to 0.85 inches of water (Figure 7.10).

8 Multiple-Hood Systems Design

A. H. Khoshakhlagh, H. M. Moda, S. Yazdanirad, and A. Gruszecka-Kosowska

8.1 OBJECTIVES OF THE LESSON

Upon completing this lesson, readers should be able to:

- Describe the general principles of designing multiple hood systems
- Explain the applications of multiple hood systems
- Describe the relevant design approach
- Explain the method of static pressure balance
- Explain the balancing method of the air regulating valve

8.2 MULTIPLE-HOOD SYSTEMS DESIGN

The design method for a single hood system is simple, and a set of single hood systems provides maximum flexibility for the user. However, single-hood systems have many disadvantages, and it is usually more cost-effective to incorporate a set of hoods with a single fan into one system. Such multiple-hood systems can be designed in different ways depending on the type of application.

A primary goal in designing multiple hood systems is to create proper airflow in each hood. This is achieved by a combination of two factors. It is possible to supply the flow of the entire system by choosing the appropriate fan. Proper flow distribution among the hoods is achieved by balancing the airflow resistance in the hoods and their related ductwork network. The process of achieving such a balance is the subject of this chapter.

8.2.1 Applications of Multiple Hood Systems

The most common application for a multiple-hood system is in manufacturing facilities with workstations that release the same type of airborne contaminant. Grinding and polishing lines are examples of such application. In these systems, all workstations are equipped with hoods that share the same flow rate and ducting, while ensuring that the minimum transfer speed is maintained across the system.

Another example is a foundry shake room, where air contaminants such as metal particles and fused-silica sand are released during the process. Since the activities and equipment vary from station to station, the geometry of the hoods and the

DOI: 10.1201/9781003501091-8

required suction levels also differ. Typically, these hoods are grouped into a multiple-hood system, as each operation generates similar contaminants. An air cleaner can then be used to efficiently address the pollutants. The design of such a system will be discussed in this chapter.

Grouping hoods in a system for activities that produce different airborne contaminants can create various design and operational problems. For example, consider a hood assembly in a metalworking shop does not control non-corrosive gases and vapors. However, one of the hoods serves a stripping tank, which releases a corrosive acid mist. If a second hood is added to the hood system, the entire system must be made of corrosion-resistant materials. In this case, the solution might be to group all the hoods dealing with non-corrosive air pollutants in one system, while installing a separate hood system made from suitable materials for handling corrosive substances. Grouping hoods that handle high concentrations of incompatible or reactive materials can lead to poor performance due to the potential for violent reactions.

In the previous examples of multiple hood systems, it was assumed that all hoods operate continuously, which may not be necessary in many cases. Each process might only be conducted during a small portion of the total shift, with work time potentially spread across multiple shifts. Continuous operation of all hoods results in energy loss. A more efficient solution is to install a separate hood-fan system for each process.

Alternatively, a multiple hood system can be used where each hood is connected to an air-regulating valve or damper that only opens when the operation is in progress. The required suction capacity of the system can be determined by analyzing the duration of all operations. This approach is commonly used for the transfer of gases and vapors; however, it often fails when controlling particulates, as variable air currents can make it difficult to maintain proper transfer speeds.

The suction system of the uniformizer chamber is another method for this application. This system ensures a minimum transfer velocity in the branches, but the low-velocity main duct or equalizing chamber acts as a settling chamber, which requires periodic cleaning.

Several other considerations should be taken into account. Installing multiple hood systems to provide suction for activities located in different departments with varying regulatory conditions should be avoided if possible. Just as it is difficult to share production equipment, it is equally challenging to share local exhaust ventilation capacity.

Stability for the operating activities that require suction is critical to the successful design and operation of a large multiple hood system. Large multiple hood systems should not be used when frequent process or equipment changes are expected. Modifying existing multiple hood systems by adding hoods without expert review is one of the most common issues seen in local exhaust ventilation systems, often leading to malfunction of the entire system.

8.2.2 Balanced Design Approach

The process of achieving the proper airflow distribution between the hoods is called balancing. Balancing can be done using either the static pressure balance method or by employing air adjustment valves. In both approaches, the desired flow is achieved for each branch and subset of the system by ensuring the necessary static pressure and airflow resistance at each junction.

In the static pressure balance method (Figure 8.1), the required static pressure at each joint is determined by selecting the appropriate hood inlet losses, duct size, and other resistance components. If the design is carried out correctly and the proper fan is installed, the static pressures generated by the system will align with the required airflow at each hood. This design method is the preferred approach and will be discussed in detail in this chapter.

In the balancing process using air regulating valves, static pressures are selected to be equal to or greater than the value required to maintain the desired flows at each junction in the system. Air adjustment valves are installed on each branch and field-adjusted to provide the necessary resistance for the branch in relation to the desired airflow (Figure 8.2). An excellent review of the two equalization methods is provided by Caplan, who highlights the significant limitations of the air regulating valve method and explains why the static pressure equalization method is preferred. A brief comparison of the two balancing methods is presented in Table 8.1.

In the third design procedure (Figure 8.3), a multiple hood system can be used when separate hoods operate alternately. In this approach, air adjustment valves are not used for balancing; instead, only on-off valves are employed. The designer must determine the timing for different processes and design the system to provide the necessary airflow through a limited number of hoods. If the air control valves are open for more hoods than specified at the same time, the system will malfunction.

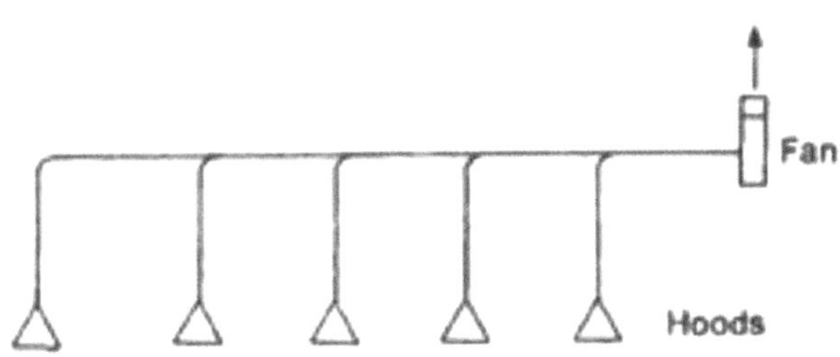

FIGURE 8.1 A multiple hood system for similar activities requires careful balancing, achieved through the proper selection of duct size and airflow.

FIGURE 8.2 Multiple hood system with adjustable air control valves is used to achieve proper air balance and ensure the correct airflow through each hood.

TABLE 8.1
Comparison of Balancing Methods

Balance without Air Adjustment Valves	Balance with Air Adjustment Valves
Air volumes cannot be easily adjusted by operators	Air volumes can be adjusted relatively easily by operators
Low flexibility to modify or expand equipment in the future	Increased flexibility to modify or expand equipment in the future
The selected exhaust air volumes for an activity may be incorrect	Incorrect estimates of exhaust air volumes can be corrected
Absence of abnormal corrosion or deposits in the duct	Partially closed air adjustment valves can lead to duct corrosion, alter airflow, and cause dust accumulation; if the air adjustment valves are tampered with, the duct may become obstructed
A more time-consuming design process compared to the air adjustment valve method	The calculations for the duct are relatively straightforward
The total air volumes are slightly higher than the design volumes due to the modified airflows	Balance can be achieved through proper air volume design
A poor branching selection is associated with the highest resistance in design calculations	The poor selection of the most resistant branch may not be immediately apparent, leading to the branch with the highest resistance being disregarded
The system placement should be carried out in full detail, and the installations must be executed precisely according to the design	Allowing changes in duct placement to avoid obstructions or interferences is not accounted for during the design phase

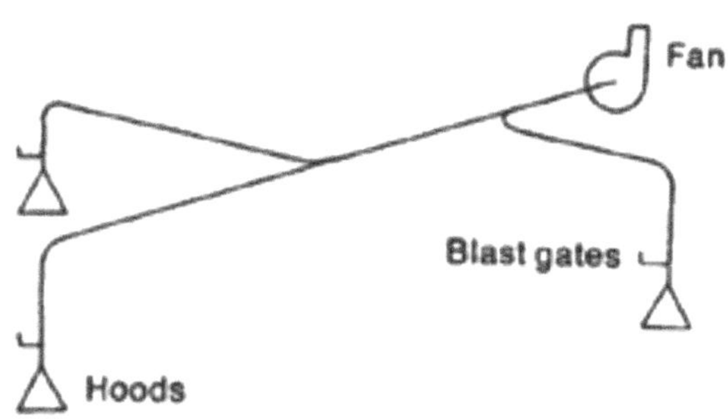

FIGURE 8.3 Multiple hood system with air adjustment valves (off-on).

The best way to avoid this issue is to interlock the operations with hood dampers and use controls to limit the number of operations that can run simultaneously. However, it may be possible to operate such systems without such strict controls if users are familiar with the operating instructions and take care to close dampers on unused hoods.

A typical multiple hood system consists of a main duct path with separate branches connected to it from various locations. The design of multiple hood systems follows the same principle as single hood systems (Chapter 7). Each branch can be treated as a single-hood system, with the main duct functioning as a network of these individual systems. The designer determines the energy losses—both frictional and due to turbulence—caused by the airflow from each branch to the junction, whether it leads to a sub-group or the main path. In this case, the additional energy loss from a single hood system, caused by the entry of the branch into the main duct, is not considered.

The goal of the static pressure balance design approach is to match the required static pressure of each branch with the available static pressure in the main duct at the point where each branch enters. If these static pressures are properly matched and a fan of appropriate capacity is selected, the designed airflow for each hood will be achieved.

To optimize the design, it is important to evaluate the results of airflow and total resistance for various duct and fan configurations. In any case, the design should begin with the branch that has the highest resistance, as this simplifies the design process. While the choice of this branch is often straightforward, the following guidelines may be helpful in selecting the appropriate starting point for the design:

- If the exhaust airflow rates of the hoods are similar, the farthest branch from the fan is likely to have the highest resistance
- A hood with a combination of high inlet losses and high airflow will likely correspond to the branch with the highest resistance
- A high-velocity branch in the duct may require the highest static pressure
- The use of long flexible duct runs for hood displacement results in maximum branching resistance

In any case, the static pressure balancing method is more flexible in this regard. If the branch with the highest resistance is not initially selected, it can still be incorporated into the system design using this method. This flexibility is not available with the air adjustment valve balancing method.

When the branch with the highest resistance is selected, friction and turbulence losses are calculated starting from the hood and progressing through the first connection in the main or sub-path. The basic procedure is shown in Figure 8.4. In the section of the system farther from the fan, two branches are connected, forming a main end (point A). The required static pressure in the first branch (1A) is calculated, followed by the required static pressure in the second branch (2A). These two static pressures are then compared. If there is a discrepancy, one of the branches will not receive the proper airflow during the system operation. This occurs because only one static pressure can exist at point A. Several design options are available to balance these static pressures and ensure the desired airflow in both branches.

Based on the example presented in Figure 8.4, hood 1 requires an airflow of 2,000 cfm and a static pressure of 0.8 inches of water at junction A. Hood 2 requires a static pressure of 0.95 inches of water at this junction to provide the required airflow of 500 cfm. During operation, the fan can only supply one static pressure value for point A. If a static pressure of 0.8 inches of water is selected, hood 1 will receive adequate airflow, but hood 2 will not achieve its design airflow of 500 cfm.

Since the speed in the duct and the airflow varies with the square root of the static pressure, the actual flow from hood 2 will be ($500 \times \left(0.8 \div 0.95\right)^{0.5} = 460$ cfm). Alternatively, the fan can be designed to supply 0.95 inches of water at point A.

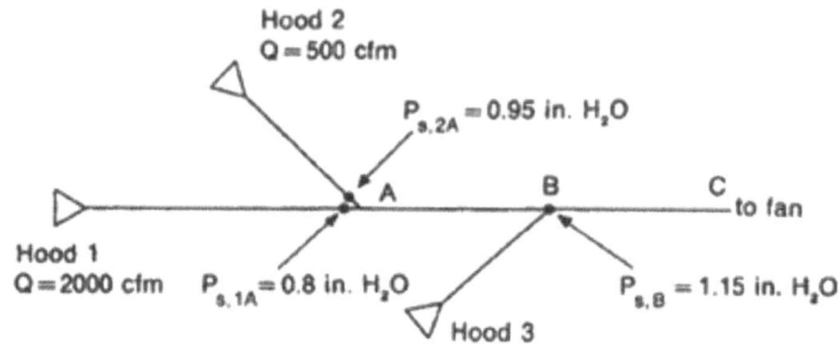

FIGURE 8.4 Branches 1A and 2A are connected at junction A. Design airflows and total losses are specified for both paths to create a balanced system.

This will ensure that hood 2 receives the correct airflow, but hood 1 will have a new airflow ((8/0 ÷ 95/0)$^{0/5}$ × 2000 = 2180 cfm). To resolve this discrepancy, there are three available options: reduce the resistance in the 2A branch, increase the resistance in the 1A branch, or increase the airflow in the 1A branch. The first step is to evaluate the design of hood 2 and its ductwork network to determine if the resistance to airflow can be reduced. For example, if the hood has a high inlet drop, it could be replaced with one that has a higher inlet coefficient. Elbows could potentially be removed, or the duct diameter could be increased if the duct path is long and the air velocity is high. This would reduce the velocity in the duct, thereby decreasing frictional losses—provided the minimum transfer velocity is not compromised. In this case, hood 2 cannot be redesigned, and the transfer rate in that branch must be maintained. Therefore, modifying the 2A branch is not a viable solution.

Now, the design of the branching for hood 1 should be considered. Since hood 2 cannot be modified, the static pressure in branch 1A must be increased to 0.95 inches of water to match the required amount of branch 2A. Increasing the static pressure in the branching of hood 1 can be achieved in several ways. If the airflow of 2,000 cfm is to remain unchanged, the airflow resistance must be increased. For example, if the hood has a high inlet drop, an air adjustment valve can be installed in the duct.

To further increase the speed in the duct and the frictional losses, the duct size can be reduced. An alternative approach is to increase the airflow to 2,180 cfm, as calculated in the previous section. While increasing airflow may improve hood performance, it will also lead to higher operating costs. If the airflow is increased, this step in the design process is referred to as modified airflow. The static pressure of 0.95 inches of water is considered the prevailing static pressure, or design point, at junction A. When calculating the friction loss in the main path (AB), the corrected flow of 2,180 cfm from hood 1 is added to the 500 cfm airflow through hood 2, yielding a total airflow of 2,680 cfm.

The duct size and friction loss coefficient are determined using standard methods, and the static pressure drop for this section is calculated as 0.2 inches of water. This value is then added to the prevailing static pressure at connection A (0.95 inches of water) to obtain the required static pressure at connection B (1.15 inches of water). The required static pressure in branch 3B is calculated and compared with the value of connection B. The design approach outlined above is then repeated for the 3B branch and the subsequent duct components (BC) until reaching the fan and outlet.

TABLE 8.2
Balancing Criteria of Multiple Hood Systems

If the Static Pressure Ratio Is from the Higher Side to the Lower Side	Action
Lower than 1.05	Acceptable: Do not alter the flow in calculations; use the higher static pressure as the reference value
Between 1.05 and 1.20	Allow the airflow to increase in the branch with the lowest static pressure, and recalculate the airflow accordingly $Q_{corrected} = Q_{design}\left(\frac{p_s \text{ duct with higher losses}}{p_s \text{ duct with lower losses}}\right)^{0.5}$
Higher than 1.20	Resize the duct with the lower static pressure accordingly

In general, static pressures at the joints will not perfectly match. So, what level of discrepancy is acceptable, and what is the preferred design approach to correct it? The method that most designers follow is to calculate the ratio of the two static pressures and refer to the recommendations provided in Table 8.2.

8.2.3 Static Pressure Balance Method

A multiple hood design is employed in a small casting vibration room. The water meter components are brass alloy casting parts that contain lead. After casting, the parts are transferred to the vibration section, where molding sand and core residues are removed. The parts are then cleaned in a high-pressure abrasive unit before being polished using rotary grinding machines, belt sanders, and base sanders. Given that the alloy contains lead, dust generated from these activities must be controlled using local exhaust ventilation.

The local exhaust ventilation system is designed using the static pressure balance method. In this example, balancing is achieved solely by adjusting the airflow. Changes to duct size as a balancing method will be discussed later in this chapter. The key points of this design are presented in the following section:

- The use of design pages is not necessary in this example, as all required design data is provided directly in Figure 8.5
- Since the system is designed to capture particles, a minimum velocity of 3500 fpm is selected; after the particles are separated by the air cleaner, the minimum velocity is reduced to 2000 fpm, as there is no longer a need to transport particles—this lower velocity allows for a larger duct diameter with reduced frictional resistance losses
- Air cleaner airflow resistance is typically provided by the manufacturer; in this example, the static pressure drop accounts for all losses related to the air cleaner, including losses due to turbulence at the filter inlet, resistance through the filter, and transfer from the air cleaner to the duct

- For simplicity, all duct paths are shown on a single page
- All branch connections to the main duct are at a 30° angle. However, before branches 2A and 3B, the angle of the elbows should be 60°
- In general, the loss due to branch connections is considered within the branch itself, not in the main duct path
- As shown in Figure 8.5, all hoods are simple hoods with a defined loss coefficient

8.2.4 Blast Gate Balance Method

In this design method, the expert chooses the hood that has the maximum resistance to flow as the starting point in calculating the losses. The design of the system is done from the hood to the first connection point. The airflow from hood 1 is combined with the airflow from hood 2. The main airflow path is determined, and losses are calculated for the section of the main path where the branch from hood 3 joins. When hood 3 joins the main path, its design airflow is added to the combined airflows of hoods 1 and 2. The main path is sized accordingly, and losses are calculated up to the next branch inlet.

In this method, the accumulation of static pressure losses ensures that if at the beginning of the calculation, the branch with the maximum resistance is correctly selected, the static pressure in the main duct at each connection will be higher than the required pressure entering the branch of the system. By adjusting the air control valves of all the hoods, the desired airflow distribution can be achieved. This allows for the required static pressure in the branch to be balanced with the static pressure in the main path.

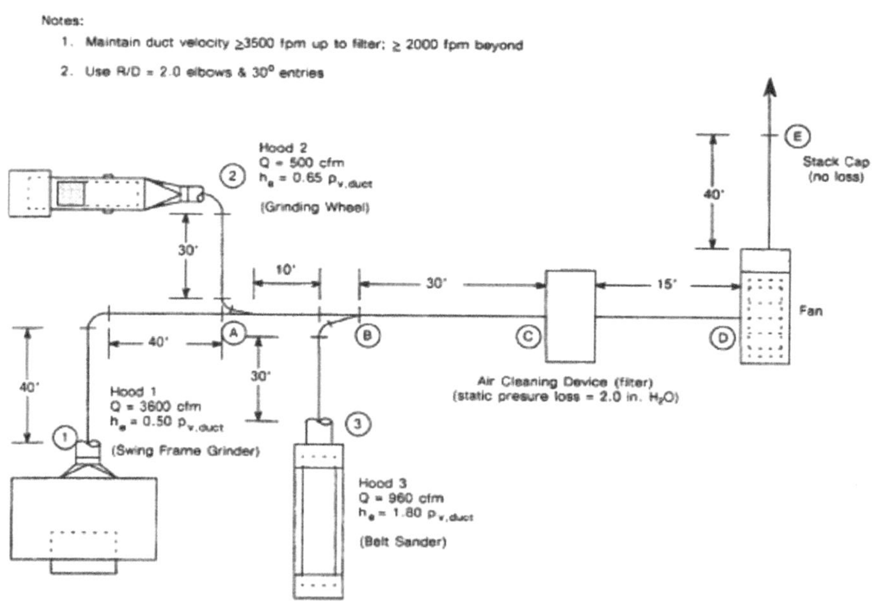

FIGURE 8.5 Layout of the ductwork for the local exhaust ventilation system in the foundry cleaning room.

This procedure is more complex than it may seem, as adjusting an air control valve in one branch can impact the airflow in adjacent branches. As a result, a trial-and-error approach is required to fine-tune the air control valves until the proper balance is achieved. Before implementing a system that relies on the balancing method of the air regulating valves, the designer should carefully assess the limitations of this approach.

9 Quantitative Evaluation of Hood Performance

A. H. Khoshakhlagh, H. M. Moda, S. Yazdanirad, and A. Gruszecka-Kosowska

9.1 OBJECTIVES OF THE LESSON

Upon completing this lesson, learners should be able to:

- Quantitatively evaluate the performance of the hood
- Explain the method of measuring hood airflow
- Calculate the capture efficiency of the hood
- Explain the factors affecting the performance of the hood
- Explain the relationship between airflow patterns and capture efficiency

9.2 QUANTITATIVE EVALUATION OF HOOD PERFORMANCE

It is important to evaluate the performance of exhaust hoods when they are installed. As explained in the following section, several different experiments can be performed in this area. Hoods are evaluated based on their ability to answer one of the following two important questions:

1. Is the airflow from the hood consistent with the designed airflow?
2. Does the hood effectively capture the contaminants generated by the process?

The answer to the first question relies on measuring both airflow and static pressure. These measurements essentially evaluate the effectiveness of the ventilation system design. In other words, the tests evaluate the designer's ability to account for and balance all static pressure drops in the system and fan. If the design has been executed correctly and the system adheres to the specifications, the measured airflow from the hood will match the design airflow.

Although measuring airflow is always essential in newly installed hoods, the second question is fundamentally more critical. The success of hood design ultimately depends on the hood's ability to effectively capture contaminants released by the process. If only half of the contaminants are captured, the design cannot be considered satisfactory. In simple terms, this indicates that the proper airflow was not selected during the design process, the hood may have been improperly installed or positioned, or there were other flaws in the design phase.

DOI: 10.1201/9781003501091-9

Typically, the success of a ventilation system in capturing contaminants is evaluated by measuring the concentrations of contaminants in the respiratory zone and comparing these results to established standards (e.g., TLV or PEL). However, if workers are moved from one location to another, the location where the sample is collected in the vicinity of the hood may come into question. The concentration of contaminants at any point outside the hood depends not only on the hood's capture efficiency but also on several other factors, such as the level of contaminant production, the general ventilation of the room, the operator's work practices (for individual samples), and exact location of the sample. In this chapter, new methods for directly measuring hood performance in capturing contaminants are explored. Ultimately, measuring a hood's capture efficiency should serve as the definite test for assessing the effectiveness of a local exhaust ventilation system.

9.2.1 Hood Airflow Measurements

The standard procedure used by ventilation engineers and occupational health professionals to evaluate hood performance involves measuring the total airflow through the hood and comparing the measured values to the design specifications. Once the hood's airflow is quantified, its performance can be periodically monitored using the hood static pressure method.

The hood airflow measurement method offers several advantages, including:

(1) It is simple and straightforward to implement; (2) It allows for easy determination of whether the performance is adequate and whether the measured flow rate matches the design flow rate; (3) It provides flexibility in adjusting the required airflow—whether by increasing or decreasing it—using methods such as adjusting the air control valve or changing the fan speed. However, the main limitation of this method as a tool for quantifying hood performance lies in the measurement of airflow speed. Knowing the airflow speed alone does not necessarily ensure proper hood operation, particularly for outdoor hoods, which depend on the capture speed at the point where contaminants are generated. Issues with selecting the appropriate capture speed and the negative effects of cross-draft air currents are examples of these limitations. For instance, an outdoor hood might have a measured airflow that matches the design valve but still fails to capture the contaminants effectively. The realization led researchers to explore direct methods for measuring the hood's capture efficiency.

9.2.2 Hood Capture Efficiency

The concept of capture efficiency means quantifying the hood performance. Such a quantitative assessment involves determining the capture efficiency of the suction system as the fraction of contaminants actually abducted by a process. This concept of capture efficiency as the best measure of hood performance has been the focus of researchers:

$$\eta = \frac{\acute{G}}{G} \tag{9.1}$$

Where, G = amount of contaminant production (g/min); $\acute{G}$ = the amount of contaminant that is captured by the exhausted system (g/min).

In practice, it may be easy to measure $\acute{G}$ by collecting a sample from the duct located downstream of the exhaust hood(s), but in many cases it is difficult to accurately measure (or estimate) G. Four approaches have been proposed to estimate G: (1) establishing a material balance to determine the emission rate; (2) measuring airborne contaminant concentrations in the workplace and relating them back to G; (3) enclosing the contaminant source in order to capture all the contaminants and then calculate G by measuring the concentration of the contaminant in the enclosure; and (4) using a tracer to simulate the emission process of contaminants.

All four methods have disadvantages. Material balance can rarely be calculated with sufficient accuracy to obtain a realistic estimate of G. The second method is not reliable due to uncertainties in the interpretation of the amount of production compared to exposure to airborne contaminants. The third method works well in the laboratory, but its implementation in the field is challenging, especially for large sources of contaminant production. The fourth method is the easiest to use both in the laboratory and in the field, but it must be ensured that the detector accurately simulates the real diffusion process.

In this field, aerosol and gas detectors are used. The most common aerosol tracer is an oil mist, which has the advantage of producing a suitable visible tracer, but it is relatively difficult to produce and measure. Gas detectors are generally easier to produce and measure than aerosols. The most common tracer gas is sulfur hexafluoride (SF_6), which is detectable at low concentrations by gas chromatography (GC).

The capture efficiency of the exhaust hood can be predicted with theoretical or experimental models. Two theoretical approaches have been developed, one using analytical solutions for fluid flow equations and the other using numerical solutions.

9.2.3 The Impact of Cross-Drafts on Hood Performance

As mentioned in the above section, the most important limitation of the hood airflow method in quantitative determination of hood performance is its lack of justification regarding the effects of cross-drafts. Based on studies conducted by various researchers, models have been developed that can be used to quantify the effects of cross-drafts on hood airflow patterns and capture efficiency. Models are provided for round, square, rectangular, and slotted hoods. The models are based on the analytical solutions of the Laplace equation in frictionless, incompressible, and non-rotating fluid flow conditions and are validated using wind tunnel tests. Programs have also been written for the PC that allow for the graphical display of model airflow predictions.

As an example of the use of these models, consider the simple slotted hood shown in Figure 9.1. This hood has a 15-inch (0.4 m) long and 1.5-inch (0.13 m) wide flange slot (10 aspect ratio) and can be used for small plating tank suction. The Flynn, Conroy, and Ellenbecker (FCE) model can be used to visualize the effects of cross-draft currents on the performance of this hood. This model is based on the high-level programming language Beginners' All-purpose Symbolic Instruction Code (BASIC), a simple computer language that uses common English words and mathematical symbols. It incorporates 850 primary keywords and sentence structures and is currently in the planning stage. An example of the model's output is shown in Figure 9.2, which presents an idealized view of this hood, including the flange, hood

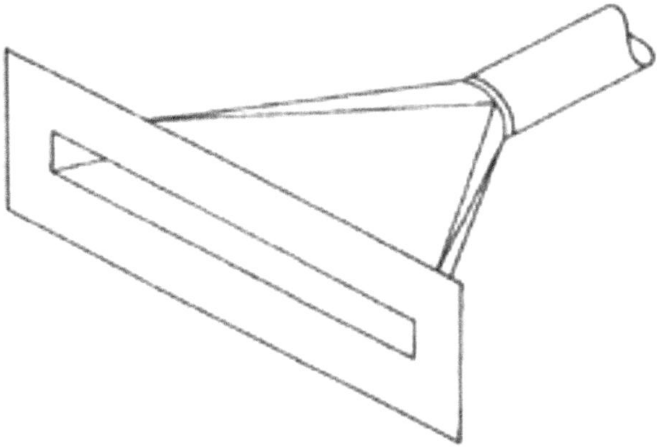

FIGURE 9.1 Basic flanged slotted hood.

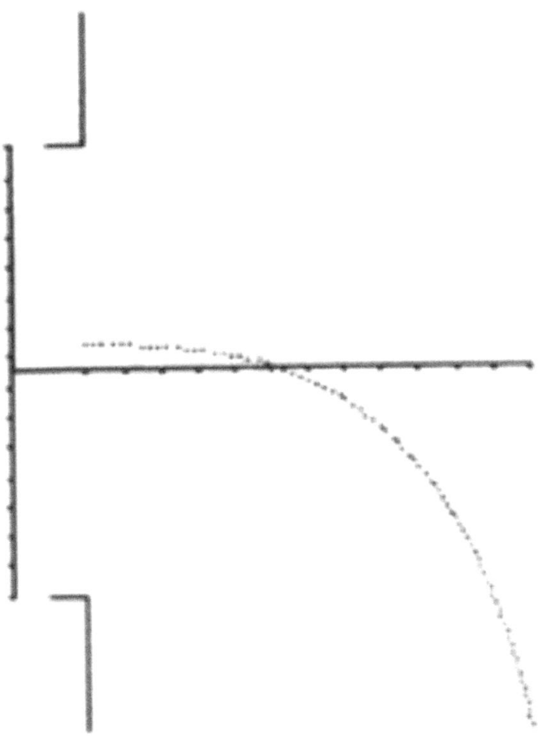

FIGURE 9.2 The hood shown using the computer model in Figure 9.1. The gap is viewed from above with vertical and horizontal distance scales at 1 in intervals, L = 15 inches, W = 1/5 inches, V_f = 25 fpm, V_d = a simple flow line for opening velocity of 1000 fpm, and cross-draft air is shown at 25 fpm.

opening, and reference scales along the opening and hood centerline. Plotted in dashed lines are the airflow lines for a particular combination of hood opening velocity V_f and draft airflow velocity V_d. In this case, the speed at the hood opening is 1000 fpm (5 m/s) and the cross-draft airflow is relatively low, 25 fpm (0.13 m/s), and it blows straight along the hood opening from the bottom to the top of the figure. A simple flow line is shown, starting at an area upstream and 12 inches (0.3 m) into the opening of the hood. Such simple lines are useful in visualizing the direction of airflow from any point in front of the hood and are important features of new models of capture efficiency. Figure 9.2 and subsequent figures in this chapter show a two-dimensional view of the airflow pattern as it exists in the centerline plane. Any other page can also be modeled and displayed, so that the entire flow of the context can be visualized.

Figures 9.3a–9.3d shows the effect of increasing cross-draft airflow on hood performance. In all cases, the cross-draft flow is blown along the hood opening (from the bottom of the figure to the top). As the cross-draft flow speed increases, fewer and

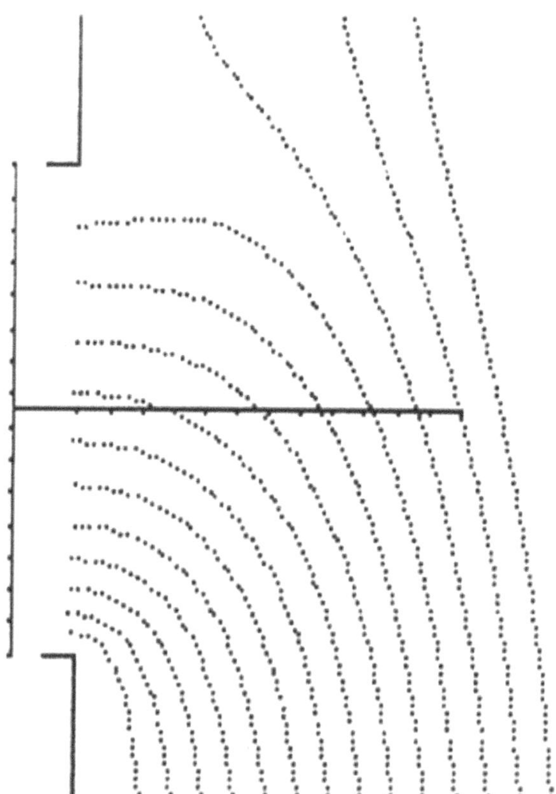

FIGURE 9.3a The hood shown in Figures 9.1 and 9.2, showing the effect of increasing the cross-draft flow; (a) V_d = 50 fpm (0.25 m/s).

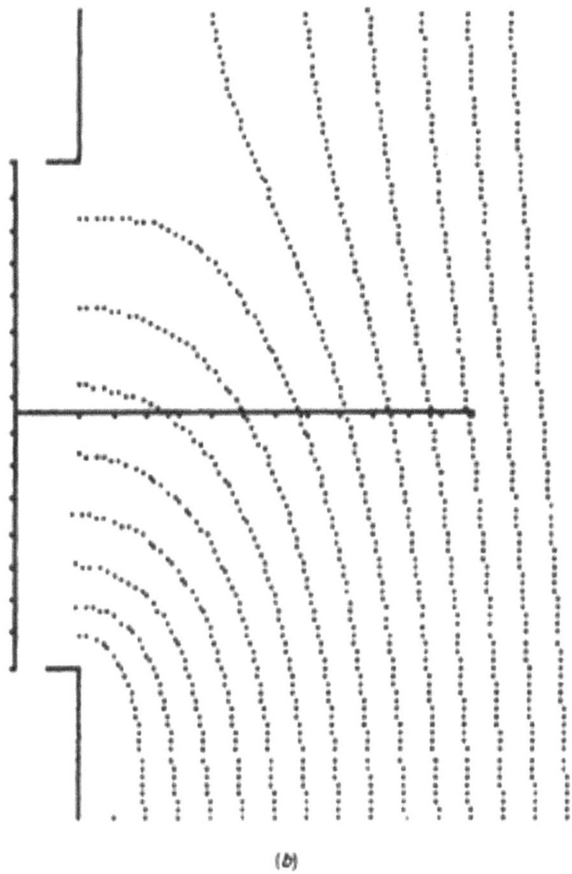

FIGURE 9.3b V_d = 100 fpm (0.5 m/s).

fewer streamlines enter the hood, so that contaminants are captured on a smaller and smaller surface. When cross-draft flow is limited (e.g., Figure 9.3a), to a velocity of 50 fpm (0.25 m/s), the hood's operating span is about 10 inches (0.25 m) along the centerline. The hood expands; when the cross-draft flow is increased to 400 fpm (2.03 m/s) (Figure 9.3d), the operating range is only about 4 inches (0.1 m).

9.3 RELATIONSHIP BETWEEN AIRFLOW PATTERNS AND ABDUCTION EFFICIENCY

Knowledge of the airflow patterns created by an exhaust hood in cross-draft airflow can be used to estimate hood capture efficiency. Consider the streamlines shown in Figure 9.3c. Note that the flow only curves around itself and enters the hood at the top edge. This flow is redrawn in Figure 9.4, where it is labeled "critical streamline" and all streamlines closer to the hood enter the hood, and those further away do not.

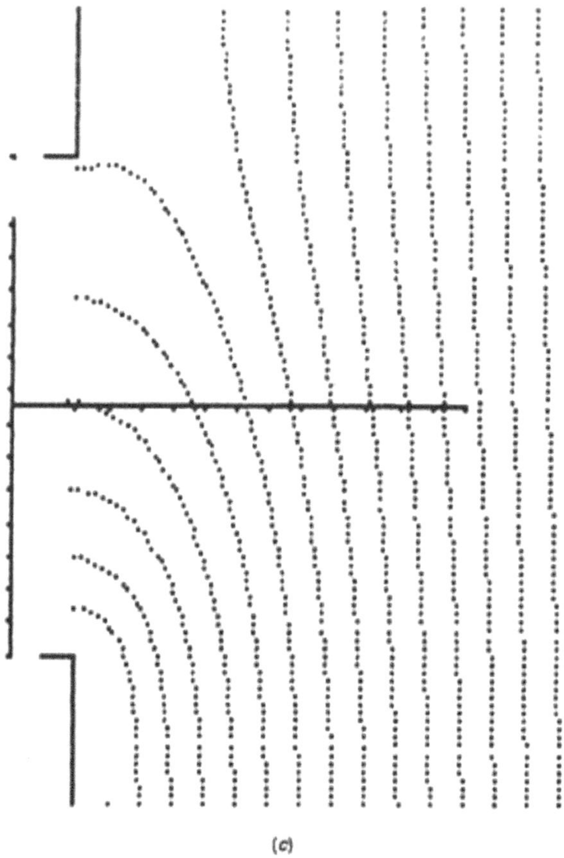

FIGURE 9.3c V_d = 200 fpm (1.01 m/s).

Air passing through the crosshatched area of Figure 9.4 illustrates the airflow entering the hood. It is expected that the contaminants transported in this vicinity will enter the hood and be captured, while those existing this level will escape from the hood.

The predicted performance of the hood at any position in the flow field can be described. Using the simple model of Figure 9.4, the capture efficiency is a step function. Inside the checkered plane it is equal to 1, and outside the checkered plane it is equal to 0. If the capture efficiency is plotted as a function of distance along the centerline x of the hood, the curve of Figure 9.5a is obtained. A point intersects the critical flow line in the central line of the hood called the critical distance (xc), at which point the efficiency of capture decreases from 1 to 0.

In real conditions, there is turbulence in the flow field, which acts to disperse the contaminants outside the main flow line. This turbulent diffusion should result in a capture efficiency curve that is sigmoidal (Figure 9.5b). If the diffusion in the main flow line is symmetric, the proper capture efficiency should occur at a critical distance of 50%.

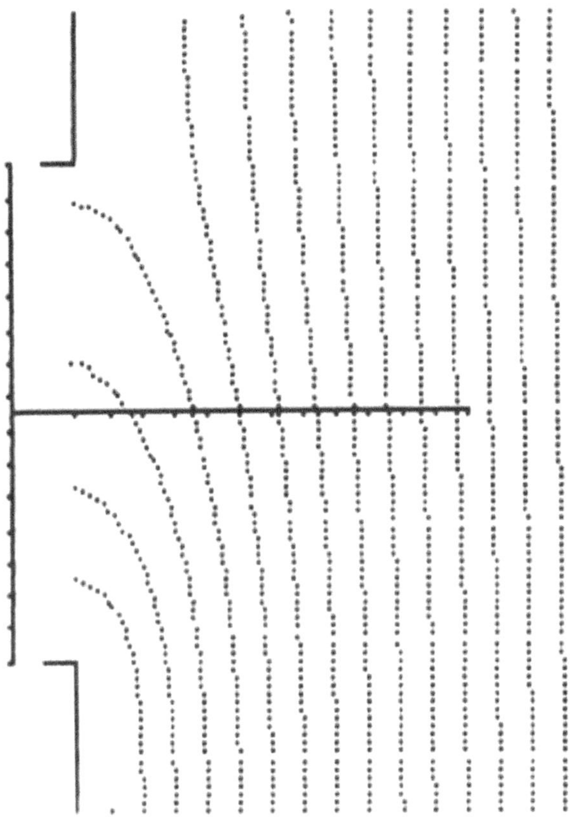

(d)

FIGURE 9.3d $V_d = 400$ fpm (2.03 m/s).

Flynn and Ellen Becker investigated the capture efficiency of several flanged circular hoods, and Conroy and Ellen Becker did the same for flanged slit hoods. They developed models for capture efficiency as a function of hood dimensions (length (L) and width (W) or diameter (D)), velocity at the hood opening (V_f), cross-draft air velocity (V_d), flow field position, and turbulence intensity (ω). Point sources for gaseous pollution were used in the experiments so that the behavior of the contaminants in the flow lines was monitored. Then, the developed models for point sources were also verified for surface sources.

The experimental results showed that the capture efficiency is best described by a logistic function, which leads to crescent-shaped capture efficiency curves similar to those in Figure 9.5b:

$$\eta = \frac{\exp\left(\dfrac{x-\mu}{\omega}\right)}{1+\exp\left(\dfrac{x-\mu}{\omega}\right)} \tag{9.2}$$

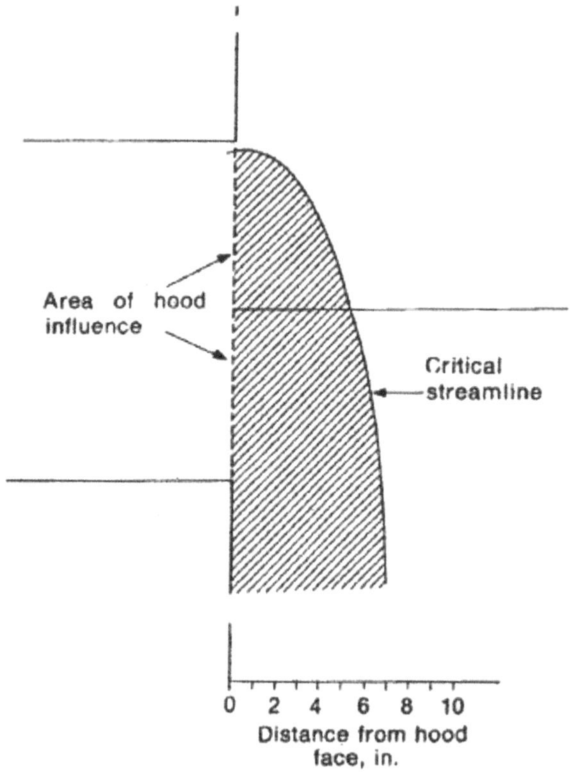

FIGURE 9.4 The hood of Figure 9.3 with $V_d = 200$ fpm (1.01 m/s), which shows the critical flow line and the influence surface of the hood.

Where μ is the critical distance measured in the laboratory, which was found to be very close to the theoretical value of x_c. μ and ω are the position in the flow field and cross-drafts.

Typical results using this equation are shown in Figure 9.6. The four available charts present efficiency curves along the centerline versus a 6-inch (0.15 m) flanged circular hood for various values of V_f / V_d. The first curve (Figure 9.6a) presents the results of the lowest tested ratio (0.92). Here, the cross drag flow of the air overcomes the velocity, and the capture efficiency reaches zero at a short distance in front of the hood. The remaining plots show the change in performance as the ratio increases. As the velocity at the hood opening is higher than the cross-draft velocity of the air, the hood has a larger "reach," and the capture efficiency is maintained at a high level at distances farther from the opening surface.

The slope of the curve changes as the interval increases due to the increased time required for the contaminant to enter the hood, which in turn results in more turbulent dispersion around the ideal contaminant flow line. As the contaminant moves further, the capture efficiency curve seems to decrease, like the step function in Figure 9.5a.

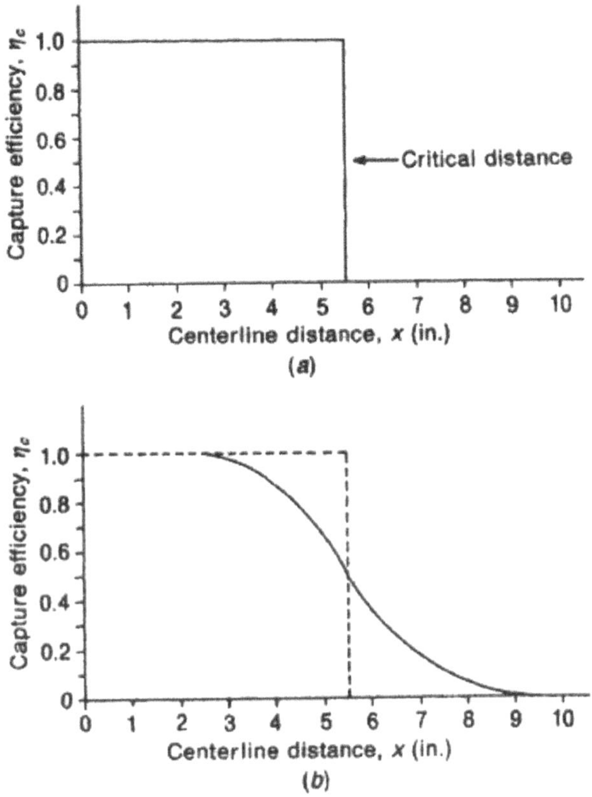

FIGURE 9.5 Hood capture efficiency models: (a) without turbulent dispersion of the contaminant, the capture efficiency is a step function that decreases from 100% to 0% at the critical distance; (b) with turbulent dispersion, the capture efficiency curve becomes crescent-shaped and gradually decreases further, passing 50% at the critical distance.

Flynn and Ellen Becker found that for the range of hoods and velocities tested, the critical distance along the hood centerline is given by the following empirical relationship:

$$\frac{X}{D} = 0.42\left(\frac{V_f}{V_d}\right)^{0.6} - 0.08 \tag{9.3}$$

Where D is the diameter of the hood. Critical distances can also be calculated at positions in the flow field that are not located on the centerline, but no simple empirical relationship is available and a computer model must be used (Conroy and Ellen Becker).

For cross-draft airflow and a given hood diamcter, Equation 9.3 shows that the critical distance is approximately proportional to the square of the velocity at the hood opening. The discussion of the capture efficiency vs. entrainment rate approach

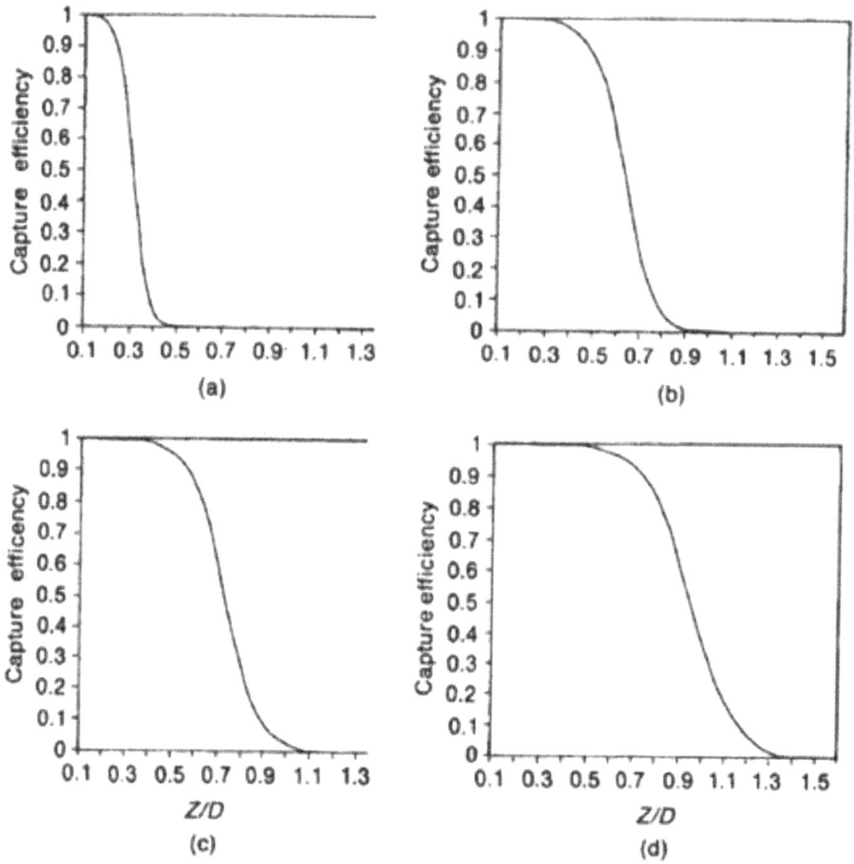

FIGURE 9.6 Theoretical capture efficiencies for a 6-inch flanged circular hood (0.15 m) in four ratios of the velocity at the opening to the cross-draft flow velocity (a) $V_f / V_d = 0.9$; (b) $V_f / V_d = 2.4$; (c) $V_f / V_d = 3$; (d) $V_f / V_d = 4.5$. Note that the critical distance increases, and the shape of the curve becomes flatter with the increase of the cross-draft current ratio.

described here is now used to describe the performance of the outer hood, i.e., Equation 9.2, which explicitly and quantitatively evaluates the effects of cross-draft flow. In other words, if the size and direction of the cross-draft flow speed is known, the flow required by the hood to overcome this speed can be calculated.

Equation 9.2 can be used to predict the capture efficiency for a point source of contaminant located anywhere in front of an external flanged hood. For surface sources of contaminant emissions, such as open surface impoundments, point source efficiency predictions can be integrated at the surface to calculate the overall capture efficiency value for the source. An example of this procedure is shown in Figure 9.7. A 1.5-inch × 15-inch (0.04 × 0.38 m) flanged hood as previously used here to ventilate a 15-inch square tank (0.38 m) open area and conditions similar to Figures 9.3a–9.3d (V_f = 200 fpm (1.01 m/s) V_d = 1000 fpm (5.08 m/s)) is used. The airflow patterns

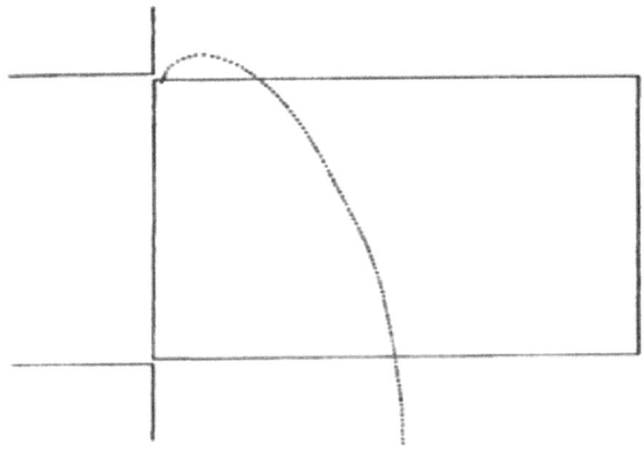

FIGURE 9.7 Surface view of a 15-inch square open surface tank and a 15-inch × 1.5-inch (0.04 × 0.38 m) flanged slotted hood showing the critical flow line with V_f = 1000 fpm (5.08 m/s), V_d = 200 fpm (1.01 m/s). Capture efficiency of surface source in this case is 0.41.

shown in Figures 9.3a–9.3d must be modified to account for the tank, which obstructs airflow down the centerline plate. The correction can be done by inserting a symmetrical image of the suction source located below the surface of the tank. The effect is to increase the range of the hood and move the critical flow line to the outside of the hood opening. The critical flow line shown in Figure 9.7 reflects this effect. The integrated model of the surface source for this case predicts the capture efficiency of only about 40% due to the high speed of the cross-draft flow and the resulting critical flow line.

9.4 LIMITATIONS OF THE CENTERLINE SPEED APPROACH

The current design method of outdoor hoods is based on choosing a suitable capture velocity and then using an equation to predict this velocity as a function of the hood airflow and the distance along the center line of the hood. The shortcomings of this procedure in the presence of significant cross tension currents can be shown using Figure 9.3c.

Figure 9.3c shows that the critical flow line crosses the centerline of the hood approximately 5.5 inches (0.14 m) from the hood opening. Since the width of the slit is 1.5 inches (0.04 m), the ratio of the critical distance (x) to the width (w)— is x / w = 5.5 inches ÷ 1.5 inches = 3.7. Therefore, calculating the capture velocity at this distance is worth considering.

Garrison's empirical model predicts that:

$$V = 0.29\,V_f \left(\frac{x}{w}\right)^{-1.1} \tag{9.4}$$

Silverman's equation predicts that:

$$V = 0.36\,V_f \left(\frac{x}{w}\right)^{-1} = 0.36 \times 1000 \left(\frac{5.5}{1.5}\right)^{-1} = 98\,\text{fpm}\,(0.5\,\text{m/s}) \tag{9.5}$$

Whereas, Flynn, Conroy, and Ellen Becker (FCE) predict that:

$$\begin{aligned} V &= \frac{V_f}{2\pi\left[0\cdot25+(x/w)^2\right]^{1/2}\left[0\cdot25+(x/l)^2\right]^{1/2}} \\ &= \frac{V_f}{2\pi\left[0\cdot25+(5\cdot5/1\cdot5)^2\right]^{1/2}\left[0\cdot25+(5\cdot5/1\cdot5)^2\right]^{1/2}} \\ &= 69\,\text{fpm}\,(0.35\,\text{m}/\text{s}) \end{aligned} \tag{9.6}$$

The authoritative Grayson and FCE models predict the same capture velocity, while the much older and accepted Silverman model predicts a much higher value of about 40%. Grayson's empirical equation (Equation 9.4) can only be used for x/w ratio less than or equal to 4, while it can be used in this example, but it cannot be used in cases like Figure 9.3b, which cross-draft currents are lower and the ratio of the critical distance to the width of the hood is more than 4. The theoretical FCE model is validated for all x/w values and experimentally for x/w values up to 7.5.

The actual velocity at the point where the critical flow line crosses the center line is the vector sum of the hood capture velocity and the cross-draft flow velocity. Using the Silverman equation in this case predicts hood overspeed, and the design based on the Silverman equation severely overestimates hood performance.

Existing design practice for open surface tank hoods also highlights the shortcomings of the capture velocity approach. The design procedure includes determining the risk potential and the release rate of the contaminant from the process in the tank. These are used along with the tank's aspect ratio and whether the tank is equipped with a baffle or not, to determine the required airflow. The specified airflow qualitatively includes only the cross-draft flow (whether or not the tank is equipped with a baffle) and includes a safety factor for contaminant toxicity.

The maximum specified airflow rate is 250 cfm/ft^2 (1.27 m^3/s). However, the airflow rate at 250 cfm/ft^2 (1.27 m^3/s) may be controlled at a speed of 0.75 m/s (approximately 150 fpm), which does not produce 150 fpm in all aspect ratios. Nevertheless, an airflow rate of 250 cfm/ft2 (1.27 m^3/s) is considered appropriate for proper control. A surface capture efficiency model is used to evaluate the adequacy of this airflow for the 15-inch (0.4 m) square source described in the section above. For the recommended gap velocity of 2000 fpm (10.2 m/s) and airflow rate of 250 cfm/ft^2 (1.27 m^3/s), the required gap width is 1.88 inches (0.05 m). If the cross-draft flow velocity (0.25 m/s) is 50 fpm, the FCE model predicts a capture efficiency of 91%. This value increases to 200 fpm (1.01 m/s) if the cross-draft flow is reduced to 61%. Usually, the recommended airflow is 150 cfm/ft^2. This airflow provides a predicted efficiency of 84% when the cross-draft velocity is 50 fpm (0.25 m/s). However, when the velocity increases to 200 fpm (1.01 m/s), the efficiency drops to only 48%. For toxic air contaminants, efficiencies of 84% or 91% are not sufficient to create safe conditions in workers' breathing zones, so efficiencies of 48% or 61% are certainly unacceptable.

10 Methods of Controlling Air Contaminants

A. H. Khoshakhlagh, H. M. Moda, S. Yazdanirad, and A. Gruszecka-Kosowska

10.1 OBJECTIVES OF THE LESSON

Upon completing this lesson, readers should be able to:

- Explain different methods of controlling air contaminants
- Describe the principles of selection and testing of filters
- Explain the mechanisms of collecting suspended particles by filters
- Explain the effective factors in choosing air pollution control equipment
- State design considerations

10.2 AN OVERVIEW OF DIFFERENT METHODS OF CONTROLLING AIR CONTAMINANTS

10.2.1 Measurement of Gas Flows and Contaminants

Sampling from stack: It is often necessary to obtain permits to install gas cleaning devices, establish compliance with regulations, and start up new equipment to meet operational needs. Also, it can be used to determine the specifications of gas cleaning equipment and verify the correct operation of the equipment. These tests determine the composition and amount of gases and suspended particles in selected locations along the process flow. The following general principles apply to the stack sampling program:

- Sampling location(s) must be acceptable to all parties that will use the results
- Sampling location(s) must have acceptable criteria regarding temperature, flow distribution, and turbulence and distance from the source of disturbances to the process flow; exceptions based on physical limitations should be identified and reported
- Samples taken from a duct or stack should represent the usual conditions during the process; appropriate chimney sweeps should be established and particulate samples taken iso-kinetic
- Sampling from stack should be done in accordance with the approved methods and established protocols at any time possible

DOI: 10.1201/9781003501091-10

- Changes in volume flow, temperature, and particulate or gaseous contaminant emissions, along with unfavorable conditions, should be identified
- A sampling report from stack should include a summary of the process, identifying any deviations from the normal course of operation that occurred during the test; a summary of the work process should be prepared during the test phase and approved by the process operator at the end of the tests
- The methods of disposal of waste produced during the test should be specified before the start of the test; this issue is particularly important in cases where test equipment is being investigated in industry because new types of waste are often generated
- Monitoring basis for experiments should be established so that results can be presented in terms of process mass, raw material consumption, energy use, etc.

Sample analysis can provide the following types of information about emission rate:

- Physical characteristics of the contaminant: solid dust, liquid mist or "smoke," waxy solids, or a viscous mixture of liquid and solid; particle size distribution: optical, physical, aerodynamic, etc.
- Concentration of suspended particles in gases, including medium and high values and a profile of concentrations in the duct or stack
- Volumetric flow of gases, including medium and large amounts and a profile of the flow in the duct or stack; volumetric flow is usually expressed in real conditions and in different standard conditions of temperature, pressure, humidity, and process state
- Chemical composition of gases and suspended particles, including recovery value, toxicity, solubility, acid dew point, etc.
- Particle densities and volume of suspended particles
- Handling characteristics of suspended particles, including the qualities of erosion, corrosiveness, abrasiveness, flocculative, or adhesion
- Flammable or explosive limit
- Electric resistance of sediments of suspended particles under laboratory and stack conditions; these data are useful for evaluating electrostatic precipitators and other electrostatically enhanced technologies

Reference methods USEPA: The US Environmental Protection Agency (USEPA) has provided methods for measuring particulate and gaseous contaminants emitted from many industrial processes. The USEPA provides fine particulate standards for ambient air quality, known as the PM_{10} and $PM_{2.5}$ standards; these revised standards focus on reducing particulate matter concentrations through the collection and control of airborne particles in the order of smaller than 10 μm (2.5 μm are concentrated fine particles). Fine particles (particle sizes with an aerodynamic diameter smaller than 2.5 μm) are of general concern because they can penetrate deep into the lungs. With the formulation of these standards, concerns were raised about the efficiency of industrial gas cleaners in different particle sizes.

TABLE 10.1
Maximum Removal Concentration of Gas Cleaning Equipment

Type	Maximum Concentration
Air cleaners	< gr/ft^3 0.002
Clean rooms	
Residential/Commercial buildings	
Factory with air recirculation	
Gas industrial cleaners	< gr/ft^3 35
Product capture in pneumatic conveyor	< gr/ft^3 3500

10.2.2 ASTM Methods

The USEPA has also approved the ASTM test methods mentioned in National Security and Public Safety Summits (NSPSS) or USEPA reference methods.

A wide range of equipment is available for the separation of suspended particles from gas streams. Typical concentration ranges for this equipment are summarized in Table 10.1.

Devices used to separate solid particles from a moving air system may be classified as: (1) air filters or (2) dust collectors.

Generally, air filters are used in common with hot air heating systems, air conditioning and ventilation systems, and at the inlet of process air systems. Suitable applications for air filters are usually limited to concentrations of contaminants that are naturally present in the atmosphere. These concentrations typically cover a range from a minimum of 0.03 grain (7000 grains = 1 lb) per 1000 ft^3 of air to a maximum of 0.2 grain/1000 ft^3 of air. It is obvious that a system that operates continuously and injects air with a concentration of 0.1 grain/ft^3 with a flow rate of 50,000 cfm into the work environment releases about 30 lbs of airborne pollutants per month into the work environment.

Dust collectors are often used at the outlet of dust control systems and processes that result in the release of contaminants into the ambient air. These systems are primarily used to capture dust to provide comfort in the work environment for workers, which are classified as dust control systems, while those to prevent escape contaminants are considered to be in the atmosphere, classified as air pollution control systems. Dust collectors are used to capture contaminant concentrations from a low amount of 0.05 grains/ft^3 of exhaust air to 20 grains/ft^3. A simple statistical example is that contaminant concentrations controlled by dust collectors may be approximately 100,000 times greater than air filters.

10.2.3 Air Filters

Three main types of filters are commonly used today: viscous impingement (sticky), dry, and electronic. The principles used by these filters in the removal of airborne solids include viscous collision, blocking (interruption), collision, diffusion, and

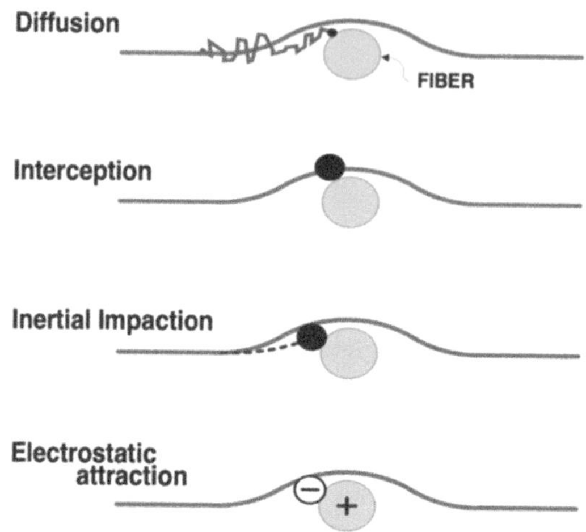

FIGURE 10.1 Particle capture mechanisms by filters.

electrostatic sediment and straining (Figure 10.1). Some filters use only one of these principles; some others work in combination.

For heavy-duty applications where high concentrations of dust are present, inertial separators are currently used as the first stage of cleaning.

10.3 AN OVERVIEW OF THE MECHANISMS INVOLVED IN PARTICLE CAPTURE BY FILTERS

10.3.1 Impaction

The trapping mechanism due to impaction applies to coarse suspended particles. The moment of inertia of coarse particles is higher than the limit that they can move along with the airflow and pass through the filter fibers, so these particles are separated from the airflow path and stop wherever they hit the fibers. The smaller the diameter of the fibers, the higher the speed of the airflow—or the larger the size of the suspended particles, the greater the inertia and the better the abduction process.

10.3.2 Interception

This process occurs when the particles are small and the distance between two adjacent filter fibers is slightly greater than the diameter of the particles. As a result, the particles can follow the flow lines, but while passing by the fibers, they become trapped. The larger the size of the suspended particle and the smaller the size of the fiber particles and the space between them, the easier the particle will be trapped and, as a result, the process will be more effective. Airflow speed has no effect on the interception process as long as it does not change the shape of the particle.

10.3.3 Diffusion or Brownian Motion

The smaller the particle size, the greater the irregularity in its movement. Very small particles (smaller than 1 micron) collide with air molecules in the airflow, resulting in create random movements similar to Brownian motion. This irregularity enhances the particle capture mechanism because the increased likelihood of particle collisions causes their paths to continuously change, leading them to collide with fibers and become trapped. Additionally, the smaller the suspended particle and the lower the air density, the longer these particles exhibit zigzag movements. Consequently, these random motions increase the probability of suspended particles striking and adhering to the filter fibers. The capacity of filters to absorb suspended particles in this mechanism increases by reducing air speed and particle size and fiber particle size.

10.3.4 The Effect of Straining

This mechanism is effective for large particle sizes. If a particle is larger than the distance between two adjacent filter fibers, it becomes trapped in the gaps between the fibers and is captured by the filter.

10.3.5 Electrostatic Attraction

In some filters, coarse fibers (where the distance between the fiber particles is large) are used to increase the amount of airflow and also to make the filter cheaper. Even if we do not charge the fibers or particles, naturally, the particles and fibers themselves can be charged due to static electricity, and the particles are removed due to the attraction between the opposite charges between the fibers and the particles, and as a result, the particles attach to the fibers. When the filter fibers are charged with static electricity, they can absorb suspended particles.

In this mechanism, the smaller charged particles are absorbed by the filter fibers and gradually form a nucleus that continuously absorbs more particles and finally, relatively large masses of suspended particles are formed inside the filter. These masses fill the empty spaces between the filter fibers and thus increase the ability of the filter to absorb more suspended particles. The mechanism of particle abduction due to electrostatic precipitation is only possible for filters with synthetic media (synthetic materials).

Filters with fiberglass media use only four mechanisms (blockage, diffusion, sieve absorption, and collision) in capturing suspended particles. The efficiency of these filters improves with the increase in the number of fiber particles. The smaller the size of the fiber particles, the number of these particles per filter surface unit increases, so the path that the suspended particle must travel through the fiber particles to exit the filter becomes more complicated. As a result, the chance that the suspended particle will be captured by the filter by one of the mentioned mechanisms increases. In filters with synthetic media, due to having larger fibers and more empty spaces between fiber particles, the possibility of suspended particles hitting the fibers and being absorbed by them is low. Therefore, the electrostatic effect and the electrostatic attraction mechanism help to increase the efficiency of these filters in capturing particles.

10.3.6 Viscous Impingement Filters

Viscous impingement filters are made of a relatively free, ordered media, usually spun glass fibers, metal plates, or layers of crimped expanded metal. The media surfaces are covered with tacky oil, which is generally called glue. The mechanism of the filter bed is such that the airflow is forced to constantly change its direction while passing through the filter. Solid particles, due to their momentum, are thrown and adsorbed to viscous coating surfaces. Larger, more massive airborne particles are filtered out by this method, while small particles tend to follow the path of the airflow and avoid being trapped. The operating characteristics of viscous impingement filters are presented in Table 10.2 as Category I.

TABLE 10.2
Filter Selection Guide

Filter Types			
A	Disposable (I)	G	Cartridge, 50–55% NBS (IIa)
B	Erasable (I)	H	Cartridge, 80–85% NBS (IIa)
C	Replaceable media (I)	J	Cartridge, 90–97% NBS (IIa)
D	Automatic renewable media (I)	K	Electronic concentrator with automatic storage of renewable media (III)
E	Cartridge 30% NBS: National bureau of standards (I or II)	L	Electronic concentrator with alternative media storage section, 97–93% NBS (III)
F	Cartridge 40% NBS (I or II)	M	Electronic air cleaner, washable plates (III)

Filter Type Measurement

The System Selection Factor	Usual Efficiency	Medium Efficiency	High Efficiency
	Air losses frictional		
Constant	D		K, M
Variable	A, B, C	E, F, G	H, J, L
	Initial cost		
Low	A, B, C	E, F, G	
Medium	D		H, J
High			K, L, M
	Operating costs		
Low	D	F	K, M
	B, C	E, G	L
High	A		H, J
	Repair and maintenance skills required		
At least	A, B, C		
Medium	D	E, F, G	H
Experienced			J, K, L, M

10.3.7 Dry Filters

Dry air filters are the broadest category of air filters in terms of the variety of designs, sizes, and shapes produced. Glass fibers are the most common media filter. Other materials used include cellulose, paper, cotton, and polyurethane and other synthetic materials. Glass fibers are widely used due to their relatively low cost and the unique ability to control the fiber diameter during manufacture (in general, smaller fiber diameters increase the air scavenging efficiency of the media).

Dry filters use direct impingement principles where very large particles are literally filtered out of the air stream as they pass through the filter openings. Impingement occurs when particles collide with and adhere to the surface of glass fibers due to natural adhesive forces, even in fibers not coated with filter adhesive. Diffusion involves very small particles behaving like gases, moving toward the filter fibers to replace particles that have impacted the surface near the fibers. Through the diffusion process, a filter can separate particles much smaller than the openings in the bed from the airflow.

Operational characteristics of dry type filters are presented in Table 10.2 in categories IIa and IIb.

10.3.8 Electronic Air Cleaners

Primarily used in applications requiring high efficiency of air cleaning, these devices are based on the principle of airflow passing through an ionization field where a potential of 12–14 kV puts a positive charge on all particles. Then, the ionized particles are alternately passed between the aluminum plates, the plates are grounded and connected to the 6–8 kV source, and finally, the particles settle on the grounded plates.

In the original design of the electronic air cleaner, a water-soluble adhesive coating is used on the plates, which retains the contaminated deposits until the plates need to be cleaned. Then, the filter is energized, and the dirt and glue layer is washed off the plates. Fresh glue is used before the electric force is applied again. Newer versions of electronic air cleaners act as agglomerators; Smaller particle masses are allowed to be removed from the plates and trapped by dry type filters or viscous impingement downstream of the electronic unit. The operating characteristics of electronic air cleaners as category III are presented in Tables 10.3a–10.3c.

Testing and measurement: Filter testing and measurement was standardized for the first time in an industry on a large scale with the implementation of the ASHRAE standard 52–76 test, which can be used to determine the nominal efficiency of filters. The ASHRAE procedure requires reporting the initial (clean) efficiency and dust holding capacity using a standard dust test, while simultaneously determining the average efficiency and dust holding capacity. To ensure the accuracy of published data, laboratory certification should be provided independent of manufacturers' performance claims. For high efficiency particulate air (HEPA) filters, such as filters used for special purpose applications including protection against radioactive contamination, clean rooms, etc., the Dioctylphthalate test (DOP) is used. In this test, a special smoke with uniform particles of 0.3 microns is utilized.

Filter selection: Unlike other mechanical components of ventilation and air conditioning systems, selecting the right air filter for a specific application requires

TABLE 10.3a
Characteristics of Air Filters

Filter Classification	Operating Principles	Type of Filter Surface	Bed Speed vs. Surface Speed fpm	Air friction loss	Efficiency Test Methods	Efficiency %	Adequate Dust Holding Capacity American Filter Institute (AFI) Grains/ft²
I. Cover Viscosity	Sieve & collision properties	Coarse fibers, metal plates, or free spaced	300–600 vs. 300–600	Variable, low, 0.50–0.10	Arrestance (ASHRAE)	Low 70–85	High 75–150
IIa. Dry	Screening & blocking feature	A media with fine fibers with a relatively close distance	20–50 vs. 250–625	Variable, Medium, 0.2–1	Dust spot (ASHRAE)	Medium 85–95 to high 85–95	Medium 5–15
IIb. HEPA	Blocking & publishing	A media with very closely spaced fine fibers	5–6 vs. 250–300	Variable, high, 2–1	DOP	Very much 99.7	Low 1–2
III. Electrostatic	Sediment & accumulation	Pages	400–600 vs. 400–600	Constant & low, or variable & medium 0.45 or 0.5–1	Dust spot (ASHRAE)	High 85–97	

TABLE 10.3b
Filter Application Instructions

Particle Size	Controlled Common Contaminant	Common Use	Common Type of Air Cleaner/Filter
> 0.3 μm	Virus (not connected) Carbon dust Tobacco smoke All the combustions	Production of electronics Production of pharmaceuticals Carcinogenic substances	HEPA/ULPA filters
0.3–1 μm	Bacteria droplet nuclei (sneeze) Cooking oil Most smokes Dust from insecticides Most face powders Most colors	Luxury commercial buildings, hospital care rooms, general surgery	Bag Filters: Microfiber or synthetic (non-flexible) non-supporting fiberglass media, 12 to 36 in deep; box filters: firmly cartridge type, 6 to 12 inches deep
1–3 μm	Legionella bacteria Humidifier Dust Lead dust Auto emission particles Drops from the nebulizer	Luxurious residential buildings Buildings with high commercial quality Hospital laboratories	Regular Pleated Filters: Wide surface with a cotton or polyester media or both, 1- to 6-inch thick box filters: firmly cartridge type, 6 to 12 inches deep
3–10 μm	Mold spores Mite feces Dog and cat hair Hair spray Fabric protector	Luxurious residential buildings Commercial buildings Industrial environments	Regular Pleated Filters: Wide surface with a cotton or polyester media or both, 1- to 6-inch thick cartridge filters: cubic or pack viscose filters Disposable: synthetic media plate filters
< 10 μm		Minimal filtration Residential gas coolers	Disposable: synthetic or fiberglass media plate, 1-inch thickness Washable: aluminum mesh, foam leather panel Electrostatic: self-charged (passive) polycarbonate plate

judgment beyond relying on well-defined parameters. Efficiency requirements, factors related to initial and operating costs, features related to repairs and maintenance, the effect of air friction loss on the system, and equipment space requirements are important factors in selection of filters.

A comprehensive analysis of how these guidelines apply to common types of air filters is provided in Table 10.2.

Dust collectors can be divided into four main categories:

1. Dry mechanical collectors
2. Wet collectors and scrubbers
3. Fabric collectors
4. Electrostatic precipitators

TABLE 10.3c
Minimum Efficiency Reporting Value (MERV)

	ASHRAE 52-2 Standard			Standard 52-1 ASHRAE
	Removal Efficiency Based on Particle Size %			Dust Spot Efficiency
MERV	0.3–1 μm	1–3 μm	3–10 μm	
16	95<	95<	95<	-
15	85–95	90<	90<	95<
14	75–85	90<	90<	90–95
13	75>	90<	90<	80–90
12	-	80<	90<	70–75
11	-	65–80	85<	60–65
10	-	50–65	85<	50–55
9	-	50>	85<	40–45
8	-	-	70<	30–35
7	-	-	50–70	25–30
6	-	-	35–50	20>
5	-	-	20–35	20>
1–4	-	-	20>	20 >

Note: HEPA filters (formerly known as MERV 17–20 filters) are not considered part of this standard.

Industrial dust originates from three main sources:

1. Abrading
2. Combustion
3. Material handling

In a typical industrial process, such as metal grinding, metal dusts in the range of 10 to 100 microns are produced. The differences in the size and concentration of suspended particles have created a significant difference in the performance requirements of dust collectors in comparison to air filters.

The usual method of rating dust collectors, even those with a high efficiency rating, is on a weight basis. For comparison, only low-efficiency air filters are measured by the weight method, whereas high efficiency filters are always measured using the dust point method or DOP.

The major differences between measuring air filters and dust collectors in the gravimetric method are that a synthetic dust test is used to measure air filters, while in dust collectors, the measurement is by the actual performance determined by the specific dust to be collected.

10.4 COLLECTOR FUNCTION

Particle collectors may be evaluated in terms of their ability to remove suspended particles from the gas stream and their ability to reduce emissions of selected particle sizes. The degree to which suspended particles are separated from the gas stream is known as the efficiency of a collector; The fraction of material released from the collection process is called infiltration. The reduction of suspended particles in a selected particle size range is known as the fractional efficiency of the particle collector.

The efficiency of a collector η is generally expressed as a percentage of the mass flow rate of the input and output materials to the collector.

$$\eta = 100\left(w_i - w_o\right) / w_i = 100 w_c / w_i \tag{10.1}$$

Where, η = collector efficiency %; w_i = mass flow rate of contaminant flow in gases entering the collector; w_o = mass flow rate of contaminant flow in exhaust gases from the collector; w_c = mass flow rate of the contaminant flow in the gases captured by the collector (Table 10.4).

Alternatively, the efficiency of a collector can be expressed in terms of the concentration of suspended particles at the inlet and outlet of the equipment. This procedure can be inconvenient because of changes in gas flow due to air leaks in the system, condensation, and changes in temperature and pressure.

Permeation P is usually measured and the efficiency for a collector is calculated using the following equations:

$$\mathrm{P} = 100 w_o - w_i \tag{10.2}$$

$$\eta = 100 - \mathrm{P} \tag{10.3}$$

TABLE 10.4
The Main Types of Particle Control Equipment

Impulse and gravity collectors	Settling chambers Chambers with shutters and baffles
Centrifugal collectors	Cyclones and multicyclones Centrifugal rotary mist collectors
Electrostatic precipitators	High-voltage (single-stage) precipitators, dry or wet plate or tube type Smoke and mist precipitators (two-stage) low voltage, dry or wet plate type
Fabric filters	Bag houses; fabric collectors; cartridge filters Disposable media filters (for dust and/or mist)
Granular media filters	Fixed filters Animated filters
Scrubbers for particles	Spray scrubbers Collision scrubbers Centrifugal type scrubbers Orifice type scrubbers Venturi scrubbers Clogged scrubbers Moving media scrubbers Electrostatic reinforced scrubbers

The fractional efficiency of a particle collector is determined by measuring the mass flow rate of incoming and outgoing contaminants to the collector in the selected particle size range. Methods for measuring the fractional efficiency of particle collectors in industrial applications are currently used due to the need to compare the performance of fine particle collectors utilized under $PM_{2.5}$ and PM_{10} guidelines for ambient air quality.

Measurements other than efficiency should also be considered in the design of industrial gas cleaning systems. Table 10.5 compares some of these factors for typical equipment. Note that this table contains only nominal values and is not a substitute for laboratory studies and engineering evaluation.

10.5 MECHANICAL COLLECTORS

10.5.1 Settling Chambers

If the particles are larger than 40 μm, they will leave the suspended state and settle within a certain period of time. Uniformizer chambers, dropout boxes, or gravity settling chambers are used to separate coarse or abrasive particles from gas streams.

Settling chambers are often used in conjunction with fabric filters or electrostatic precipitators to reduce overall system cost. Settling chambers act as a pre-collector to remove coarse particles from the gas stream.

Settling chambers sometimes contain baffles to distribute the gas flow and act as a surface for coarse particles to collide. In other designs, baffles are used to change the direction of the gas flow; as a result, the coarse particles are removed from the gas flow by inertial forces. The fractional efficiency for a settling chamber with uniform gas flow may be calculated by the following equation:

$$\eta = 100u_t\, L / HV \tag{10.4}$$

Where, η = collector efficiency for particles with settling velocity ut, %; u_t = settling velocity for selected particles, fps; L = chamber length, ft; H = chamber height, ft; V = surface velocity of gases from the chamber, fps.

The surface velocity of gases from the chamber is determined by measuring the volume flow of gases entering and exiting the chamber and the cross-sectional area of the chamber. This average speed should be low enough to prevent the re-entry of settled dust. A surface speed lower than 60 fpm is suitable for many materials.

The typical data related to settlement can be extracted from Table 10.6. Due to the air pockets within in the particles, the density of a dust particle can be significantly lower than the actual density of the material from which it is composed.

10.5.2 Inertia Collectors

Baffle collectors: baffle chambers are widely used to control particles larger than 10 μm in diameter. Baffles cause a sudden change in the direction of the gas flow. Due to their inertia, the particles move away from the gases at high speed and are either trapped in a hopper or removed in a dense side stream. The side stream is filtered by a high efficiency cyclone or collector or discharged to the atmosphere. In general, the

TABLE 10.5
Performance Measurements Related to Gas Cleaning Equipment

Type of Particle Collector	Particle Diameter, μm	Maximum Loading, gr/ft³	Collector Efficiency, % in Mass	Pressure Drop: Gas, in. water	Pressure Drop: liquid, psi	Energies at 1000 cfm (gas)	Relative Energy Requirement	Surface Velocity, fpm	Capacity Limits, 1000 cfm	Required Space (relative)
Inertial dry collectors										
Settling chamber	>40	>5	50	0.1–0.5	-	-	1	300–600	-	High
Baffle chamber	>20	>5	50	0.5–1.5	-	-	1.5	1000–2000	-	Medium
Continuous removal chamber	>20	>1	70	<1	-	-	3	2000–4000	50	low
Cyclone	>15	>1	85	0.5–3	-	-	1.5–9	2000–4000	50	Medium
Multicyclone	>5	>1	95	2–10	-	-	6–20	2000–4000	200	low
Collision	>10	>1	90	1–2	-	-	3–6	2000–4000	-	low
Dynamic	>10	>1	90	Provides pressure	-	1–2 hp	10–20	-	50	-
Electrostatic precipitator										
High voltage	>0.01	>0.1	99	0.2–1 -	0.1–0.6 Kw	0.8–20	60–400	10–2000		High
Low voltage	>0.001	0.5	90–99	0.2–0.5 -	0.03–0.06 Kw	0.5–1	200–700	0.1–100		Medium
Fabric filters										
Bag houses	>0.08	>0.5	99	2–6 -	-	6–20	1–20	200		High
Cartridge filters	>0.05	>0.1	99+	2–8 -			0.5–5	40–50		Medium

					Wet scrubbers					
Gravity spray	>10	>1	70	0.1–1	20–100	0.5–2 gpm	5	100–200	100	Medium
Centrifugal	>5	>1	90	2–8	20–100	1–10 gpm	12–26	2000–4000	100	Medium
Collision	>5	>1	95	2–8	20–100	1–5 gpm	9–31	3000–6000	100	Medium
Packed media	>5	>0.1	90	0.5–10	5–30	5–15 gpm	4–34	100–300	50	Medium
Dynamic	>2	>1	95	Provides pressure	5–30	1–5 gpm 3–20 hp	30–200	3000–4000	50 50	Low
Immersed orifice	>2	>0.1	90	2–6	-	Without pumping	9–21	3000	50	Medium
Jet	>2	>0.1	90	Provides pressure	50100	50–100 gpm	15–30	2000–20,000	100	Low
Venturi	>0.1	>0.1	95–99	10–60	1030	3–10 gpm	30–300	12,000–42,000	100	Low

TABLE 10.6
The Final Velocities of Particle Settling, fps

Particle Density Relative to Water	Particle Diameter μm									
	1	2	5	10	20	50	100	200	500	1000
0.05	5.9×10^{-6}	2.3×10^{-5}	1.3×10^{-4}	5.2×10^{-4}	2.3×10^{-3}	1.3×10^{-3}	5.2×10^{-2}	0.18	0.75	1.7
0.1	1.2×10^{-5}	4.6×10^{-5}	2.6×10^{-4}	1.0×10^{-3}	4.6×10^{-3}	2.6×10^{-3}	9.8×10^{-2}	0.36	1.3	2.7
0.2	2.4×10^{-5}	9.2×10^{-5}	5.2×10^{-4}	2.1×10^{-3}	9.2×10^{-3}	5.2×10^{-2}	0.19	0.62	2.1	4.3
0.5	5.9×10^{-5}	2.3×10^{-4}	1.3×10^{-3}	5.2×10^{-3}	2.3×10^{-2}	0.13	0.46	1.4	4.0	8.2
1	1.2×10^{-4}	4.6×10^{-4}	2.6×10^{-3}	1.0×10^{-2}	3.9×10^{-2}	0.25	0.82	2.3	6.4	12.8
2	2.4×10^{-4}	9.2×10^{-4}	5.2×10^{-3}	2.1×10^{-2}	8.2×10^{-2}	0.46	1.5	3.7	10.2	20.5
5	5.9×10^{-4}	2.3×10^{-3}	1.3×10^{-2}	4.9×10^{-2}	0.21	1.1	3.2	7.3	18.9	36.1
10	**1.2×10^{-3}**	**4.6×10^{-3}**	**2.6×10^{-2}**	**0.1**	**0.39**	**2.0**	**5.4**	**11.5**	**29.2**	**56.8**

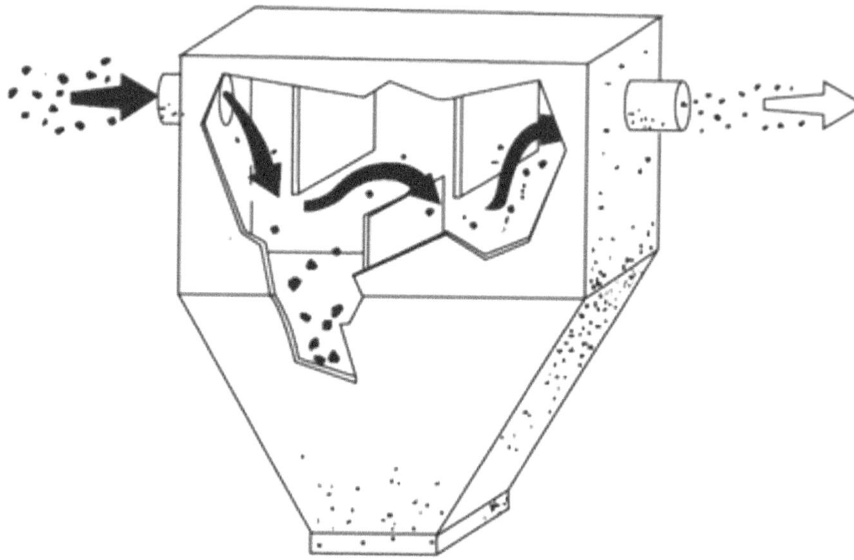

FIGURE 10.2 Baffle collector.

pressure drop of inertial collectors with baffles is greater than that of settling chambers, but this loss is balanced by higher collection efficiency and smaller equipment.

Inertial collectors are also sometimes used for mist control. In some applications, the interior of the collector may be irrigated to prevent re-entry of dry dust and remove dissolved sediments.

A common baffle collector is shown in Figure 10.2.

10.5.3 Centrifugal Dry Collectors

Centrifugal dry collectors have the lowest manufacturing cost and have been used as common dust collection equipment for many years. They are effective for almost all sizes of medium to coarse granular dusts and are usually used as primary collectors or as pre-cleaners in conjunction with higher efficiency final collectors. Their advantages include low cost, cheapness in operation and maintenance, and relatively small size.

Due to stringent emission standards, the use of centrifugal dry collectors is almost exclusively for product recovery applications upstream of more efficient secondary collectors.

Centrifugal dry collectors are commercially available in a wide variety of designs, but all rely on centrifugal force to separate dust particles from the air stream.

The amount of centrifugal force that can be exerted on a particle is a function of the particle's mass and angular acceleration (which depends on the air speeds in the collector). With increasing speed, energy losses due to friction and turbulence also increase, and pressure drop along the collector increases. In general, the dust collection efficiency increases with the increase of the pressure drop of the collector.

Unfortunately, the collection efficiency cannot be improved sufficiently by increasing the energy input and collector pressure drop. Although higher velocities result in a larger centrifugal force vector that causes the particle to deviate from the airflow path, the aerodynamic forces that create resistance to the deviation path also increase. This effect is more pronounced especially for smaller particles because they have a larger cross-sectional area per unit mass. Therefore, any centrifugal dry collector has an upper performance limit (which can be expressed as pressure drop for simplicity). Beyond that, the collection efficiency for a particular dust type cannot be increased for practical purposes. Determining the size and selection of the collector is largely dependent on the selection of the optimal operating point. The selected point should have the highest efficiency in terms of collection efficiency for the input power.

Figure 10.3 shows some common cyclone collectors.

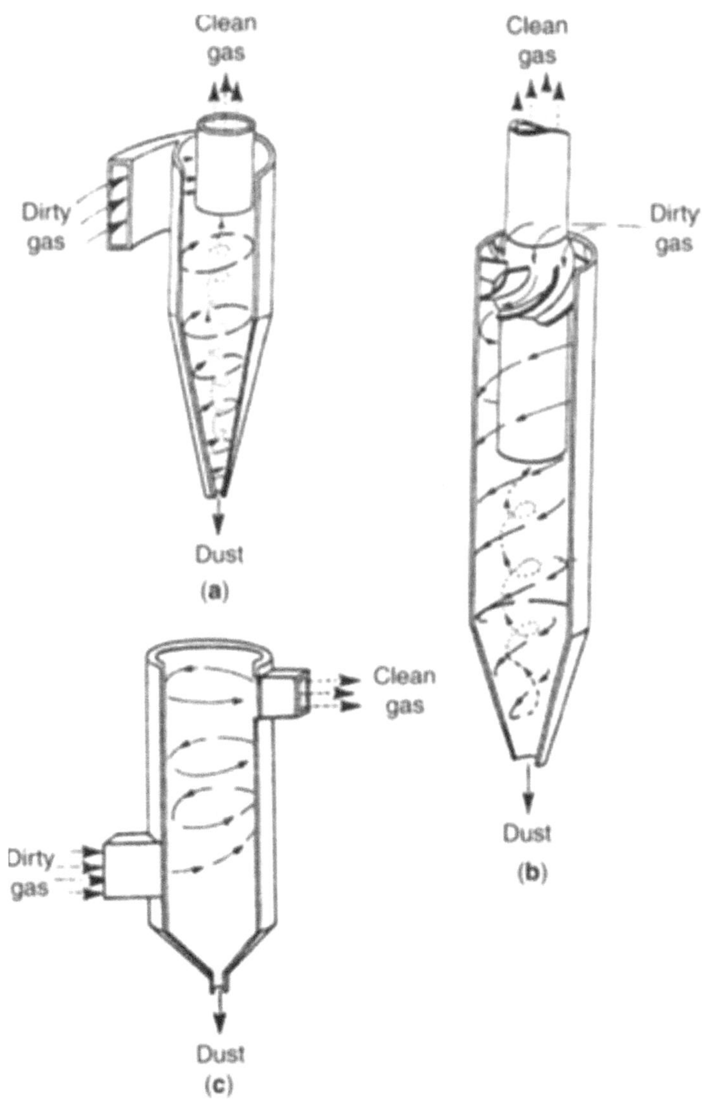

FIGURE 10.3 Types of cyclones: (a) top inlet; (b) axial inlet; (c) bottom inlet.

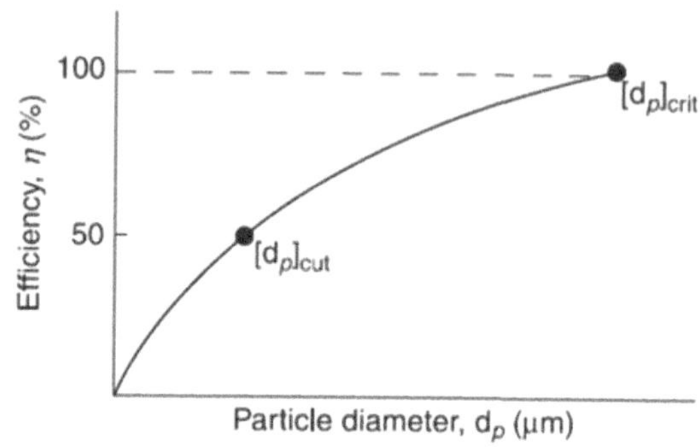

FIGURE 10.4 Cyclone efficiency.

A low-efficiency cyclone operates with a static pressure drop of 1 to 1.5 inches of water between the inlet and outlet and can remove 50% of particles 5 to 10 micrometers in diameter. High-efficiency cyclones operate with a static pressure drop of 3 to 8 inches of water between inlet and outlet and can remove 70% of particles with a diameter of 5 micrometers.

The efficiency of a cyclone depends on the particle density, shape, and size (the aerodynamic size D_p, which is the mean of the particle size range). Cyclone efficiency may be estimated from Figure 10.4.

The D_{pc} parameter, known as cut-off diameter, is defined as the diameter of particles collected with 50% efficiency. The cut-off diameter can be estimated using the following equation:

$$D_{pc} = \sqrt{\frac{9\mu b}{2N_e V_i (\rho_p - \rho_g)\pi}} \tag{10.5}$$

Where, D_{pc} = cut-off diameter, ft; μ = absolute gas viscosity, centipoise; b = cyclone inlet width, ft; N_e = number of effective rotations in the cyclone. It is approximately 5 for a high efficiency cyclone and varies from 0.5 to 10 for other cyclones; V_i = inlet gas velocity, fpm; ρ_p = particle density, lb/ft^3; ρ_g = gas density, lb/ft^3.

10.5.4 Electrostatic Precipitators

The equipment is able to achieve dust collection efficiency equal to a fabric collector. However, due to the high initial cost of installation, the control of only small and medium volumes of air and the need for experienced personnel to properly maintain them have limited their use. High-voltage electrostatic precipitators should not be confused with the low-voltage, low dust holding capacity designs used for filtration in air conditioning systems.

Electrostatic precipitators use the forces on charged particles passing through an electric field to separate the particles from the airflow in which they are suspended. In each precipitator, three specific functions must be performed:

Ionization: Charging contaminant particles
Collection: The target particles move towards the collection electrodes based on electrical force
Purification: Removal of accumulated contaminants from the precipitator

The principle of collection relies on the ability to apply a negative charge to particles in the gas stream, which causes them to move and attach to grounded or positively charged collection plates. Most precipitators are designed for horizontal airflow at velocities of 100 to 600 fpm. Collector plates are typically parallel components on 9-inch centers and are made in a variety of ways, including corrugated or perforated plates, or rod curtains.

Rigid electrodes or weighted or hard wire are located in the center of the collection plates. The voltage difference between the electrode and the plate in most designs is between 60,000 and 75,000 volts. Cylindrical forms of collection plates around the electrode rod are used where water is collected for washing material and where gas flow is under high pressure or vacuum.

The removal of collected materials is done through vibration or shaking mechanism in predetermined or continuous intervals—or draining is usually done without stopping airflow from the precipitator, although a slight drop in exhaust air has been observed in most applications during the purge cycle.

The pressure drop can be negligible. The collection efficiency is high and almost uniform, regardless of particle size, including sub-micron particles. Relatively large space is required, and operating cost increases where low gas volumes (below 50,000 cfm) are used because high-voltage equipment is expensive. By increasing the humidity in the airflow as a change in the dielectric properties of the dust, the efficiency improves. On the other hand, heavy concentrations lead to a decrease in the collection efficiency by creating a space charge on the particles and weakening the corona effect of the electrodes.

Under these conditions, a pre-cleaner, often of the centrifugal dry type, is used to reduce the concentration entering the precipitator.

Electrostatic precipitators are widely utilized in high temperature gas cleaning in equipment such as blast furnaces, open hearth furnaces, and central station of pulverized fuel boilers. Black liquor papermaking furnaces are also used in chemical industries, including sulfuric acid, carbon black, cement furnaces, and soda ash factories. Because the voltage setting is close to the sparkover potential, application is limited to non-explosive and non-combustible materials unless the carrier gas stream is an inert gas.

Units that perform ionization and collection simultaneously in a single structure are called single-stage precipitators (Figure 10.5); they have widely spaced electrodes (3–6.15 inches) and typically high voltages (20–60 kV) but relatively low field gradients (rarely up to 11 kV/in). In two-stage precipitators, ionization and collection occur independently in the yielding and sedimentation structures (Figure 10.6). Because the ionization and collection electrodes are closely spaced (0.7–1.5 inches), two-stage precipitators typically operate at high field gradients (usually greater than 10 kV/in) but at relatively low voltages (typically 10 kV or less, and up to 14 kV in some cases). Because of fundamental differences in ionization processes and how

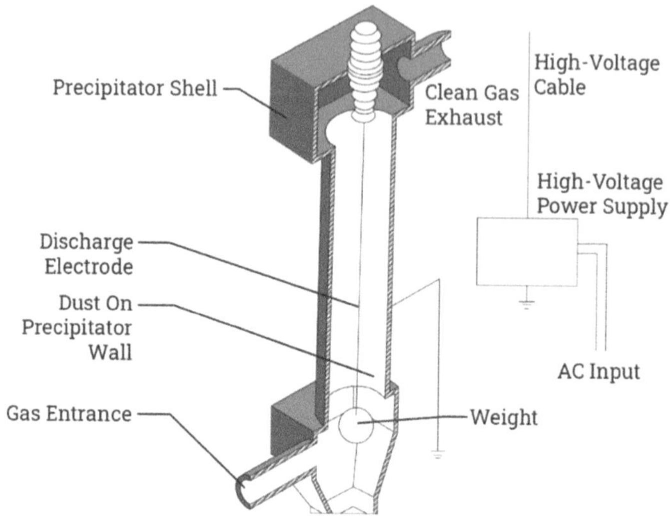

FIGURE 10.5 Common single-stage electrostatic precipitator.

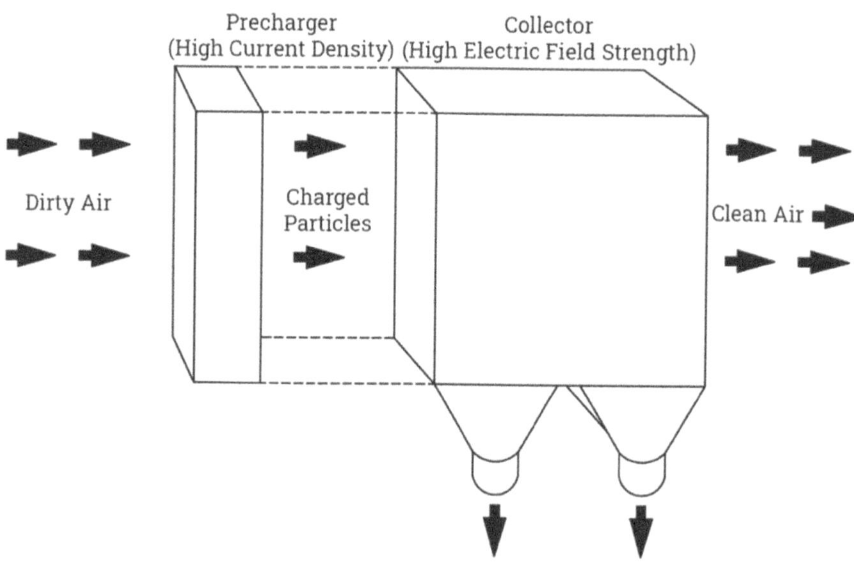

FIGURE 10.6 Common two-stage electrostatic precipitator.

they are constructed, high-voltage (single-stage) and low-voltage (two-stage) precipitators are chosen for completely different air purification needs.

10.5.4.1 Single-Stage Designs

Figure 10.5 shows a single-stage precipitator. The yield electrodes are located between the parallel collecting plates, and gas flows through the precipitators horizontally. High-voltage precipitators collect larger particles better than small particles (they are less efficient in collecting pollutants smaller than 1–2 μm). Their deposition efficiency depends to some extent on the relative electrical resistance of the contaminant being collected. Most single-stage precipitators are very efficient at collecting conductive or dielectric contaminants.

High-voltage single-stage precipitators can easily control heavy loads of dry dust. Most are configured for continuous operation (using vibratory cleaning or online vibration). They can continuously remove large quantities (hundreds of pounds per hour) of airborne materials such as foundry vibratory processes, cement, ceramics, chemical dusts, fly ash, dust and blast furnace (blow furnace), and boiler emissions of the paper production industry.

Although they are rarely used to clean exhaust airflows much less than 50,000 cfm, single-stage precipitators can be used to control airflows up to 2,000,000 cfm. The gas velocity from the electrostatic field is typically 60 to 400 fpm with a treatment time in the range of 2–10 seconds. Only single-stage "wet" precipitators are used to collect liquid contaminants.

Single-stage industrial electrostatic precipitators are distinguished from low-voltage two-stage precipitator designs by several features:

- The distance between the electrodes is increased to accept the collection and cleaning of the electrode under the unfavorable conditions of industrial gas flows and high dust load
- The structure is designed to operate at 850 °F and a pressure drop of 30+ inches of water
- They are usually suitable for industrial applications that produce ozone; as a result, they may operate with negative and polarized ionization, providing maximum electric field strength between the electrodes
- They are usually custom made and assembled based on the specific application on-site

Single-stage precipitators can be designed for collection efficiencies higher than 99.9% for precisely specified conditions. Dust characteristics such as particle size distribution and deposit electrical resistivity, along with changes in gas composition and flow speed, can significantly affect performance.

10.5.4.2 Two-Stage Designs

Two-stage precipitators are produced in two general forms: 1) electronic air cleaners, typically used for light dust loads in commercial/residential ventilation and air conditioning; and 2) heavy-duty industrial electrostatic precipitators used to control heavy loads of sub-micron liquid particles emitted from hot industrial processes and machining operations. Two-stage industrial clarifiers often include sumps and drains that continuously feed the collected liquids through gravity runoff into the tanks.

Two-stage precipitators are frequently used in industry to collect submicron contaminants that are difficult (or impossible) to collect by other types of equipment. Two-stage precipitators are often used for flows with flow rates less than 1000 cfm and are rarely used for contaminated airflows with flow rates greater than 50,000–75,000 cfm.

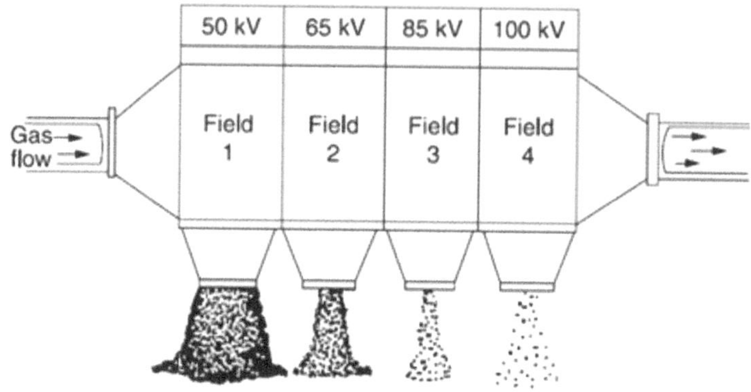

FIGURE 10.7 Stage or field sectionalization.

The gas velocity from the collection fields usually ranges from 200–700 fpm, with purification times of 0.015 to 0.05 seconds in the ionizing fields and 0.06–0.25 seconds in the collection fields.

Because liquid drainage, fume, or sedimentation mist is a reliable collector cleaning mechanism, two-stage low-voltage precipitators are often recommended for liquid contaminant collection where few solids are present.

The performance of the precipitator depends on the number of sections or fields installed (Figure 10.7).

Since the early 1970s, an important application for low-voltage two-stage electrostatic precipitators has been as gas scrubbers in condensing precipitator systems. The design of these air pollution control systems (and similar condensing filtration systems) is based on this principle: Although hot gases or smokes emitted by many processes are not easily filtered or precipitated (because they are in the vapor phase), condensation of fumes or aerosol mists formed by cooling of vaporized contaminants can be effectively collected by filtration or precipitation (Figure 10.8).

Many dense aerosol fumes are composed of submicron liquid droplets, making them suitable for collection by low-voltage two-stage precipitators.

Low-voltage precipitators can be effective in collecting aerosol particles smaller than 1 μm (0.001–0.01 μm), which are responsible for most blue smoke emissions and plume opacity. Condensation precipitator systems may be a good choice for removing residual blue smoke turbidity formed by condensation of aerosol smoke plumes from hot processes, dryers, ovens, furnaces, or other exhaust air cleaning devices.

When dense aerosol contaminants are odorous, deposition of odorant submicron droplets can prevent long-range drift of odorous substances.

10.5.5 Fabric Collectors

Fabric collectors are one of the most successful and old methods of dust collection.

They are usually used when small particle sizes and a high degree of cleaning efficiency are required. They can also recover maximum amounts of valuable process

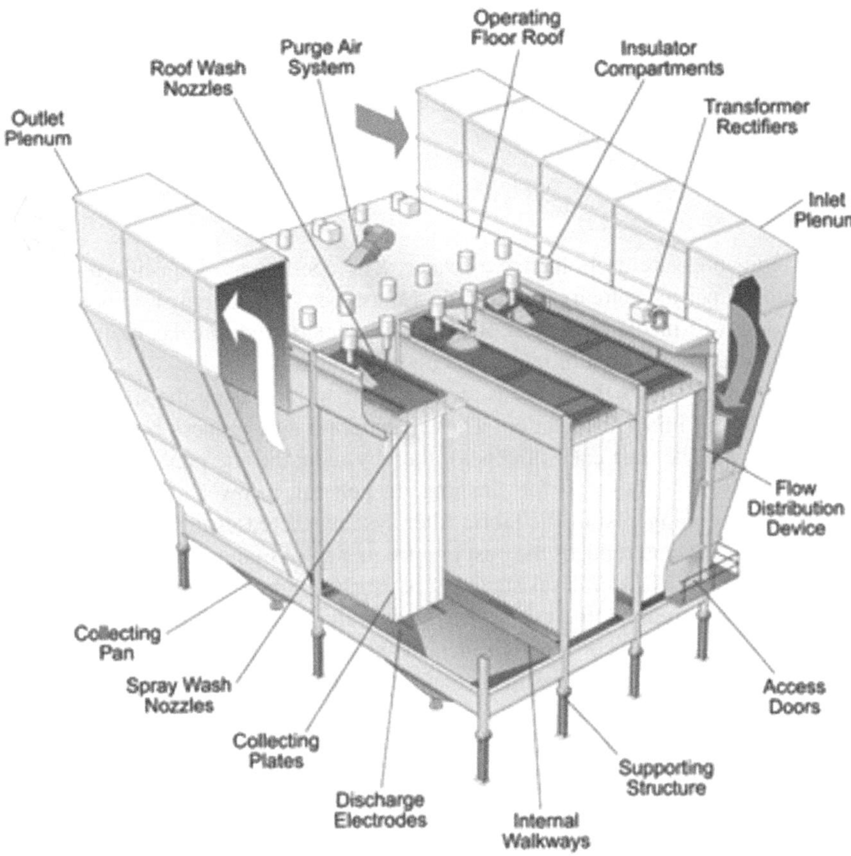

FIGURE 10.8 Wet electrostatic precipitators.

materials in the dry state. More recently, stricter regulations have been imposed on the use of fabric collectors due to process emissions limitations.

All fabric collectors use a similar method of separating suspended particles from the airflow. Air laden with dust passes through a tube or fabric package, where particles, larger than the interstices of the fabric, are settled by the effect of simple sieve properties. A coating or cake of dust quickly forms on the surface of the air inlet to the fabric; the dust cake acts as a highly efficient filter, capable of removing fumes and sub-micron dust, while the fabric acts as a support structure for the cake.

The size of a fabric filter is based on the available experimental data related to the amount of fabric media needed to clean the volume flow of the desired gas with an acceptable static pressure drop along the bed. The suitable conditions and media for its use are selected by pilot testing or through experience working with similar media in full scale. These tests provide a suggested range of approach speeds for a particular application.

In practical applications, approach velocity is expressed as gas-to-media ratio or filtration speed (that is, the speed at which the gas passes through the media filter). The gas-to-media ratio is the ratio of the volume flow of gas from the fabric filter to the surface of the fabric that participates in the filtration—in fact, it is the average speed of gas approach to the surface of the media filter.

It is difficult to estimate the relationship between pressure drop and gas-to-media ratio. However, when the flow entering the dust cake filter is smooth and uniform, the pressure drop will be proportional to the convergence rate:

$$\Delta p = KV_iW = KQWA \tag{10.6}$$

Where, Δp = pressure drop, inches of water; K = coefficient of specific resistance, pressure drop per air velocity unit and dust mass per surface unit; V_i = is approach velocity or gas-to-media ratio, fpm; W = surface density of dust cake, oz/ft^2; Q = gas volume flow, cfm; A = area of fabric that abducts gases, ft^2.

Equation 10.6 shows that there is an advantage to increasing the surface area of the fabric during initial installation. The larger fabric area reduces both the gas-to-media ratio and the dust cake thickness, thereby reducing the pressure drop in the collector as well as the need for cleaning. In general, a lower gas-to-media ratio reduces the operating cost of the fabric filter system, increases the useful life of the filter components, and reduces the cost and frequency of maintenance.

The coefficient of specific resistance K is usually higher for fine dust. Using the main collector to remove the coarse fraction rarely causes a significant change in pressure drop along the collector. In fact, the coarse dust fraction helps reduce the operating pressure, because it results in a more porous dust cake that allows for better dust cake release.

Figure 10.9 shows the dependence of pressure drop versus time for a single-chamber fabric filter operating on a purge cycle.

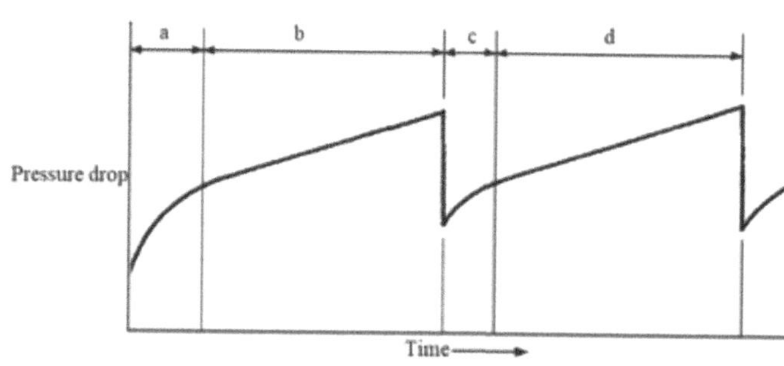

a- Initial formation of dust cake

b- Settling during the homogeneous formation of dust cake

c- Removal of cake and initial formation of cake

d- Interval between cleaning processes

FIGURE 10.9 Dependence on the time of pressure drop along the fabric filter.

It is assumed that the volume flow, particle concentration, and particle size distribution remain constant. In practice, these assumptions are usually not acceptable. However, the intervals between cleaning processes are usually long enough that these changes are negligible for most systems. Between the cleaning processes, the dependence of the pressure drop on time is almost linear.

The volume flow of gases in a process often varies in response to changes in the pressure drop across the fabric filter, which depends on the operating point of the process and the fan. There are many different designs for fabric collectors. In fact, collectors differ in the method used to remove the settled material from the fabric surface. Common methods include mechanical shaker, reverse air collapse, and pulse-jet (Figures 10.10–10.13).

Choosing the fabric reconditioning method (removing dust accumulation from the fabric surface) is primarily a function of the nature of the dust (continuous or intermittent process) and fabric selection. For example, fiberglass cloth is often chosen for higher temperature applications (up to 500°F). However, they are generally limited to reverse air equipment because the glass fibers are degraded when subjected to the flexure imposed by the shaking mechanism. Shaker collectors are often used for intermittent process applications, while the pulse jet type is a more common choice for continuous processes.

Fabric choice options include all organic and synthetic fibers. The choice of fabric is usually a function of the temperature and fume content of the exhaust gases and the chemical composition of the captured dust. Table 10.7 provides a summary of the temperature ranges and characteristics of fabric filter media.

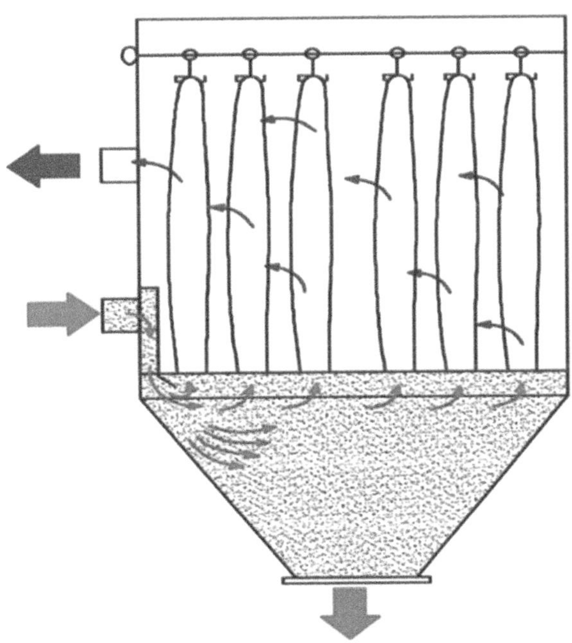

FIGURE 10.10 Shaker baghouse.

FIGURE 10.11 Shaker dust collector.

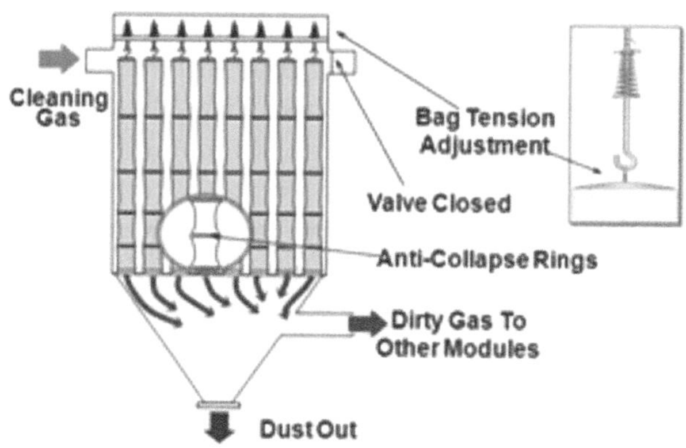

FIGURE 10.12 Reverse air dust collector.

10.5.6 Wet Collectors

Wet dust collectors have provided a relatively simple and moderate cost solution to many dust and air pollution control problems. In general, they have less space requirements than other collectors. Because the size of the equipment is small compared to the air cleaning capacity, most collectors can be obtained from the manufacturer fully assembled or assembled from the main parts, with simple installation.

The particle sizes, which can be controlled by the wet scrubber device, range from 0.3–50 micrometers or more. Wet collectors can be classified into three categories: (1) low energy (1 W/cfm max., 1–6 inches of water); (2) moderate energy (1–3 W/cfm, 6–18 inches of water); and (3) high energy > 3 W/cfm, >18 inches of water). Typical wet scrubber operation is summarized in Table 10.5.

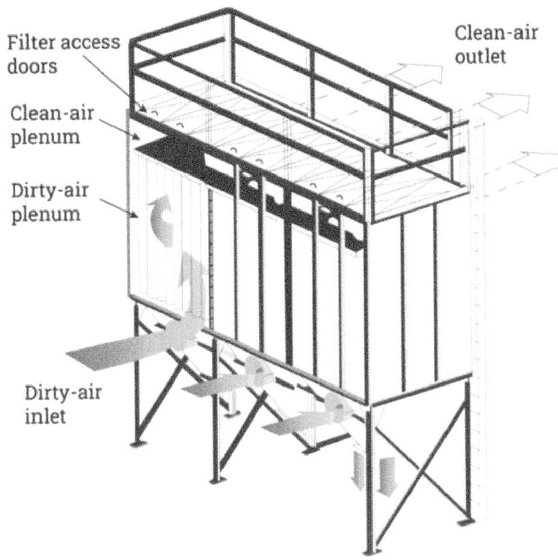

FIGURE 10.13 Pulse jet dust collector.

Wet collectors can clean hot and humid gases that are difficult or even impossible to control with other collectors. Wet collectors are often able to eliminate or reduce the risks associated with the collection of explosive or highly flammable materials. In addition, since the solids are collected wet, secondary dust problems during material disposal are eliminated. They are commercially available in a wide variety of designs, shapes, and sizes; the collection principles employed include centrifugal force, impact, and inertial impact, either individually or in combination. Researchers who study the performance of wet collectors focus mostly on contact power theory, which states that for properly designed equipment, the collection efficiency is a function of the energy expended in the process of contacting air with water and is independent of the design of the collector. Accordingly, collectors of suitable design operate at or near similar pressure drops where the same performance can be expected.

All wet collectors have a fractional efficiency characteristic. That is, their cleaning efficiency varies directly with the size of the collected particles. In general, collectors that operate at very low pressure drop only remove medium- to large-sized particles. High collection efficiency of fine particles requires an increase in energy input, which is observed in collectors with higher pressure drop.

It is possible to collect sub-micron particles, fume, and smoke with high efficiency by a Venturi type collector with high energy (Figure 10.14). Currently, Venturi designs are used in many applications previously limited to fabric or electrostatic collectors. According to contact power theory, Venturi-type collectors require significant input energy to achieve high collection efficiency of submicron particles. To solve the long-term problem of using a contactor moving media to wash SO_2 fumes from the combustion products of electricity production and large industrial boilers, water collectors are a suitable solution.

TABLE 10.7
Temperature Ranges and Fabric Filter Media Characteristics

Fabric Type	Maximum Continuous Operating Temperature, °F	Acid Resistance	Alkali Resistance	Flex Abrasion	Cost Relative to Cotton
Cotton	180	weak	very good	very good	1
wool	200	good	weak	average to good	2.75
Nylon[a,b]	200	weak	great	great	2.50
Nomax[a,b]	400	average	very good	very good	8
Acrylic	260	good	average	good	3
Polypropylene	180	great	great	very good	1.75
Polyethylene	145	great	great	very good	2
Teflon[b]	425	great	great	good	30
Glass fibers	500	average to good	average to good	weak	5
Polyester[a]	275	good	good	very good	2.50
Cellulose	180	weak	good	good	-

[a] When these fibers are exposed to hot and humid atmosphere, they undergo a hydrolysis reaction.
[b] DuPont trademark.

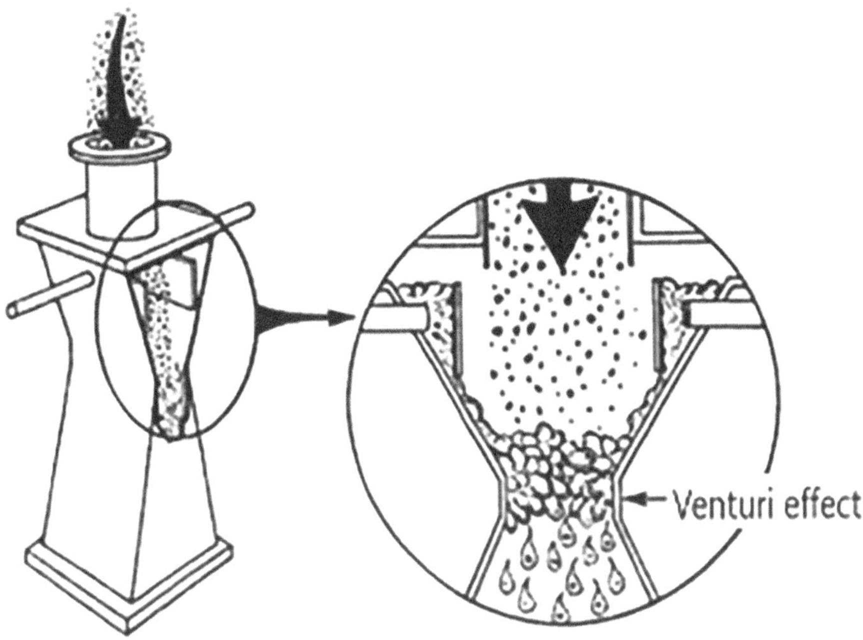

FIGURE 10.14 Venturi scrubber.

The water requirements of the collector represent ongoing operating costs that must be evaluated in the selection of specific equipment. Additional considerations are necessary when using wet collectors to wash some process fumes—as a water solution and a chemical additive may be necessary to achieve the desired goal. When the amount of water required is high, in order to save, it is necessary to use the water return system. Such systems usually utilize a tank or settling pond to separate the collected material by gravity.

Since the water returned to the collector will invariably contain some solids, it is best for these applications to choose a collector that does not require spray nozzles or small water orifices. Corrosive substances are often present in the collector. Advanced structural materials can provide satisfactory protection against almost all corrosive agents, but chemical compositions must be properly predicted and identified to select the appropriate structural materials.

10.5.7 Adsorption

Adsorption is a fluid-solid mass transfer technology in which the contaminant is absorbed, which is called an adsorbed substance, and the solid absorbing substance is called an adsorbent. In gas purification, the fluid is a gas. The adsorbed substance adheres to the solid surface through physical adsorption or chemical adsorption. In physical surface adsorption, weak physical bonding forces are created, mainly van der Waals forces, while stronger chemical bonds are formed in chemical adsorption.

The most common adsorbent in air or gas pollution control is activated carbon (AC). In a study conducted by Khoshkhlagh et al. (2021) on activated carbon in the adsorption of toluene, the results showed that the adsorption capacity of activated carbon (110.39 mg g^{-1}) is suitable for trapping the said contaminant. In other studies, the adsorption of toluene by activated carbon composites—copper and iron metal-organic frameworks—were investigated, and the results of the studies showed that hybridization of activated carbon with metal-organic frameworks increased the adsorption capacity of activated carbon. In general, activated carbon is in the form of balls or granules in sizes usually less than 1 mm to several millimeters. AC is quite efficient at removing many VOCs, although not all contaminants will be removed with the same efficiency.

In the following section, the results of the studies of Khoshkhlagh and colleagues in the field of air pollution control in the workplace by the adsorption mechanism are presented. These results and the method used in it can be useful for graduate students in master's and doctoral degrees in research matters.

10.5.7.1 Investigating the Characteristics of the Studied Adsorbents

The specific surface area of commercial-activated carbon used in the study was 815 m^2 g^{-1}. The activated carbon isotherm was of the first type and placed in the category of microporous. Based on BET analysis, the specific surface area and the total volume of functionalized activated carbon cavities were 632 m^2 g^{-1} and 0.277 cm^3 g^{-1}, respectively, and the average diameter of the cavities was 1.79 nm.

Fourier-transform infrared spectroscopy (FT-IR) of commercial and functionalized activated carbon showed that after oxidation of commercial active carbon surface by nitric acid, the absorption peaks showed a good change in the surface. The absorbed band at 1185 cm^{-1} was related to C-O stretching vibration band and OH bending groups in phenolic and carboxyl groups. The band observed at 1715 cm^{-1} corresponded to the stretching vibrational band of aldehydes, ketones, or carboxyl groups, although the 3400 cm^{-1} band corresponding to the carboxylic OH group was not well defined due to water absorption. Also, the 1577 cm^{-1} band belonged to the aromatic ring. This means that with surface oxidation, many carbonyl groups have been created on the active carbon surface.

The well-known method for determining the surface chemistry of carbons is Boehm titration. Due to the neutralization of various species present on the carbon surface, hydrochloric acid was used to determine the content of the main groups. Titration results for commercial activated carbon and carbon oxidized by hydrochloride and different bases with different strengths were done. Boehm's titration showed that there are amounts of phenolic, lactonic, and carboxylic groups in the structure of the samples.

The chemical nature of surface oxygen complexes on the samples after oxidation was investigated by both the FTIR test and titration method. Scanning electron microscopy analysis of the Fe-MIL-101 metal-organic framework showed a cubic single-crystal structure and entangled crystals, which is consistent with the X-ray diffraction test results. The isotherm of metal-organic framework Fe-MIL-101 was of the first type and was placed in the category of microporous. Based on BET analysis, the specific surface area and the total pore volume of the Fe-MIL-101 metal-organic framework were found to be 681 m^2 g^{-1} and 0.363 cm^3 g^{-1}, respectively, and the average pore diameter was 2.29 nm. According to the FT-IR spectrum of the synthesized Fe-MIL-101 metal-organic framework, the peak in the band gap of 1000–1200 cm^{-1} was related to the stretching C-O bond.

The absorption band 1596 cm^{-1} was a sign of asymmetric stretching absorption; the stretching band 1390 cm^{-1} belonged to the symmetric stretching absorption of the carboxylic acid group, and the two absorption bands 1506 and 741 cm^{-1} were related to the vibrations of the phenyl ring. The absorption peak at 543 cm^{-1} was related to Fe–O vibration.

The X-ray diffraction pattern of the metal-organic framework Fe-MIL-101 showed main peaks at $2\theta = 1.9°$, 4.4°, 9.7°, and 18.8°, indicating the successful synthesis of the Fe-MIL-101 structure. The peak at $2\theta = 16.4°$ also corresponds to a characteristic feature of the Fe-MIL-101 crystal structure.

Scanning electron microscopy analysis of the Cu-BDC metal-organic framework showed a cubic single crystal structure and entangled crystals, which is consistent with X-ray diffraction test results.

The isotherm of the Cu-BDC metal-organic framework was classified as Type I, indicating it is microporous. Based on BET analysis, the specific surface area and total pore volume of the Cu-BDC metal-organic framework were determined to be 686 $m^2\ g^{-1}$ and 0.335 $cm^3\ g^{-1}$, respectively, with an average pore diameter of 1.95 nm.

The absorption band of 1601 cm^{-1} was a sign of asymmetric stretching absorption, and the stretching band of 1390 cm^{-1} belonged to the symmetric stretching absorption of carboxylic acid group. The absorption peak at 1700 cm^{-1} was related to the stretching C-O bond in free carboxylic acids (H_2BDC). Two absorption bands, 1514 cm^{-1} and 741 cm^{-1}, were related to aromatic stretching CH groups related to phenyl ring and absorption peaks 468 cm^{-1} and 554 cm^{-1} related to Cu–O vibration.

Scanning electron microscope analysis for Fe-MIL-101@OAC sample showed that this nanocomposite has a crystalline structure, and the corresponding Fe-MIL-101 structures are homogeneously distributed inside the functionalized activated carbon.

Adsorption/desorption graph was done on Fe-MIL-101@OAC nanocomposite. The obtained isotherm was of the first type and was placed in the category of microporous tending to mesoporous. Based on BET analysis, the specific surface area and the total pore volume of Fe-MIL-101@OAC sample were 578 $m^2\ g^{-1}$ and 0.347 $cm^3\ g^{-1}$, respectively, and the average diameter of the pores was 2.38 nm.

The XRD results of Fe-MIL-101@OAC sample were similar to the peaks observed for the Fe-MIL-101 metal-organic framework, but the peaks were around (20–25°) $2\theta = 24°$ corresponding to the 002 plane of the amorphous carbon graphitic structure, slightly wide.

The SEM image of the said nanocomposite showed that according to the scanning electron microscope image, this nanocomposite has a crystalline structure and Cu-BDC structures are homogeneously distributed inside the functionalized activated carbon.

Adsorption/desorption graph was performed at 77 K temperature on Cu-BDC@OAC nanocomposite. The obtained isotherm was of the first type and was placed in the category of microporous tending to mesoporous. Based on BET analysis, the specific surface area and the total volume of pores of Cu-BDC@OAC sample were obtained as 712 $m^2\ g^{-1}$ and 0.336 $cm^3\ g^{-1}$, respectively, and the average diameter of pores was 1.82 nm.

All the peaks observed in the Cu-BDC metal-organic framework were also observed without any change in the Cu-BDC@OAC sample. The 1700 cm^{-1} peak for the functionalized activated carbon shifted to a lower wavenumber or disappeared, which could be due to the interaction between Cu ions in the metal-organic Cu-BDC framework and the carboxylic functional groups ($-COO^-$) formed on the surface of the functionalized activated carbon.

In the next step, the direct-reading instrument was evaluated. For this purpose, after making a certain concentration in the setup designed to determine the efficiency of the direct-reading instrument, the outlet air for different tests in four concentrations of 20, 100, 200, and 400 ppm; three temperature values of 21°, 30°, and 40°C; and three values of relative humidity, 30, 50, and 80%; were set. Each experiment was repeated four times. The overall average response factor of the device was found to be 1.09±0.02, which means that the actual concentration of 100 ppm is proportional to the concentration of 109 ppm read by PID.

In the next step, the adsorption capacity and breakthrough time of the adsorbents were investigated, which are presented in the following section:

1. The maximum breakthrough time of activated carbon adsorbent: 67.18 min
2. The maximum breakthrough time of non-functionalized activated carbon adsorbent: 32/29 min
3. The maximum adsorption rate of Fe-MIL-101@OAC adsorbent: 155/17 mg g^{-1}
4. The maximum breakthrough time of the Fe-MIL-101@OAC adsorbent: 101/52 min
5. The maximum adsorption rate of Cu-BDC@OAC adsorbent: 220/79 mg g^{-1}
6. The maximum breakthrough time of Cu-BDC@OAC adsorbent: 152/83 min

After the reaction, the surface of activated carbon functionalized with HNO_3, the number of acidic functional groups, especially phenolic and carboxylic groups, increased (Boehm titration), which indicates that more metal-ligand complexes are deposited on the surface of activated carbon. Also, some researchers have reported the adsorption mechanism of benzene ring of VOCs in terms of π-π electron acceptor-donor complex, π-π interaction, and electrostatic interaction of functionalized activated carbon. The adsorption mechanisms of Cu-BDC@OAC composite were adsorption of activated carbon and Cu-BDC pore fillers, cation π bonding, π-π interaction, and hydrophobic interaction from the phenyl ring in the wall of the MOF structure.

Also, the adsorption process of toluene by adsorbents was also modeled, which is mentioned in the following section.

10.5.7.2 Analysis and Review of the Central Composite Design

The response surface method can be used as the best technique for experimental modeling. Data fitting is done by various models such as linear, quadratic, and cubic models to obtain regression equations.

After obtaining the data related to the adsorption capacity of toluene on the synthesized adsorbents, the selection of the appropriate model was performed to determine the optimal adsorption conditions by comparing the results of the following three tests.

- The test to examine the linear relationship (first degree) of independent variables with the dependent variable
- The test to examine the linear relationship (first degree) of independent variables with the dependent variable and to examine the interference relationship or the simultaneous effect of independent variables two by two on the dependent variable
- Investigating the linear relationship (quadratic) of independent variables with the dependent variable

The values obtained from the four parameters—R-squared, adjusted R-squared, p-value, and lack of fit—serve as criteria for selecting the appropriate test to continue the modeling process. In this case, the R-squared and adjusted R-squared values are equal to each other and to 1, indicating optimal conditions for modeling. Also, the values of p-value and lack of fit of the selected model should be smaller and larger than 0.05, respectively. The F-value is a measure of the deviation of the data from the mean value. Sufficient accuracy, when it is higher than 4, indicates the suitability of the model.

After selecting the appropriate test, regression analysis was performed on the data related to the toluene adsorption process. Using the coefficients of the independent variables and the constant terms from the regression equations, the optimal values for the parameters affecting toluene adsorption on the synthesized adsorbents were determined using the Solver add-in in Excel.

Therefore, according to the values obtained from the mentioned parameters, the test model for examining the linear relationship (quadratic) of independent variables with the dependent variable was selected to continue the modeling process for adsorbents.

10.5.7.3 Modeling Related to Functionalized Activated Carbon

The value of F for this model is 49.75, which means that the variance of each variable is significant compared to the error variance, and all the main parameters play an important role as a response. The F-value determined that the effect of different parameters on the process follows the following sequence:

Humidity > temperature > concentration > adsorbent mass > flow rate

Humidity with F=2289.62 is the most effective factor in the process of toluene adsorption by functionalized activated carbon. In addition, the adjusted coefficient of determination is equal to $R^2_{Adj} = 0.9531$ in the case of high toluene adsorption, and this fully indicates the accuracy of the statistical model.

10.5.7.4 Modeling of Functionalized Activated Carbon – Fe Metal-Organic Framework (MOFs) Nanocomposites

The value of F for this model is 57.29, which means that the variance of each variable is significant compared to the error variance, and all the main parameters play an important role as a response. The F-value determined that the effect of different parameters on the process follows the following sequence:

Humidity > temperature > concentration > adsorbent mass > flow rate

Humidity with F = 2403.43 is the most influential factor in the toluene adsorption process. In addition, the adjusted coefficient of determination is equal to R^2_{Adj} = 0.9591 in the case of high toluene adsorption, and this fully indicates the accuracy of the statistical model.

10.5.7.5 Modeling of Functionalized Activated Carbon - Copper Metal-Organic Framework Nanocomposites

The value of F for this model is 48.01, which means that the variance of each variable is significant compared to the error variance, and all the main parameters play an important role as a response. The F-value determined the effect of different parameters on the process in the following order:

Humidity > concentration > temperature > adsorbent mass > flow rate

Humidity with F=2422.38 is the most influential factor in the toluene adsorption process. In addition, the adjusted coefficient of determination is equal to R^2_{Adj} = 0.9514 in the case of high toluene adsorption, and this fully indicates the accuracy of the statistical model.

10.5.7.6 Determining the Optimal Conditions of the Adsorption Process

In order to achieve the maximum amount of toluene removal from the gas flow by the adsorbents, the optimal conditions for this process were determined. Therefore, by using the Design-Expert 11 software and the optimization section by Derringer's Desirability Function method, the optimal value for each of the operating conditions (adsorbent mass, flow rate, concentration, humidity, and temperature) was selected "in the range," while the answer (adsorption capacity) was chosen to achieve the maximum amount of adsorption. In the aforementioned method, the range of performance is examined in the range of 0 to 1, where the number 0 indicates unfavorable conditions and the number 1 indicates completely favorable and optimal conditions from an operational point of view.

The optimal conditions were obtained as follows: adsorbent mass was 60 mg, flow rate was 90 ml/min, concentration was 500 ppm, humidity was 30%, and temperature was 26°C. The highest amount of adsorption was related to Cu-BDC@OAC nanocomposite. The relative standard deviation for the optimal adsorbent was 2.3%, which indicates the acceptable performance of the toluene adsorption optimization method.

In the next step, the effect of each operational parameter on toluene adsorption by Cu-BDC@OAC nanocomposite (selective adsorbent) was investigated, which are presented in the following section.

10.5.7.7 The Effect of Adsorbent Mass in Toluene Adsorption

In this experiment, the input gas flow rate was 150 mL min^{-1}, the concentration was 350 ppm, the temperature was 40°C, and the humidity was 50%. Increasing the amount of adsorbent mass had an effect on the breakthrough curve. The dynamic adsorption capacity in masses of 0.1, 0.3, and 0.5 grams were 195.04, 183.14, and 149.57 mg g^{-1}, respectively.

10.5.7.8 The Effect of Toluene Concentration in the Inlet Flow to the Adsorbent Media

The effect of concentrations of 100, 200, and 400 ppm on the dynamic adsorption parameters of toluene were investigated. In this experiment, the amount of adsorbent was 0.1 g, the flow rate of the inlet gas was 150 mL min^{-1}, the temperature was 40°C, and the humidity was 50%. Increasing the amount of toluene concentration had an effect on the breakthrough curve. The dynamic adsorption capacity was obtained at concentrations of 100, 200, and 400 ppm; 178.65, 183.82, and 198.92 mg g^{-1}, respectively.

10.5.7.9 The Effect of Gas Flow Rate Passing through the Adsorbent Media

In this experiment, the amount of adsorbent was 0.1 g, the concentration was 350 ppm, the temperature was 40°C, and the humidity was 50%. Increasing the flow rate had an effect on the breakthrough curve. The dynamic adsorption capacity in values of 100, 250, and 500 ml/min was obtained as 203.38, 187.95, and 160.25 mg g^{-1}, respectively.

The retention time of the contaminant in the adsorption media at flow rates of 100, 250, and 500 ml/min was calculated as 0.19, 0.07, and 0.03 seconds, respectively.

10.5.7.10 The Effect of Temperature on the Adsorption Parameters of Toluene by the Adsorbent

The effect of three temperature values of 21, 30, and 40° Celsius on the adsorption rate was investigated in the conditions where the amount of adsorbent was 0.1 gram, the concentration was 350 ppm, the inlet gas flow rate was 350 mL min^{-1}, and the humidity was 50%.

Dynamic adsorption capacity at temperatures of 21, 30, and 40 degrees Celsius were calculated as 203.42, 199.54, and 195.04 mg g^{-1}, respectively.

10.5.7.11 The Effect of Relative Humidity on the Adsorption Parameters of Toluene by the Adsorbent

The effect of the presence of relative humidity with percentages of 30, 50, and 80% on the dynamic adsorption of toluene was investigated. In this experiment, the amount of adsorbent was 0.1 g, the concentration was 350 ppm, the inlet gas flow rate was 350 mL min^{-1}, and the temperature was 40°C.

The dynamic adsorption capacity in relative humidity with percentages of 30, 50, and 80% was obtained as 205.92, 195.04, and 183.65 mg g^{-1}, respectively.

In the last step, the reusability of the synthesized nanocomposites was investigated, which is presented in the following section.

10.5.7.12 Investigating the Reusability of Synthesized Nanocomposites

The reusability of the synthesized nanocomposites was investigated. Thermal desorption of Cu-BDC@OAC and Fe-MIL-101@OAC nanocomposites was carried out at a temperature of 150°C in a vacuum oven for 4 h, under a pressure of 30 millibars. Adsorption capacity and failure time of both synthesized nanocomposites were calculated after three temperature desorption cycles. The toluene adsorption capacity of Fe-MIL-101@OAC and Cu-BDC@OAC nanocomposites in experiment 3 was obtained as 122.75 and 172.93 mg/g of adsorbent, respectively. Both Fe-BDC@OAC and Cu-BDC@OAC adsorbents showed significant stable and desorption capacity. The adsorption capacity of nanocomposites decreased slightly, which in fact indicates a slight effect on the structure of the material such as morphology.

10.6 INDOOR AIR QUALITY IN THE WORKPLACE

Providing suitable and healthy quality of indoor air for work environments is one of the most important concerns of HVAC engineers. Exposure to indoor contaminants can be more serious than exposure to outdoor contaminants because the concentration of indoor contaminants is generally higher than outdoor contaminants. ASHRAE 62 standards have addressed many indoor air quality concerns. Table 10.8 presents the concentrations of some contaminants in indoor spaces.

Some of the unacceptable signs and symptoms related to low indoor air quality are:

- Headaches
- Burning eyes
- Fatigue
- Breathing problems
- Asthmatic attacks
- Decreased memory
- Dizziness
- Depression
- Burning of the skin
- Sinus infections
- Cold, flu, and viruses

ASHRAE 62 standard provides two options for maintaining acceptable air quality:

- Ventilation procedure
- Indoor space air quality method

TABLE 10.8
Sources, Possible Concentrations, and Indoor-to-Outdoor Concentration Ratios of Some Contaminants in the Indoor Environment

Contaminant	Sources of Pollution in Indoor Space	Probable Concentration in Indoor Space	Concentration Ratios in the Interior Space to the Open Space	The Desired Location
Carbon monoxide	Defective combustion equipment, engines, heating systems	100 ppm	1<<	Skating fields, administrative offices, houses, cars, shops
Respirable particles	Gas stoves, fireplaces, cigarettes, volatile condensation, aerosol sprays, chemical reactions, cooking	100–500 μg/m^3	1<<	Homes, Administrative offices, cars, public places, restaurants
Organic vapors	Combustion, solvents, resin products, pesticides, sprays	Not reported	1<	Homes, restaurants, public places, offices, hospitals
Nitrogen dioxide	Combustion, gas stoves, water heaters, dryers, cigarettes, engines	200–1000 μg/m^3	1<<	Houses, skating fields
Sulfur dioxide	Heating systems	20 μg/m^3	1>	
Total particulate matter without smoke	Combustion, heating systems	100 μg/m^3	1	Homes, administrative offices, transportation, restaurants
Sulphate	Matches, gas stoves	5 μg/m^3	1>	Houses, administrative offices
Formaldehyde	Insulation, product fasteners, particle board	0.05–1 ppm	1<	Houses, administrative offices
Radon and its family	Building materials, underground water, soil	0.1–200 nCi/m^3	1<<	Houses, buildings

(*Continued*)

TABLE 10.8 (Continued)

Contaminant	Sources of Pollution in Indoor Space	Probable Concentration in Indoor Space	Concentration Ratios in the Interior Space to the Open Space	The Desired Location
Asbestos	Fireproof insulation	>10^6 fiber/m^3	1	Homes, administrative offices, schools
Synthetic and mineral fibers	Fabrics, carpets, wall boards	Not reported	……..	Homes, administrative offices, schools
Carbon dioxide	Combustion, humans, pets	3000 ppm	1<<	Homes, administrative offices, schools
Living organisms	Humans, pets, rodents, insects, plants, fungi, humidifiers, air conditioners	Not reported	1<	Homes, administrative offices, schools, hospitals, public places
Ozone	Electric arc sources of ultraviolet radiation	20 ppb 200 ppb	1> 1<	Airplanes Administrative offices

10.6.1 Ventilation Procedure

In the ventilation procedure, a minimum acceptable amount of outdoor air to the conditioned space should be provided to maintain acceptable concentrations of indoor contaminants such as carbon dioxide (CO_2), odor, tobacco smoke produced by residents, and other polluting sources. In Table 10.11, some indoor air contaminants and their sources are listed. The acceptable concentration of outdoor air contaminants is presented in Table 10.8, and the minimum required amount of outdoor air is presented in Table 10.9. In fact, a minimum airflow of 15 cfm per person in outdoor air is required to maintain the CO_2 concentration below the acceptable concentration of 0.1% (1000 parts per million, ppm), assuming relative CO_2 production per person (Tables 10.10 and 10.11).

Concentration of air contaminants: The concentration of indoor air contaminants depends on

1. type of contaminant and
2. the amount of contaminants removal

The pollutant removal process can be done by diluting the air in the interior and using local suction ventilation. It is assumed that the air inside is completely mixed and the concentrations are stable. The steady state concentration of contaminants in the indoor space can be expressed as Equation 10.7:

TABLE 10.9
Ventilation Flow Rates Based on ASHRAE 62 Standard

Application	Maximum Estimated Space Occupancy 1000 ft² or 100 m²	Outdoor Air Requirements				Comments
		cfm/ person	L/s: Liters per second. person	ft²/cfm	m². L/s	
		Laundry houses, dry cleaning				
Commercial laundry	10	25	13			
Commercial dry cleaning	20	30	15			
Warehouse, sorting clothes	30	35	18			Dry cleaning processes may require more airflow.
Coin-operated public laundries	20	15	8			
Coin-operated public dry cleaning	20	15	8			
		Drinks and foods				
Dining halls	70	20	10			
Attendance dining, self-service restaurant	100	20	10			Additional smoke removal equipment may be required. Compensating air for the exhaust hood may be required to provide additional ventilation air. The sum of outdoor air and transfer air with acceptable quality from adjacent spaces should be 1.5 cfm/ft² (7.5 L/s. m²) to provide the minimum amount of suction.
Kitchens	20	15	8			

		Service stations, repairs, parking spaces				
Covered parking spaces	1.50			1.50	7.50	In the distribution of air between people, the worker's place and the contaminant concentration from running engines should be considered; suction systems should be used in stations where engines work. Contaminant sensors may be used in ventilation control.
Car repair shops	1.50			1.50	7.50	
	Hotels, motels, resorts, dormitories			cfm/room	Room L/s	Regardless of room size.
Bedrooms				30	15	
Reception rooms				30	15	
Bathrooms				35	18	Capacity provided for intermittent use.
Lobbies	30	15	8			
Meeting rooms	50	20	10			Additional smoke removal equipment may be required.
Conference rooms	120	15	8			
Dormitories	20	15	8			
Coffee shops	120	30	15			

(*Continued*)

TABLE 10.9 (Continued)

Application	Maximum Estimated Space Occupancy 1000 ft² or 100 m²	Outdoor Air Requirements				Comments
		cfm/ person	L/s: Liters per second. person	ft²/cfm	m². L/s	
		Administrative offices				
Office space	7	20	10			Some office equipment may require local suction.
Reception places	60	15	8			
Information entry points and telecommunication centers	60	20	10			Additional smoke removal equipment may be required.
Meeting rooms	50	20	10			
Public spaces				ft²/ cfm	m² . L/s	
Corridors and installations				0.05	0.25	Mechanical suction without recirculation is recommended.
Public sanitary facilities, toilets	50	25				
Locker and dressing rooms				0.5	2.5	Typically, local mechanical suction is provided without recirculation through transfer air.
Smoking lounge	70	60	30			
Elevators				1	5	It is normally supplied through transfer air.
		Retail stores and showrooms				
Underground and street	30			1.50	0.30	
Upper floors of the building	50			1	0.20	
Warehouse	15			0.75	0.15	Localized mechanical suction is typically provided via transfer air; suction without recirculation is recommended.
Dress changing rooms				0.20	1	
Malls and arcades	20			0.20	1	
Shipping and receiving goods	10			0.20	1	
Wholesale stores	5			0.05	0.25	
Smoking lounge	70	30	60			

		Specialized activities				
Hairdressers	25	15	8			
Beauty salons	25	25	13			
Massage salons	20	15	8			
Flower shops	8	15	8			Ventilation requirements may be determined based on plant growth conditions.
Clothiers, furniture stores						
Fabric shops, pharmacies, iron market	8	15	8			
Supermarkets	8	15	8			
Pet shops				1	5	
		Sports and entertainment				
Spectator areas	150	15	8			
Game rooms	70	25	13	0.50	2.50	
Ice arenas				0.50	2.50	When internal combustion engines are used for repairs and maintenance of playing surfaces, increasing the ventilation rate may be necessary. For higher values, humidity control may be required.
Swimming pools						
Sports halls (gymnastics)	30	20	10			
Aerobic halls	100	25	13			
Bowling alleys	70	25	13			
		Theaters				
Ticket offices	60	20	10			
Lobbies	150	20	10			
Auditoriums	150	15	8			
Stages, studios	70	15	8			Special ventilation is required to eliminate scene-specific effects (e.g., dry ice vapors, mists, etc.).

(*Continued*)

TABLE 10.8 (Continued)

Application	Maximum Estimated Space Occupancy 1000 ft² or 100 m²	Outdoor Air Requirements				Comments
		cfm/ person	L/s: Liters per second. person	ft²/cfm	m². L/s	
			Transportation			
Waiting rooms	100	15	8			Ventilation inside vehicles may require special considerations.
positions	100	15	8			
Vehicles	150	15	8			
			Work rooms			
Meat processing departments	10	15	8			Spaces maintained at low temperatures (−10 to 50°F, or −23 to 10°C) are not covered by these requirements, unless the occupancy is continuous. Ventilation from adjacent spaces is allowed. When occupancy is intermittent, infiltration typically increases ventilation requirements.
Photography workshops	10	15	8			
Houses Dark	10	15	8			
pharmacies	10	15	8			
Bank trust funds	10	15	8			
Departments of printing, reproduction				0.50	2.50	The installed equipment must be equipped with control and positive suction (if necessary) for undesirable contaminants (toxic or non-toxic).

		Institutional facilities				
Training classrooms	50	15	8			For processes or activities such as animal laboratory work, special contaminant control systems may be required.
Laboratories	30	20	10			
Workshops	30	20	10			
Music rooms	50	15	8			
Libraries	20	15	8			
Locker rooms				0.50	2.50	It is normally supplied through transfer air.
Corridors				0.10	0.50	Local mechanical suction without recirculation is recommended.
Lecture halls	150	15	8			
Smoking lounges	70	60	30			
		Hospitals, nursing, and convalescent homes				
Sick rooms	10	25	13			Specific requirements or codes and pressure relationships may determine minimum ventilation flow rates and filter efficiency.
Medical activity	20	15	8			
Operating rooms	20	30	15			
Resuscitation and ICU	20	15	8			
Dissection halls (anatomy)				0.50	2.50	In cases of presence of contaminants, higher flow rates may be required.
Physiotherapy	20	15	8			The air should not be recirculated to other spaces.
		Correctional facilities				
cells	20	20	10			
Dining halls	100	15	8			
Guard stations	40	15	8			

TABLE 10.10
Workplace Air Quality Standards in the United States

Contaminants	Time Average	Primary Standard Levels	Secondary Standard Levels
Suspended particles	yearly (geometric mean) 24 h	μg/m^3 75 μg/m^3 260	μg/m^3 60 μg/m^3 150
Sulfur oxides	yearly (geometric mean) 24 h 3 h	μg/m^3 80 (ppm 0.03) 365 μg/m^3 (ppm 0.14)	μg/m^3 130(ppm 0.5)
Carbon monoxide	8 h 1 h	mg/m^3 10(9 ppm) mg/m^3 40(35 ppm)	mg/m^3 10(9 ppm) mg/m^3 40(35 ppm)
Nitrogen dioxide	yearly (geometric mean)	μg/m^3 100(0.05 ppm)	μg/m^3 100(0.05 ppm)
Ozone	1 h	μg/m^3 240(0.12 ppm)	μg/m^3 240(0.12 ppm)
Hydrocarbons (non-methane)	h 3 (am 3 or am 9)	μg/m^3 160(0.24 ppm)	μg/m^3 160(0.24 ppm)
lead	3 months	1.5 μg/m^3	1.5 μg/m^3

TABLE 10.11
Indoor and Outdoor Space Air Contaminants and Sources

Sources	Contaminants
	Open space
Ambient air	SO_2, NO, NO_2, O_3, hydrocarbons, CO, particles, and bio-aerosols
Motor vehicles	CO, Pb, particles, and bio-aerosols
Soil	Radon, organic substances
	Internal
	Building structural materials
Stone, concrete	Radon
Particle board, plywood	Formaldehyde
Insulation	Formaldehyde, fiberglass
Fire retardant fabric	Asbestos
Adhesives	Organic material
Color	Mercury, organic materials
Building contents	
Heating and cooking combustion equipment	CO, NO, NO_2, formaldehyde, particles, organic matter
Furnishings	Organic matter
Natural gas, drinking water	Radon
Residents	CO, NO_2, particles, organic matter, odors
Tobacco smoke	Fluorine carbons, vinyl chloride, organic matter
Aerosol sprays	Organic matter, odors, NH_2
Cooking and cleaning products	Organic matter
Hobbies and crafts	
Wet organic matter, stagnant water	
Coil drain pans	Bio-aerosols
Humidifiers	

$$C_i = C_o + \frac{N}{Q_{oa}} \tag{10.7}$$

Where, C_i = stable concentration of indoor space, μg/ft^3 or ppm; C_o = stable outdoor concentration, μg/ft^3 or ppm; N = total contaminant produced in indoor space, μg/min; Q_{oa} = ventilation rate, cfm.

10.6.2 The Procedure for Ensuring Optimal Indoor Air Quality

In this procedure, the minimum amount of outdoor air given in Table 10.8 can be reduced by recirculating the room air and utilizing air cleaning devices such as air filters. By using this procedure to minimize the amount of outdoor air for ventilation, energy consumption during heating and cooling operations by HVAC systems is significantly reduced. The effect of minimizing the use of outdoor air is a function of the type, efficiency, and location of the filter in the air distribution system and the amount of recirculated air.

Excessive reduction of ventilation and penetration to solve thermal comfort problems and reduce energy consumption can affect indoor air quality and may even be against the standard, so such procedures should be used with caution and under the guidance of an experienced specialist in HVAC design and analysis. HVAC design engineers and others related to building ventilation and indoor air quality should carry a copy of ASHRAE 62 standard to achieve acceptable indoor air ventilation. This standard, like all ASHRAE standards, is reviewed regularly and contains requirements for the design and assessment of ventilation for work environments.

Figure 10.15 illustrates an HVAC system with filters installed in two possible locations: A and B. Filter location A is in the recirculation airflow path, while filter location B is in the mixed airflow path. Variable air volume (VAV) systems can reduce recirculated air when the heat load is met, as determined by the flow reduction factor, F_r. For each type of ventilation system, a mass balance equation for the contaminant can be set up to determine the concentration of the contaminant. The result of the mass balance for the contaminant in different system conditions based on Figure 10.2 is presented in Table 10.12. If the pollution of space is known, the equations of Table 10.12 can be used to determine the required flow rate in the open space. If the flow rate in the open space is known, the equations of Table 10.12 can be used to determine the contaminant concentration:

10.7 AIR FILTRATION TECHNOLOGIES

The main groups of air filtration technologies are: filters (mechanical), ionizers, ozone generators, and ultraviolet rays.

10.7.1 HEPA Filters

Particle removal filters are highly efficient. HEPA filters use a powerful air blower to force air movement through a very tight membrane to achieve high efficiency particle filtration.

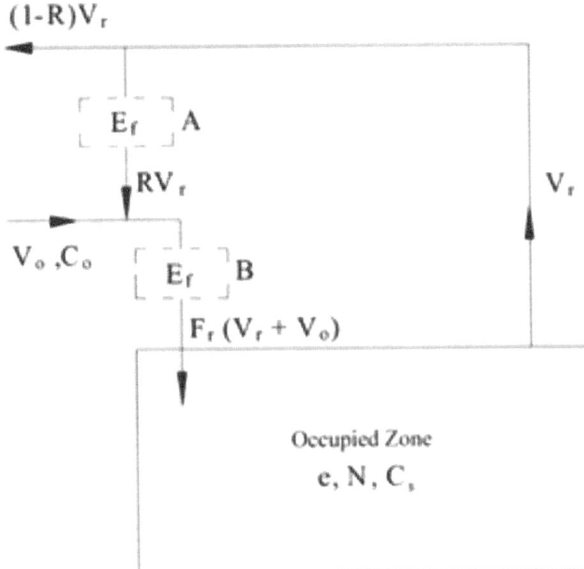

FIGURE 10.15 HVAC ventilation system schematic.

A, B = location of the filter; V = volume flow; C = contaminant concentration; E = efficiency or effectiveness; F_r = flow reduction coefficient; N = amount of contaminant production; R = recirculation flow coefficient; Subscripts: f = filter; o = open space; r = recursive; s = provision; v = ventilation

Advantages: They are very efficient in filtering the air passing through the filter and they filter particles up to 0.03 microns in diameter. Disadvantages: The filter can become a breeding ground for bacteria, molds, and fungi. They do not eliminate odors, gases, pesticides, viruses, and many bacteria. Due to the narrowness of the filter pores, the airflow is reduced. They are generally not used in central systems and are only offered as stand-alone units.

10.7.2 Carbon Media Filters

Carbon bed filters consist of filter cloth impregnated with carbon. These filters usually have a fabric or foam filter to maintain the carbon bed, which has a unique ability as a physical filter to trap particles—and the filtration process is performed on a chemical basis by reacting with some odors and some heavy gases.

Advantages: They absorb odors and certain gases, filter suspended particles, and can be installed in either separate or central rooms. Disadvantages: They must be changed frequently, can promote the growth of microorganisms, are prone to becoming blocked and ceasing to function, and reduce the amount of airflow.

10.7.3 Fiber and Foam Filters

In open-cell or fibrous foam filters, air passes through a matrix of foam cells or fibers of fiberglass, wire, plastic, or fabric; typically, these filters capture only medium-to-coarse particles.

TABLE 10.12
Contaminant Concentration in Space or Open Air with Recirculation and Filtration

	Parameters						
Class	The Filter Location	Flow	Temperature	Open Space Air	Required Open Space Air	Contaminant Concentration in Space	Amount of Recirculation Required
I	None	VAV	Constant	100%	$V_o = \frac{N}{E_v F_r (C_s - C_0)}$	$C_s = C_0 + \frac{N}{E_v F_r V_0}$	$RV_r = \frac{N + E_v V_0 (C_0 - C_s)}{E_v E_f C_s}$
II	A	Constant	Variable	Constant	$V_o = \frac{N - E_v RV_r E_f C_s}{E_v (C_s - C_0)}$	$C_s = \frac{N + E_v V_0 C_0}{E_v (V_0 + RV_r E_f)}$	$RV_r = \frac{N + E_v V_0 (C_0 - C_s)}{E_v F_r E_f C_s}$
III	A	VAV	Constant	Constant	$V_o = \frac{N - F_r E_v RV_r E_f C_s}{E_v (C_s - C_0)}$	$C_s = \frac{N + E_v V_0 C_0}{E_v (V_0 + F_r RV_r E_f)}$	$RV_r = \frac{N + E_v F_r V_0 (C_0 - C_s)}{E_v F_r E_f C_s}$
IV	A	VAV	Constant	Proportional	$V_o = \frac{N - F_r E_r RV_r E_f C_s}{E_v F_r (C_s - C_0)}$	$C_s = \frac{N + E_v V_0 C_0}{E_v (V_0 + RV_r E_f)}$	$RV_r = \frac{N + E_v V_0 (C_0 - C_s)}{E_v F_r E_f C_s}$
V	B	Constant	Variable	Constant	$V_o = \frac{N - E_v RV_r E_f C_s}{E_v (C_s - (1 - E_f) C_0)}$	$C_s = \frac{N + E_v V_0 (1 - E_f) C_0}{E_v (V_0 + RV_r E_f)}$	$RV_r = \frac{N + E_v V_0 [(1 - E_f) C_0 - C_s]}{E_v E_f C_s}$
VI	B	VAV	Constant	Constant	$V_o = \frac{N - E_v F_r RV_r E_f C_s}{E_v (C_s - (1 - E_f) C_0)}$	$C_s = \frac{N + E_v V_0 (1 - E_f) C_0}{E_v (V_0 + F_r RV_r E_f)}$	$RV_r = \frac{N + E_v V_0 [(1 - E_f) C_0 - C_s]}{E_v F_r E_f C_s}$
VII	B	VAV	Constant	Proportional	$V_o = \frac{N - E_v F_r RV_r E_f C_s}{E_v F_r (C_s - (1 - E_f) C_0)}$	$C_s = \frac{N + E_v F_r V_0 (1 - E_f) C_0}{E_v F_r (V_0 + RV_r E_f)}$	$RV_r = \frac{N + E_v F_r V_0 [(1 - E_f) C_0 - C_s]}{E_v F_r E_f C_s}$

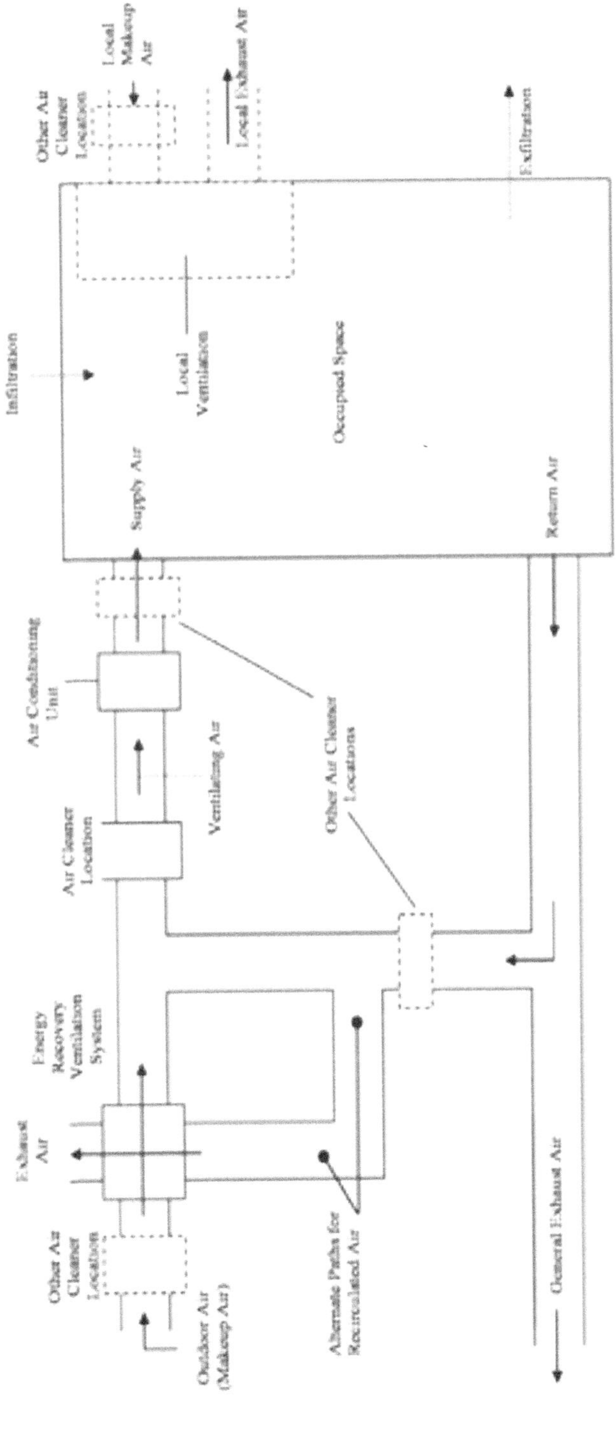

FIGURE 10.16 Ventilation system based on ANSI/ASHRAE standard.

Advantages: They are low cost, have little resistance to air and can be installed in central units. Disadvantages: Due to the accumulation of particles, they become a place for bacteria production and filter only large particles.

10.7.4 Ozone

Ozone is like oxygen but with an extra atom. This extra atom is known as a loose radical, which looks for organic substances to bind to so that the oxidation process can occur. Ozone is known as a "compatible" oxidizer.

Advantages: Ozone is an oxidizing gas that moves throughout the room and oxidizes all organic substances. Ozone can neutralize odors and gases. Ozone destroys microorganisms and does not reduce airflow. Ozone units can be installed in central units or in any room. Disadvantages: Ozone has no effect on solid particles, and the level of ozone exposure must be controlled.

10.7.5 Ultraviolet Light

Ultraviolet rays have been used in the medical profession as disinfectants for many years. Ultraviolet light can clean the air that passes directly through it.

Advantages: UV light can kill bacteria, fungi, mold, and some gases, but it does not reduce airflow. It can be installed in a separate or central room unit. Disadvantages: UV light has no effect on suspended particles, it requires direct contact, and humans must be protected from exposure to UV rays.

Filters differ from each other in terms of size, shape, initial cost, and operating cost. Filters can be classified based on efficiency (ability to remove particles), resistance to airflow (pressure drop), and useful life (holding capacity). In general, the filter allows contaminants to pass through the air stream. Higher efficiency or density of a filter means more restriction against airflow, and more restriction requires a blower in an air conditioning system. In general, filters with higher efficiency and lower air resistance should be considered in conditions of presence of particles.

Figure 10.16 shows a ventilation system based on ANSI/ASHRAE 2019-1.62.

11 Fans

A. H. Khoshakhlagh, H. M. Moda, S. Yazdanirad, and A. Gruszecka-Kosowska

11.1 OBJECTIVES OF THE LESSON

Upon completing this lesson, readers should be able to:

- Identify the types of fans
- Explain the design procedure of fans
- Explain the topics related to the curve of fans
- Explain the fan laws
- State the selection criteria of fans

11.2 FANS

The fan and motor system provides mechanical energy to move the contaminated air from the dust generating source to a dust collector.

11.2.1 Types of Fans

The two main types of industrial fans are:

- Centrifugal fans
- Axial-flow fans

11.2.1.1 Centrifugal Fans

Centrifugal fans consist of a wheel or motor mounted on a shaft that rotates in a scroll-shaped housing. The air enters the motor opening, performs a right angle rotation, and enters the helical chamber through the motor blades using centrifugal force. Centrifugal force applies static pressure to the air. Also, the spiral divergent shape converts part of the velocity pressure into static pressure.

The three main types of centrifugal fans are (Figures 11.1 and 11.2):

11.2.1.1.1 Radial-Blade Fans

They are suitable for handling heavy dust and operate efficiently at low to medium flow rates under high pressure. Their smooth, radial blades resist clogging by materials and exhibit considerable wear resistance. These fans operate at medium rotational speeds and produce moderate noise levels.

DOI: 10.1201/9781003501091-11

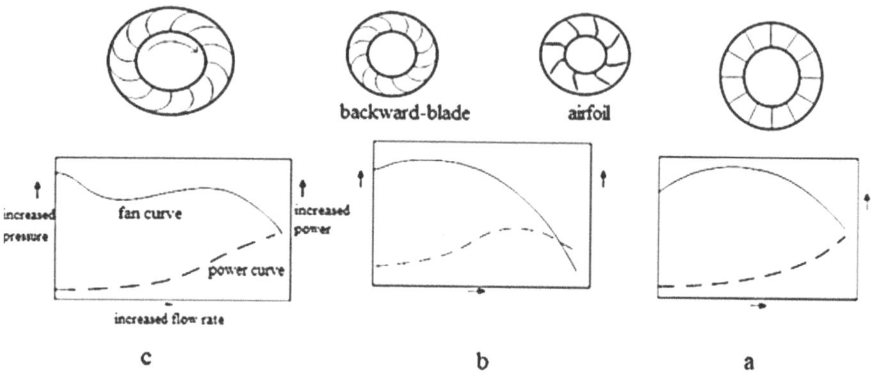

FIGURE 11.1 Centrifugal fan: (a) radial-blade fan; (b) backward-curved blade fan, airfoil fan; (c) forward-curved-blade fan.

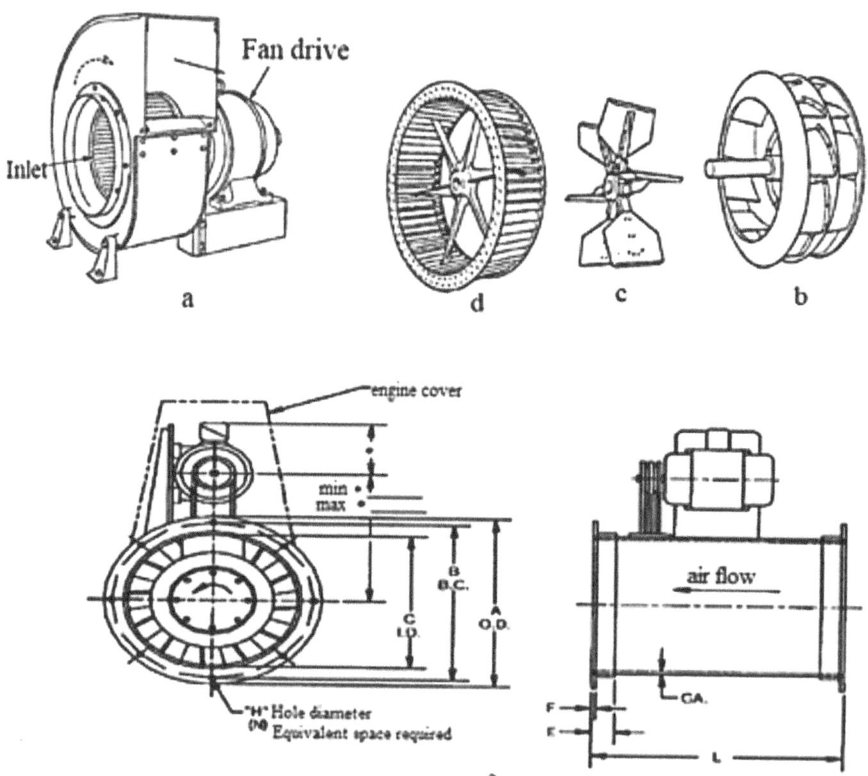

FIGURE 11.2 Centrifugal fans: (a) General view of centrifugal fan; (b) backward-curved blade fan; (c) radial-blade fan; (d) forward-curved-blade fan; (e) Centrifugal fan in the operating line.

11.2.1.1.2 Backward-Curved Blade Fans

Backward-curved blade fans operate at higher speeds and are therefore more efficient. Since material may accumulate on the blades, these aerators should be used downstream of the dust collector. At high flow rates, their performance decreases. Backward-curved blade fans are noisier than radial blade fans; however, due to their higher efficiency, they are typically used in high-volume dust collection systems. They are also commonly employed in heating, ventilation, and air handling units.

11.2.1.1.3 Airfoil Fans

Airfoil fans have the highest efficiency among centrifugal fans and are less noisy than other centrifugal fans. Due to their high mechanical strength, they have a very suitable capability in producing high pressures. In cases where the airflow is not polluted and due to the characteristic of these fans in not imposing an overload on the electric motor, they are often used in cases where a high flow rate is required. These fans are suitable for systems with variable resistance due to the absence of a stagnation zone on the curve.

11.2.1.1.4 Forward-Curved-Blade Fans

Forward-curved-blade fans have curved blades that move in the direction of rotation. They have less space requirements, lower rotational speeds, and therefore less noise. They are usually used against low to moderate static pressure and low to moderate flow rate at low pressure. They are used in heating, fan, and air handling units.

11.2.1.2 Axial Flow Fan

Axial flow fans are used in systems that have low resistance levels. These fans move the air parallel to the axis of rotation of the fan. The rotation of the propellers moves the air in a parallel path and creates a spiral flow pattern.

The three main types of axial fans are shown below (Figure 11.3):

11.2.1.2.1 Propeller Fans

This fan moves a large amount of air at very low static pressures. They are usually used for general ventilation or dilution ventilation and work well at pressure drops

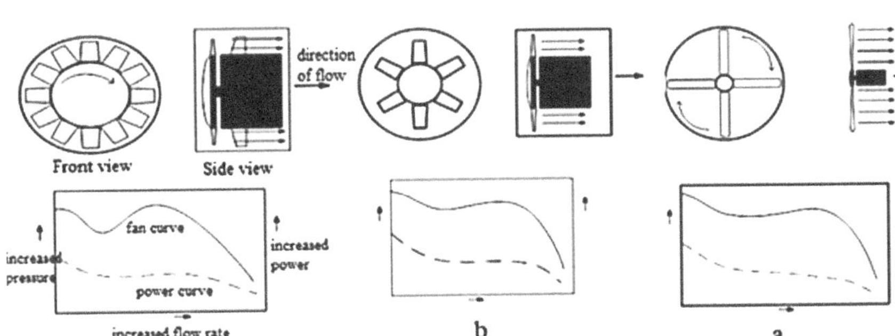

FIGURE 11.3 Axial flow fan: (a) propeller; (b) tube-axial; (c) vane axial.

up to 0.5 inches of water (124.4 pascals). Among the characteristics of this fan are that it is relatively noisy, its power consumption decreases as the flow rate increases, it generally operates at low efficiency, but achieves maximum efficiency at minimum pressure. Its main use is in exhausting the ambient air.

11.2.1.2.2 Tube-Axial Fans

Tube-axial fans are similar to propeller fans, except that they are placed on a tube or cylinder. They are suitable for applications with medium pressure and high flow rate. Therefore, they are more efficient than propeller fans and can perform well at a pressure drop of 3–4 inches of water (743–3.995 pascals). They are suitable for conveying air containing substances such as condensing fumes or pigments. The characteristics of these fans include higher efficiency and better performance at higher pressures compared to propeller fans, as well as moderate noise levels relative to other axial fans. Due to their high rotational speed, they can be operated at speeds below 1100 revolutions per minute when connected via a belt pulley system. The main application of these fans is in ventilation, heating, and cooling of places where short length ducts are used.

11.2.1.2.3 Vane Axial Fans

Vane axial fans are similar to tube axial fans, the difference being that they have exit vanes to guide the flow directly and convert its kinetic energy into static pressure—plus they have non-uniform flow characteristics. If airfoil propellers are used, the efficiency of this type of fan reaches 85%. They are usually connected directly to the electric motor and have medium to high pressures. They are easily adapted to multi-stage applications and can perform well at static pressures as low as 14–16 inches of water (3.98–3.483 pascals) and are generally only applicable to clean air. Their main use is air suction to discharge from boilers.

11.3 AIR EJECTORS

Air ejectors (also called injectors or jet pumps) are a means of moving air without passing through a fan. Typical designs are shown in Figure 11.4a (a simple ejector) and Figure 11.4b (a Venturi type ejector). Airflow (secondary air) in the large duct is induced by the action of the primary fluid jet exiting the nozzle at high speed. The outlet fluid can be water, steam, compressed air, or exhaust air from a fan; Thus, since exhausting the first three involves excessive energy consumption, the usual exhaust fluid is air.

Ejectors are inherently inefficient devices because most of the jet's energy is spent overcoming friction. The maximum efficiency that can be achieved under optimal conditions is only 40%, while the typical efficiency in field conditions varies from 15–25%. Such low efficiency exhausters are limited to applications where air cannot contact the fan; Very hot, sticky, corrosive, or explosive gas flows are examples of it.

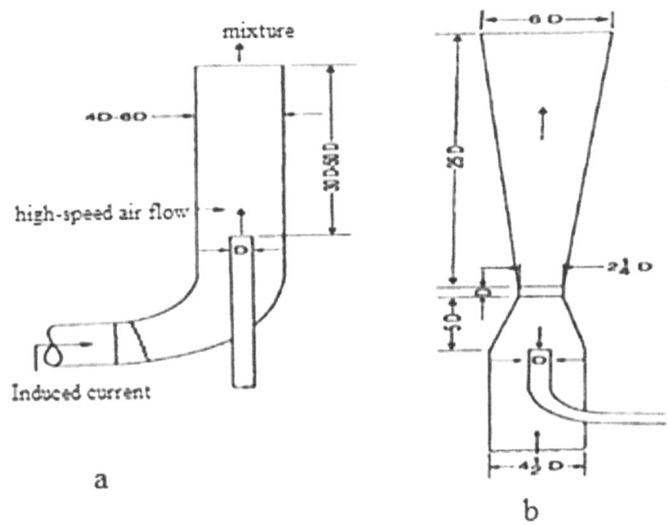

FIGURE 11.4 Ejectors: (a) simple design; (b) Venturi design.

11.4 FAN CURVES

The performance characteristics of a fan can be described by referring to several important variables. Variables include the amount of air transferred by the fan (Q, cfm or m^3/s), the static pressure created by the fan) P_s, in. H_2O or Pa(, fan efficiency (η), fan rotation speed (rpm, N), and power) H, W, or hp) required by the fan. Empirical formulas have been developed in relation to these variables in relation to each other. These formulas can be presented in the form of fan tables or as an illustration known as a fan curve.

Manufacturers specify fan performance using fan tables, but it is easier to visualize the performance of fan types using fan curves; in this section, we will start with fan curves and then relate the curves to fan tables. Since the end result of using the fan is to deliver a certain amount of airflow to the desired location; therefore, the performance of the fan is expressed according to the airflow. After that, based on the fan curves, other fan variables are related to the airflow. Thus, in such curves, airflow is plotted on the horizontal axis and other desired variables on the vertical axis.

11.4.1 Static Pressure Curve

The basic operation of a typical radial blade fan is shown in Figure 11.5a, where the amount of air transferred by the fan is plotted against the resistance to airflow provided by any system connected to it. As expected, the maximum airflow occurs when there is no resistance in the fan, which is called the free delivery no pressure point (FDNP). On the other hand, in case of infinite resistance (i.e., when the inlet or outlet of the fan is completely blocked), no airflow is formed. In terminology, this is called the static no delivery point (SND). In this condition, the blower develops

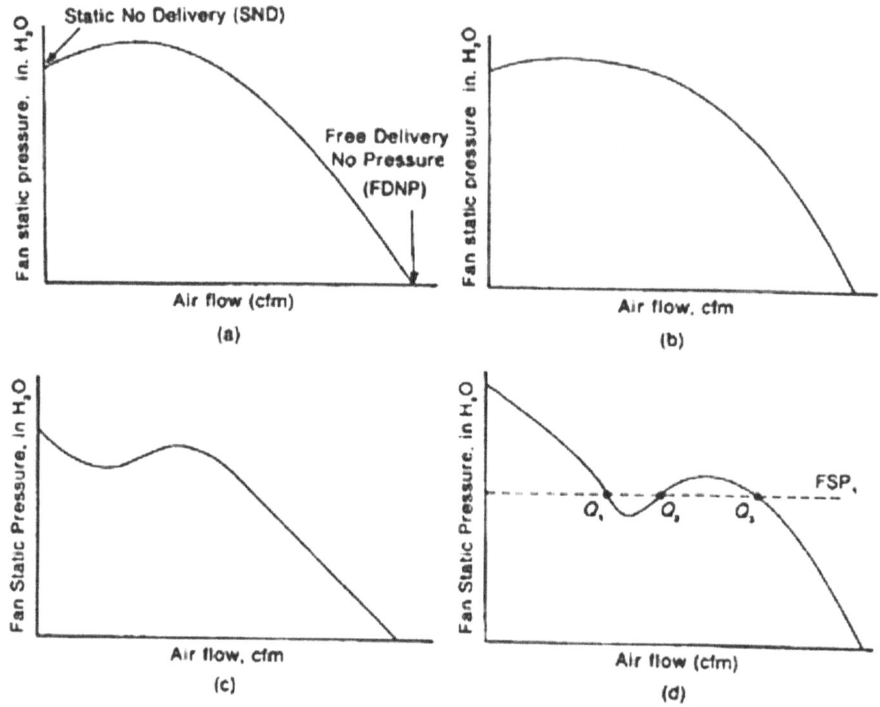

FIGURE 11.5 Static pressure curve of fan: (a) Radial blade; (b) backward-curved-blade; (c) forward-curved-blade; (d) vane axial.

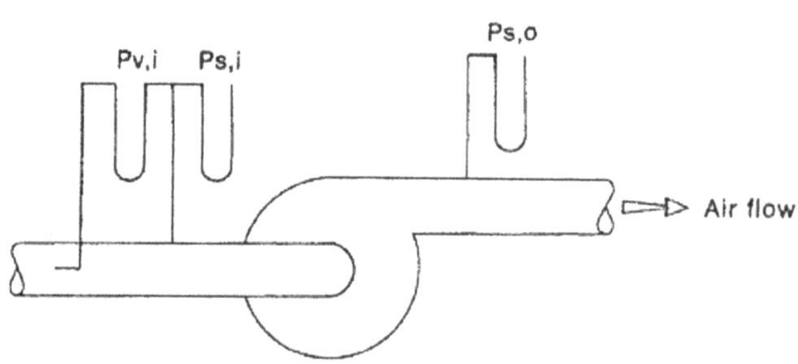

FIGURE 11.6 Fan pressures.

its maximum static pressure, as shown in the curve. However, the static pressure is partially reduced due to insufficient airflow, which increases because of the fan's inefficiency in this area.

The exact shape of the fan pressure curve between these two points depends on the specific design of the fan, but the curve shown is of the radial blade fan type. The static pressure plotted on this curve is called fan static pressure (P_s). This quantity can be expressed with the help of the simple suction system shown in Figure 11.6. To suck air from this system, the fan must create a certain negative static pressure at the inlet ($P_{s,i}$) and positive static pressure at the outlet of the fan ($P_{s,o}$). As a result, the total amount of static pressure that must be exerted by the fan is:

$$P_s = |P_{s.i}| + |P_{s.o}| \tag{11.1}$$

The static pressure of the fan is slightly different from the total static pressure required to move air from the system, which is expressed as follows:

$$P_s = p_s - p_{v,I} \tag{11.2}$$

Where $p_{v,i}$ is the pressure of the air at the inlet to the fan. This is a somewhat arbitrary definition of the static pressure of the fan, differing from the standard definition of the total pressure of the fan (P_t), which is derived from Equation 11.2 presented by Jorgensen.

The static pressure curve for the backward-curved-blade centrifugal fan, forward-curved-blade centrifugal fan, and vane axial fans are shown in Figure 11.5b–d, respectively. Note that the fan static pressure curve "slope" is shown for vane axial fans and to a lesser extent for forward-curved-blade centrifugal fans. This region is called "quiet," which is the part where the unstable operation of the fan is caused by the aerodynamic conditions when these fans operate in airflows much lower than the maximum design airflows. Fan operation in or near the stagnation zone is strictly prohibited, because the flow will be unstable. This instability is shown in Figure 11.6; at static pressure P_1, three different airflows (Q_1, Q_2, and Q_3) can be transferred. Such instability can lead to air oscillation and significant mechanical vibration. These fans must always work at a sufficiently high airflow rate to avoid the stagnation zone in the lower slope of fan static pressure curve.

11.4.2 Power Curve

To develop the real power curve (brake power curve) and mechanical efficiency presented in the following section, we must first define the amount of moving air power (H_a). Air power is a measure of the work done by the fan and is expressed based on the amount of energy added to the moving air per unit time. In general, the work (W) done in moving an airflow Q at a pressure drop Δp will be:

$$W = (\Delta p)Q \tag{11.3}$$

If the pressure drop is expressed in pounds per square foot (*psf*) and airflow is expressed in cfm, work is calculated in engineering units of ft-lb/min. In calculating the fan air power, the pressure change that should be used in the equation is the total pressure required to receive air from the system, which is usually expressed in inches of water. The following transformations can be used together with Equation 11.3 to obtain the desired relationship:

$$H_a(\text{hp}) = \frac{p_t(\text{in H}_2\text{O})Q(\text{cfm})}{33{,}000\left[\frac{\left(\text{ft}-\frac{\text{lb}}{\text{min}}\right)}{\text{hp}}\right]0\cdot192\left[\text{in}\frac{\text{H}_2\text{O}}{\left(\frac{\text{lb}}{\text{ft}^2}\right)}\right]} = \frac{p_tQ}{6343} \tag{11.4a}$$

In the metric system, the units are compatible, so the equation is easily done. In the metric system we will have:

$$H_a = p_t Q \tag{11.4b}$$

This expression expresses the amount of energy transferred to the moving air per unit of time. However, in the fan design, the power requirements of the fan driving motor should be clear. If the engine operates at 100% power in transferring energy to the fan blades, and if the fan is 100% efficient in transferring rotational energy to moving air energy, the required power of the motor (H_m) will be equal to the power of the air.

The transfer efficiency of both energies is less than 100% so that the power required by the engine is always significantly higher than the air power. The efficiency of the motor in the conversion of electrical energy to rotational mechanical energy is expressed by η_m, and the efficiency of the fan in the conversion of rotational energy to moving air is expressed by η_T (fan standard terminology for total mechanical efficiency); the energy conversion process can be visualized by Figure 11.7. The work done by the rotary fan is called the real power of the fan (H_b). Mathematically, H_a and H_b are related by the simple equation:

$$H_b = \frac{H_a}{\eta_T} = \frac{p_tQ}{6343\eta_T} \tag{11.5a}$$

In metric units, it will be:

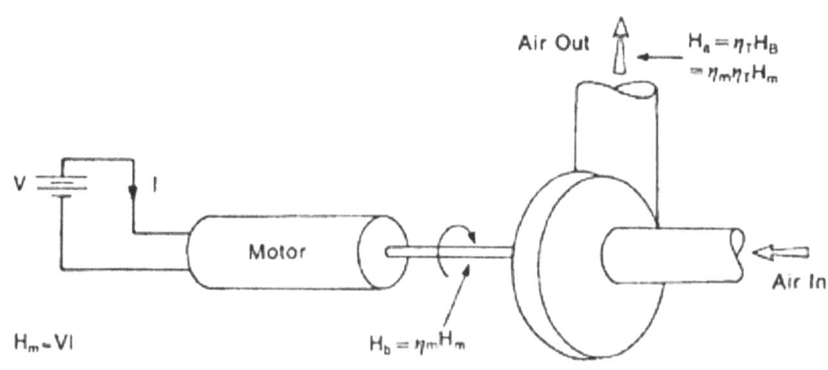

FIGURE 11.7 Fan-motor horsepower relationships.

$$H_b = \frac{p_t Q}{\eta_T} \tag{11.5b}$$

The electric power required by the motor will be:

$$H_m = \frac{H_b}{\eta_m} \tag{11.6}$$

The combination of Equations 11.5 and 11.6 shows the relationship between the electric power supplied to the fan and the work done by the moving air:

$$H_a = H_m \eta_m \eta_T \tag{11.7}$$

The fan motor efficiency η_m is usually around 95%, so the real power and motor power are almost equal. Fan power requirements are usually specified in terms of reduced power, and the user of fan tables should remember that the motor power used will be slightly higher than the specified value.

This introduction in fan power terminology is a necessary prerequisite to understanding fan power curves. A typical derating curve for a radial blade fan is shown in Figure 11.8a. The curve does not intersect the origin, as predicted by Equation 11.5, because the fan uses some energy to rotate the impeller even when it is off when no

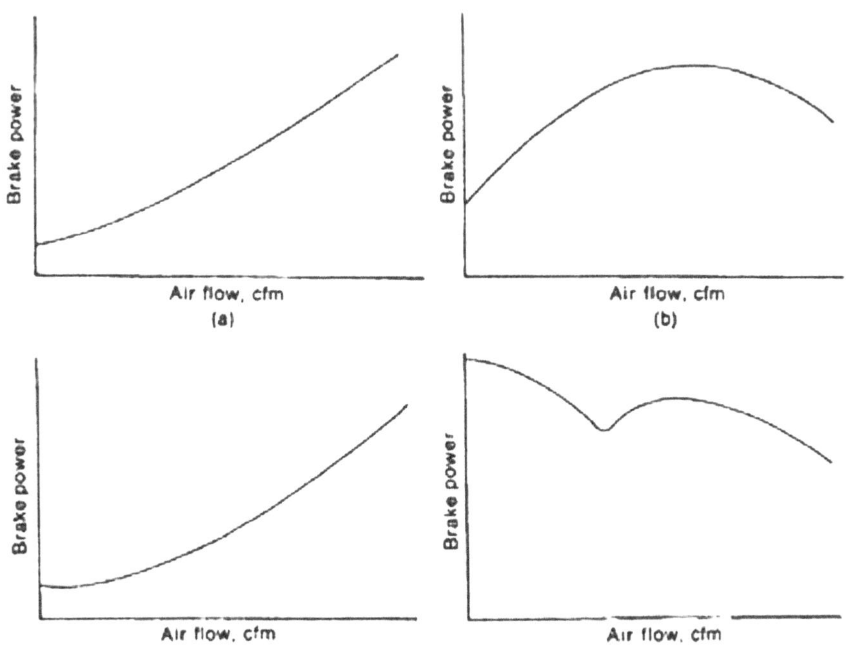

FIGURE 11.8 Common curves of fan power: (a) radial blade; (b) backwards-curved-blade; (c) forward-curved-blade; (d) vane axial.

air is flowing. The power curve rises steadily with increasing current and reaches its maximum at freewheeling (resulting in the possibility that the motor will stall if such a fan is operated under no-load conditions).

Similar curves are shown in Figure 11.8b for a backward-curved-blade fan, in Figure 11.8c for a forward-curved-blade fan, and in Figure 11.8d for a vane axial fan. Note that the downward curve is the real power curve at high airflows for a backward-curved-blade fan. This "load-limiting" feature is a desirable feature of this type of fan, since the power requirement does not exceed the defined levels when the fan is operating under low resistance conditions.

11.4.3 Mechanical Efficiency Curve

Mechanical efficiency refers to the fan's efficiency in transferring rotational mechanical energy to flowing air energy. Two different efficiencies, called total efficiency and static efficiency, are used in standard fan terminology. The total efficiency η_T was expressed in Equation 11.8. Here, it will be reset:

$$\eta_T = \frac{H_a}{H_b} \tag{11.8}$$

By substituting Equation 11.4 in Equation 11.5, we will have:

$$\eta_T = \frac{p_t Q}{6343\,H_b} \tag{11.9a}$$

Note that Equation 11.6 is its simplified form, which is simply used to calculate total efficiency. In metric units, it will be:

$$\eta_T = \frac{p_t Q}{H_b} \tag{11.9b}$$

Note that the total efficiency is defined in terms of the total pressure required of the fan. If static pressure is used instead of total pressure, a new efficiency called static efficiency (η_s) is expressed as follows:

$$\eta_s = \frac{p_s Q}{6343\,H_b} \tag{11.10a}$$

In metric units, it will be:

$$\eta_T = \frac{p_s Q}{H_b} \tag{11.10b}$$

Five parameters: Static pressure, total pressure, horsepower, total efficiency, and static efficiency can be plotted as functions of airflow for any particular fan. Fan curves for several types of blowers are shown in Figure 11.9.

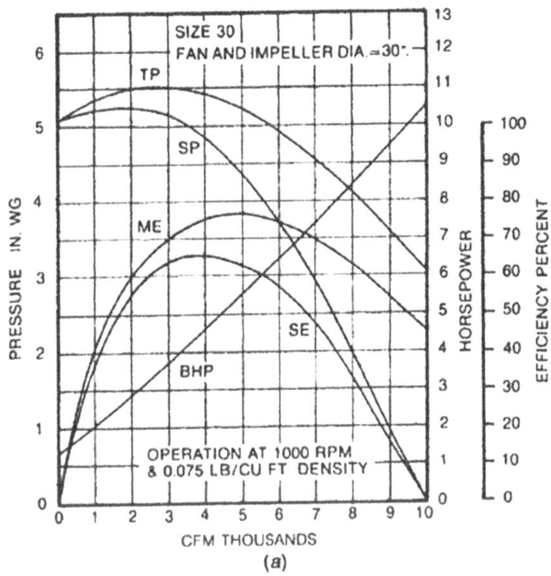

FIGURE 11.9 Common performance curves for different fan designs: (a) performance curve for radial blade centrifugal fan.

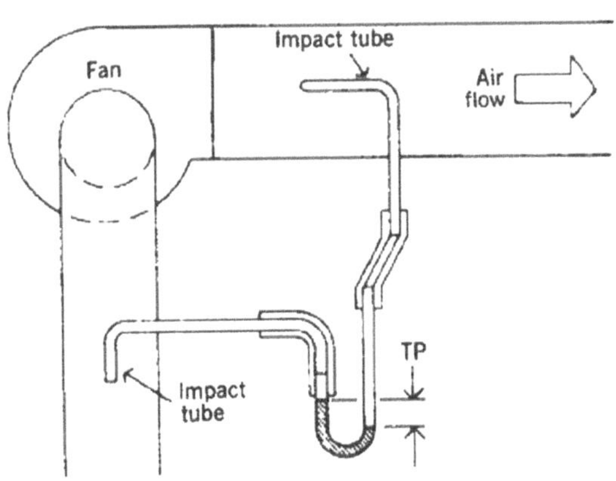

FIGURE 11.10 Total pressure fan.

11.5 CLASSIFICATION OF FANS BASED ON PRESSURE

11.5.1 Total Pressure Fan (TP)

The total pressure of the fan is the total pressure that increases along the fan, or the difference between the total pressure at the outlet of the fan and the total pressure at the inlet of the fan. Total fan pressure is a measure of total mechanical energy added to air or gas. How to measure is shown in Figure 11.10.

11.5.2 Static Pressure Fan (SP)

In the static pressure fan (Figure 11.11), the static pressure increases along the fan, that is, the difference between the static pressure at the outlet and the static pressure at the inlet. Fan static pressure is provided by manufacturers as a fan curve (Figure 11.12) that relates static pressure to cfm. In cases where the inlet and outlet velocities are the same, the static pressure is equal to the total pressure (SP = TP), provided that the air density at the inlet and outlet is also the same.

11.5.3 Velocity Pressure Fan (VP)

The fan velocity pressure (Figure 11.12) is the pressure related to the fan output velocity. In fact, the kinetic energy per unit volume is the flow.

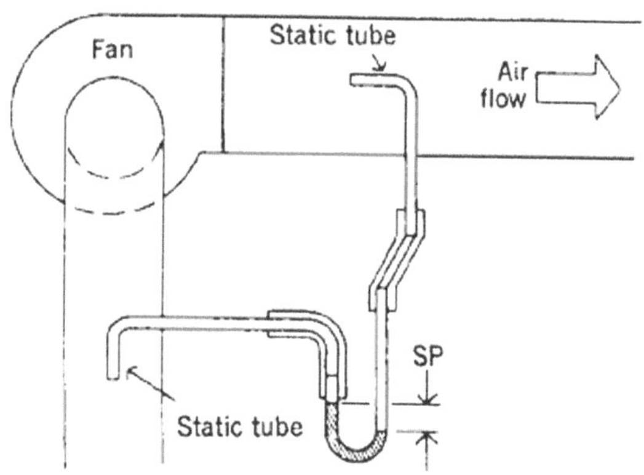

FIGURE 11.11 Static pressure fan.

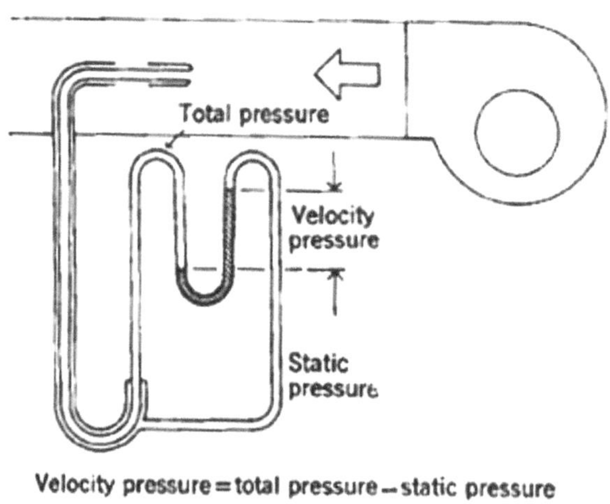

FIGURE 11.12 Velocity pressure fan.

11.5.4 Scope of Application

The scope of application is the operating scope of volumes and pressures in a satisfactorily operating fan as determined by the manufacturer (Figure 11.13).

11.5.5 Blocked Tight Static Pressure (BTSP)

The operating condition is when the fan outlet is completely closed and as a result, the air will not flow (Figure 11.14).

11.5.6 Free Transfer

Also called wide open CFM (WOCFM). Under these operating conditions, the static pressure along the fan is zero (Figure 11.15).

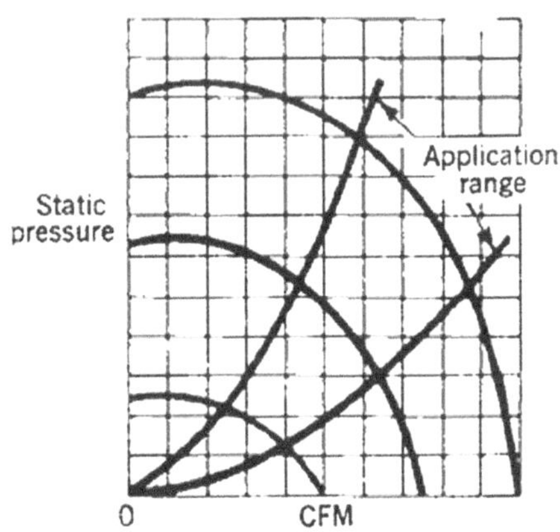

FIGURE 11.13 Fan curve.

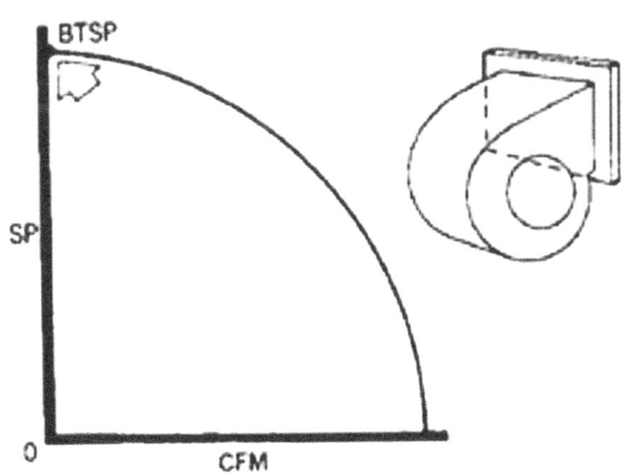

FIGURE 11.14 Fully blocked static pressure.

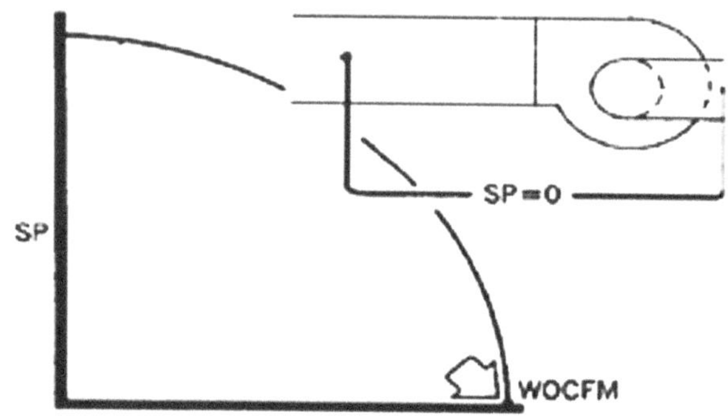

FIGURE 11.15 Free transfer.

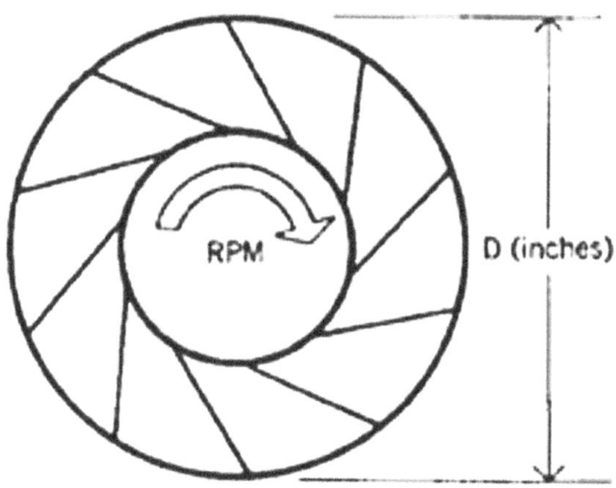

FIGURE 11.16 Tip speed.

11.5.6.1 Wide Open Brake Horsepower (WOBHP)

The power consumed is when the fan is working in free transmission. Typically, fan specifications are based on the percentage of wide-open CFM (WOCFM) specified for a fan, then related to the percentage of fully blocked tight static pressure (BTSP) and the percentage of WOBHP.

11.5.6.2 Tip Speed (TS)

Peripheral velocity, also known as the circumferential velocity of the fan wheel, is calculated using the fan's RPM and expressed in ft/min (Figure 11.16).

$$\text{TS} = \frac{\pi D \times RPM}{12} \tag{11.11}$$

Where D is expressed in inches.

11.6 FAN LAWS

Fan laws provide performance variables for a set of matched fans and can be used for both centrifugal and axial flow types. The variables involved include the size of the fan, which is related to the diameter of the wheel (D), rotational velocity in revolutions per minute (RPM), gas density (d), capacity (cfm), static pressure (SP), brake horsepower (BHP), sound power level, and efficiency. The laws for RPM, fan size, and density effects are discussed below. Fan laws for geometrically proportioned fans will be exact; however, because the accuracies in the builds are usually not proportional, the performance will be slightly better for fans with larger sizes.

11.6.1 Effect of Change in RPM

Initially, fan laws were considered to vary only in RPM (constant system) in a fan and a specific air handling system at a given density. Fan law 1 is shown in Figure 11.17.

cfm is directly related to RPM:

$$\frac{cfm_1}{cfm_2} = \frac{RPM_1}{RPM_2} \tag{11.12}$$

Static pressure is directly related to the square of RPM:

$$\frac{SP_1}{SP_2} = \left(\frac{RPM_1}{RPM_2}\right)^2 \tag{11.13}$$

The real power is directly related to the third power of RPM:

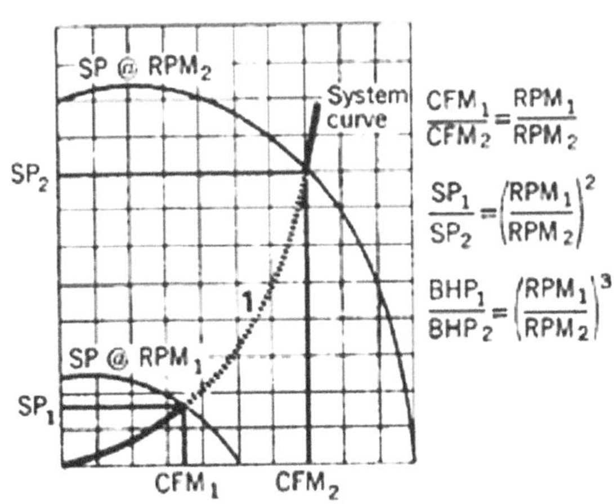

FIGURE 11.17 Fan Law No. 1: change RPM.

$$\frac{BHP_1}{BHP_2} = \left(\frac{RPM_1}{RPM_2}\right)^3 \quad (11.14)$$

In this condition, the efficiency does not change.

The operating point (percentage of fully open CFM) for a given system will not change. Changing the speed in percentage WOCFM will not change the static efficiency, etc.

Note that the SP change is more than the CFM change.

11.6.2 Effect of Change in Fan Size

Fan laws 2 and 3 change performance in proportion to changes in fan size; they are either based on a constant tip speed or a constant RPM. Law No. 2 of the fan is shown in Figure 11.18; with constant tip speed, air density, fan ratios, and constant operating point, the fan size will change.

CFM and BHP are directly related to the square of the wheel diameter.

$$\frac{CFM_1}{CFM_2} = \frac{BHP_1}{BHP_2} = \left(\frac{D_1}{D_2}\right)^2 \quad (11.15)$$

Static pressure remains constant: $SP_1 = SP_2$

Law No. 2 of fans is mostly used by fan designers and rarely applied in the field.

Law No. 3 of the fan is shown in Figure 11.19; with constant RPM, air density, fan ratios, and operating point are constant.

CFM has an inverse relationship with the third power of the fan diameter:

$$\frac{CFM_1}{CFM_2} = \left(\frac{D_2}{D_1}\right)^3 \quad (11.16)$$

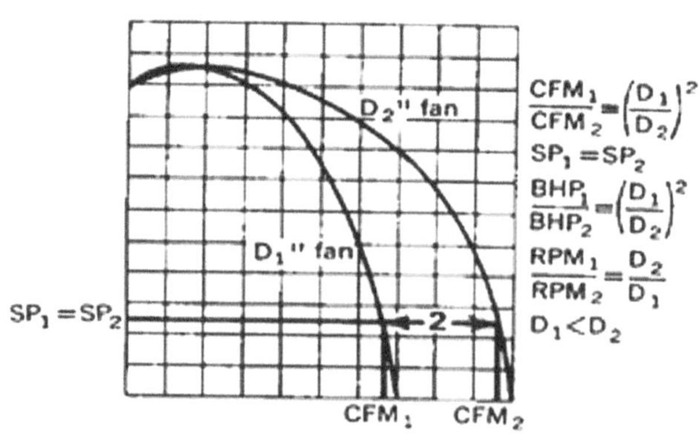

FIGURE 11.18 Law No. 2 of the fan: change in wheel diameter (constant volume).

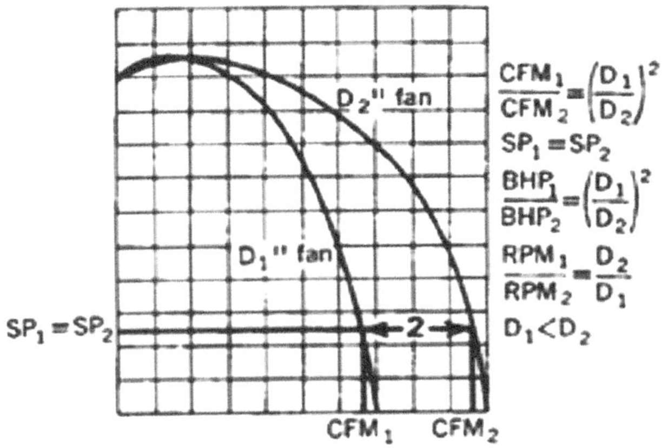

FIGURE 11.19 Law No. 3 of the fan: change in wheel diameter (constant RPM).

Static pressure has a direct relationship with the square of the wheel diameter:

$$\frac{SP_1}{SP_2} = \left(\frac{D_1}{D_2}\right)^2 \tag{11.17}$$

The tip speed has a direct relationship with the wheel diameter:

$$\frac{TS_1}{TS_2} = \frac{D_1}{D_2} \tag{11.18}$$

The real power has a direct relationship with the fifth power of the wheel diameter:

$$\frac{BHP_1}{BHP_2} = \left(\frac{D_1}{D_2}\right)^5 \tag{11.19}$$

Fan Law No. 3, with constant density and RPM, is used by fan manufacturers to provide performance data for fan families of proportional geometry.

11.6.3 Effects of Density

The next effect is the change in air density on the fan's performance. There are three fan laws for this condition.

Fan Law No. 4 is shown in Figure 11.20 with system constant volume (CFM), fan size, and RPM.

Fan volume in CFM will not change with density. A fan is a constant volume device and will produce the same CFM regardless of air density:

$$\text{CFM}_1 = \text{CFM}_2 \tag{11.20}$$

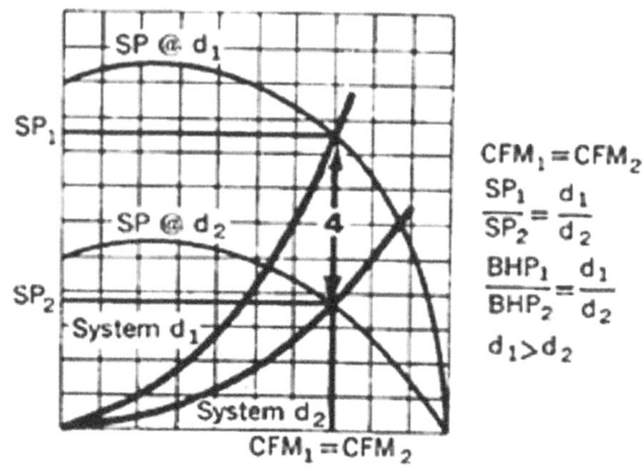

FIGURE 11.20 Law No. 4 of fan: effect of change in density (constant volume).

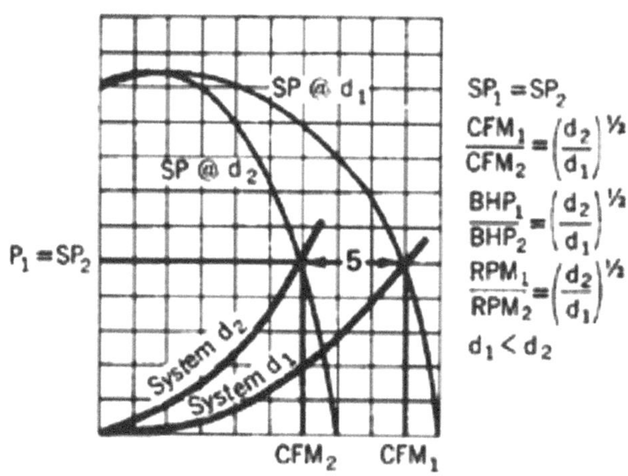

FIGURE 11.21 Law No. 5 of fan – effect of change in density (constant static pressure).

The static pressure and BHP created by the fan is directly related to the density. In other words, the air is heavier, so more pressure is produced and more real power is required.

$$\frac{SP_1}{SP_2} = \frac{BHP_1}{BHP_2} = \frac{D_1}{D_2} \tag{11.21}$$

Fan Law No. 5 is shown in Figure 11.21 with constant pressure, system, and fan size. Variable RPM, CFM, and real power have an inverse relationship with the square root of density:

$$\frac{CFM_1}{CFM_2} = \frac{RPM_1}{RPM_2} = \frac{BHP_1}{BHP_2} = \left(\frac{d_2}{d_1}\right)^{\frac{1}{2}} \tag{11.22}$$

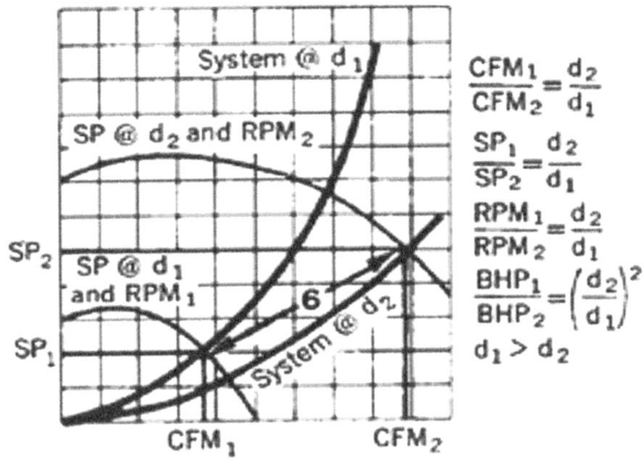

FIGURE 11.22 Law No. 6 of the fan – the effect of change in density (constant mass flow rate).

In this condition, the static pressure remains constant: $SP_1 = SP_2$

Law No. 6 of the fan is shown in Figure 11.22 – with flow constant of mass flow rate, constant system, and constant fan size.

Variable RPM, CFM, and static pressure are inversely related to density:

$$\frac{CFM_1}{CFM_2} = \frac{RPM_1}{RPM_2} = \frac{SP_1}{SP_2} = \frac{d_2}{d_1} \tag{11.23}$$

Real power has an inverse relationship with the square of the density:

$$\frac{BHP_1}{BHP_2} = \left(\frac{d_2}{d_1}\right)^{\frac{1}{2}} \tag{11.24}$$

Laws No. 4 and 6 of fans are the basis for selecting fans; for other standard air densities, selection is made using fan list tables that are provided based on standard air.

11.6.4 Relationship between Fan Curves and Fan Tables

Graphically, a change in speed in the fan causes a change in the static pressure curve, as shown in Figure 11.23. For each specific fan motor configuration, a set of static pressure curves are available, each curve corresponding to the rotational speed of a fan; each of these static pressure curves includes a set of four other curves (total pressure, horsepower, total efficiency, and static efficiency), so countless curves can be drawn for any given fan. Fan manufacturers present the data in a table rather than a large set of curves, which makes it easier to use the data.

A typical fan rating table for a backward-curved-blade centrifugal fan is given in Table 11.1. Each row of the table corresponds to an airflow, and each pair of two columns corresponds to a fan static pressure. The entries in the table show the

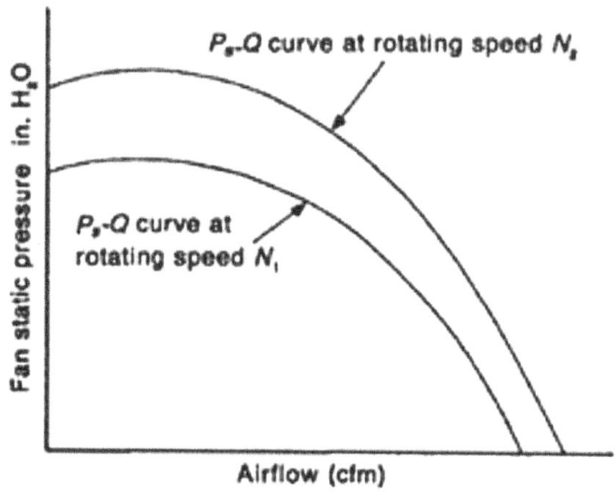

FIGURE 11.23 The effect of the change in the rotation speed of the fan N on the airflow curve – static pressure. Increasing N affects the upward movement of the curve.

necessary fan rotation speed to obtain any combination of Q's and P's and the real power required to move the fan. If the table entries are selected for a constant rotational speed, a fan curve similar to the one described in the above section can be drawn. To illustrate this procedure, the entries near 600 and 800 rpm are underlined in Table 11.1 and grouped together in the first three columns of Table 11.2. Equation 11.7 can be used to calculate the static efficiency for any set of conditions in Table 11.2; these values are shown in column 4.

Now, the values in Table 11.2 can be used to draw the fan curve (Figure 11.24) by plotting the static pressure of the fan (Figure 11.24a), the actual power (Figure 11.24b), and the static efficiency (Figure 11.24d). It can be used for two rotation speeds; the effect of changing rotational speeds is clearly evident. Note that these curves are plotted using a true fan table similar to the general fan curve for a backward-curved-bladed fan shown in Figure 11.10.

Typically, static efficiency values are not listed in the fan rating table; r instead, the point of maximum efficiency for each static pressure value is highlighted by underlining or shading the corresponding entry in the table. Therefore, in the example presented, according to Table 11.1, the point of maximum efficiency during operation is obtained at the static pressure of the fan of 4 inches of water with the rotation speed of 553 rpm and the airflow of 9000 cfm. The maximum efficiency is obtained at the static pressure of the fan of 5 inches of water with a rotation speed of 626 rpm and an airflow of 11,000 cfm. The maximum efficiency at a rotation speed of 600 rpm should be reduced in the range of 9000 cfm and 11000 cfm, as shown in Figure 11.24c.

A. The data in the first three columns from Table 11.1 have been used at the fixed rotational speed of the fan 600 and 800 rpm; the static efficiencies of the fan are calculated using these values and Equation 10.7.

TABLE 11.1
Section of the Standard Fan Rating Table

		Static Pressure 0f 1 inch		Static Pressure of 2 inches		Static Pressure of 3 inches		Static Pressure of 4 inches	
Cfm	Output Speed	RPM	hp	RPM	Hp	RPM	hp	RPM	hp
400	867	273	0.88						
5000	1084	280	1.09	383	2.24	466	3.48		
6000	1301	291	1.33	388	2.63	469	4.04	539	5.53
7000	1518	302	1.60	396	3.05	473	4.62	541	6.27
8000	1735	315	1.91	406	3.52	480	5.23	546	7.04
9000	1952	327	2.26	417	4.03	489	5.89	553	7.85
1000	2169	340	2.65	429	4.59	500	6.61	561	8.71
11,000	2386	353	3.10	441	5.21	511	7.38	571	9.64
12,000	2603	368	3.60	454	5.88	522	8.22	582	10.6
13,000	2819	383	4.18	466	6.61	534	9.12	593	11.7
14,000	3036	398	4.83	479	7.39	547	10.1	605	12.8
15,000	3253	415	5.55	492	8.25	560	11.1	617	14
16,000	3470	431	6.36	506	9.19	572	12.2	630	15.3
17,000	3687	448	7.25	521	10.2	585	13.4	642	16.7
18,000	3904	465	8.24	536	11.4	598	14.7	655	18.1
19,000	4121	483	9.32	551	12.6	612	16	668	19.6
20,000	4338	501	10.5	567	13.9	626	17.5	680	21.2
21,000	4555	519	11.8	583	15.4	640	19.1	693	22.9
22,000	5772	537	13.2	600	16.9	655	20.8	701	24.8
23,000	4989	556	14.7	616	18.6	671	22.6	721	26.7
24,000	5206	575	16.4	633	20.4	687	24.6	736	28.8

25,000	5422	593	18.2	650	22.4	703	26.7	751	31.1
26,000	5639			668	24.5	719	28.9	766	33.4
27,000	5856			686	26.7	735	31.3	782	36
28,000	6073			703	29	752	33.8	797	38.6
29,000	6290			721	31.5	769	36.5	813	41.4
30,000	6507			740	34.2	786	39.3	830	44.4
31,000	6724			758	37.1	803	42.3	846	47.6
32,000	6941			777	40.1	821	45.4	863	50.9

TABLE 11.2
Calculation of the Static Efficiency of the Fan for the Common Fan Table 11.1

P_s (in. H_2O)	Q (cfm)	H_B (hp)	η_s (%)
in 600 rpm			
1	25,000	18.2	2
2	22,000	16.9	41
3	18,000	14.7	58
4	14,000	12.8	69
5	7000	8	69
in 800 rpm			
3	31,000	42.3	35
4	28,000	38.6	46
5	25,000	35.6	55
6	22,000	33.2	63
7	19,000	30.7	68
8	15,000	26.2	72
9	9000	18.7	68

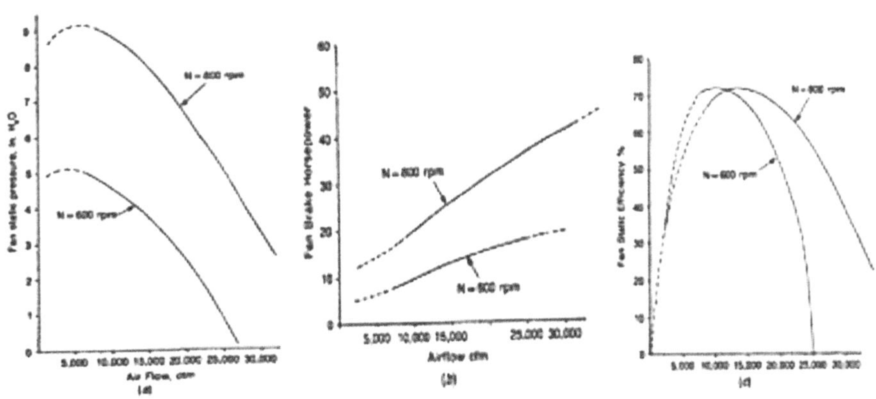

FIGURE 11.24 Fan performance curves adapted from Table 11.1: (a) fan static pressure versus airflow; (b) actual power versus airflow; (c) static efficiency versus actual power.

11.7 THE USE OF FANS IN VENTILATION SYSTEMS

The fan is always used as part of the ventilation system (in some simple progressive ventilation systems, it is one of the main components of the ventilation system). In this way, understanding the interaction between the fan and the rest of the system is considered an important principle in system design. The type, size, and appropriate fan rotation speed should be selected to produce the favorable airflow in the system.

11.7.1 General Exhaust Ventilation Systems

The most common fan for general exhaust ventilation is the propeller type axial fan. This fan is designed to move a large volume of air against low resistance, which is what is normally required for general ventilation systems. Choosing and installing general ventilation fans is relatively simple. However, for various reasons, many of these systems perform poorly; for example: if the vent size is chosen incorrectly, or if the vent is installed in the wrong location, or if adequate replacement air is not provided.

A properly designed general exhaust ventilation system will operate at static pressures very close to zero, so a fan blower can provide maximum airflow. However, the most common design flaw in such systems is the lack of a suitable replacement system for air. If the amount of air not supplied by the outlet airflow from the room is drawn in by the general exhaust ventilation system, the room will be "starved" of air and the fan suction will create negative static pressure conditions within the room.

11.7.2 Local Exhaust Ventilation Systems

The proper selection of fan for use in LEV system includes matching the fan specifications with the exhaust system specifications to achieve the desired airflow. In this regard, the exhaust system can be considered as a constant resistance to airflow. The amount of air flowing in this resistance network depends on the specifications of the fan used. It may be more clearly expressed using an electrical analogy (Figure 11.24). Applying voltage (V) to a network of resistors leads to a current passing through each branch of the network. The current passing through each branch (I2, I1, etc.) depends on the relative amount of each resistance, while the total current (I) depends on the overall resistance of the network.

In the same way, the branches of the ventilation system can be viewed as representing the network of resistance to airflow. The pressure difference created by the fan is similar to the voltage in an electric circuit; the magnitude of this pressure difference, along with the overall resistance of the ventilation system, determines the total airflow (Q). The airflow in each branch is determined by the relative resistance to the airflow provided by that branch.

This analogy may be useful to a point, but it is not complete. In an electrical system, it is usually possible to supply a constant voltage, but a fan creates a variable pressure value described by the fan curve. The first step in determining the actual operating point of the fan is to determine the resistance characteristics of the exhaust system.

11.7.3 Exhaust System Resistance Curve

The resistance to airflow in most components in the exhaust system is proportional to the average square of the air speed passing through that component; this applies to hoods, ducts, elbows, and other connections. The main exception to this general law is in the case of air cleaning devices that may be used in the air ventilation system. Many of these devices have pressure drops that vary significantly with speed and/or linearly over time.

Consider the airflow resistance from the exhaust system in the plating shop system, ignoring the air cleaning devices. The plating system requires an airflow with a flow rate of 11230 cfm with a system static pressure of 1.71 in. H_2O. The inlet speed pressure to the fan is 0.43 in. H_2O; therefore, the fan's static pressure will be:

$$P_s = p_s - p_{v,i}$$

$$= 1.71\,\text{in.}H_2O - 0.43\,\text{in.}H_2O$$

$$= 1.28\,\text{in.}H_2O$$

This system exhibits the resistance-airflow relationship shown in Figure 11.25. The system resistance curve demonstrates a quadratic relationship between airflow and the static pressure provided by the fan. Specifically, doubling the airflow rate from the design value of 11230 cfm to 22460 cfm results in a fourfold increase in the fan's static pressure, rising from 1.28 in. H_2O to 5.12 in. H_2O.

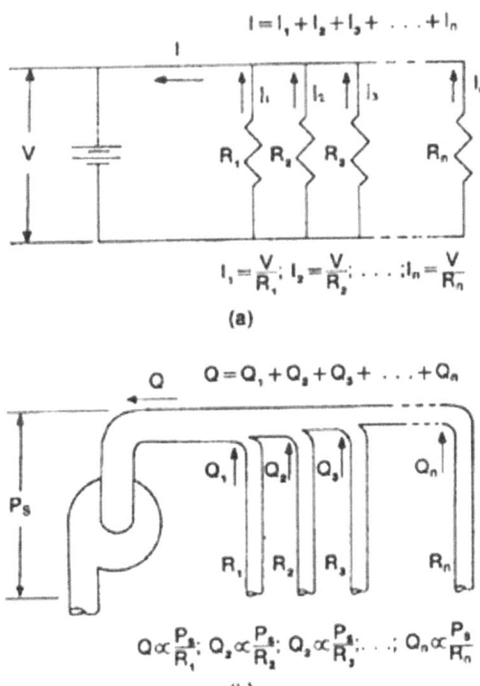

FIGURE 11.25 Similarity between an electrical network (a) and a local exhaust ventilation system, (b) The voltage is similar to the static pressure of the fan, electrical resistance versus airflow resistance, and electrical current versus airflow.

11.7.4 Matching the Fan with the System

The system resistance curve defines the airflow through the system at any value of the static pressure applied by the fan. What should be given special attention in the design process is the selection of ventilation that provides the desired static pressure (1.28 in. H_2O) to create the desired airflow (11230 cfm). The data in Table 11.1 indicate that the fan in this table may have adequate performance for this activity. At a flow rate of 11,000 cfm and a pressure drop between 1 and 2 inches of H2O, the fans are operating at approximately their maximum efficiency, as indicated by the underlined RPM values shown below. The problem involves selecting the fan's rotational speed to achieve a flow rate of 1 11,230 CFM with a pressure drop of 1.28 in. H2O.

The system resistance curve (Figure 11.26) and the fan static pressure curve at N = 600 rpm (Figure 11.24) can be drawn on the same graph; the results are shown in Figure 11.27. The fan curve illustrates the airflow required at various static pressures for the fan to overcome airflow resistance, while the system static curve represents the pressure needed to overcome the airflow resistance of that specific system. It is clear that the two curves intersect at a point where the airflow is 17500 cfm and the fan static pressure is 3.2 in. H_2O. This is the only common point between the functional characteristics of the fan and the exhaust system, so that if this fan is installed in this duct network and operates at this speed (600 rpm), the airflow will be suctioned with a flow rate of 17500 cfm and the fan static pressure will be 3.2 in. H_2O.

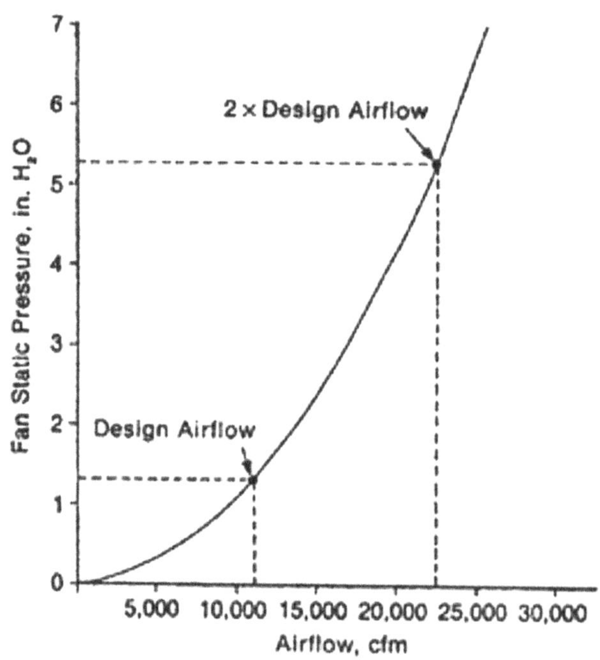

FIGURE 11.26 The static pressure required to induce different airflows for the local exhaust ventilation system of the electroplating system.

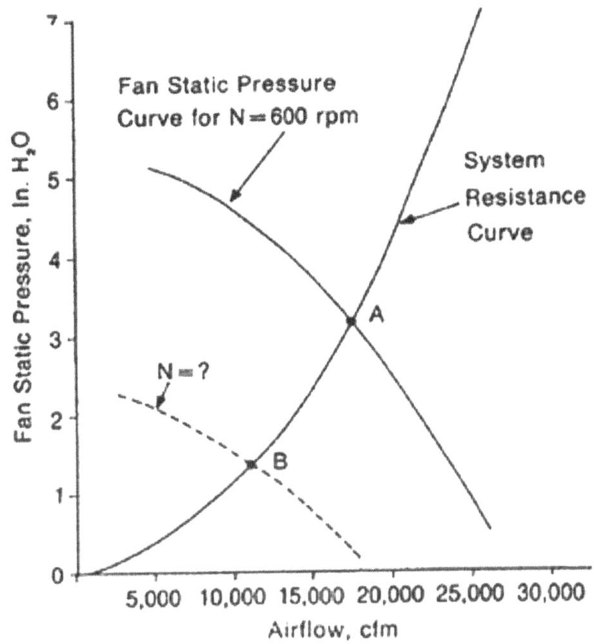

FIGURE 11.27 The fan static pressure curve at N = 600 rpm from Figure 11.24 and the system resistance curve of Figure 11.26 are drawn in one figure. Pay attention to the intersection point (a) and the desired performance point (b).

This amount of airflow is significantly higher than what is required in the electroplating system. Recall that a set of fan static pressure curves are available, which are different for each rotational speed; two of these curves (N = 800 rpm and N = 600 rpm) are considered for the fan in Figure 11.24a. One of these curves, and only one of them, must pass through the desired operating point (b) (dashed line in Figure 11.27). A key part of the fan selection procedure is to select the appropriate rotation speed of the fan to achieve the desired intersection point with the system resistance curve. This procedure is described in the next section.

11.8 FAN SELECTION

When choosing a fan, the following points should be considered:

- Required volume
- Fan static pressure
- Type of materials to be moved through the fan (for example, a radial blade fan should be used when fibrous materials or heavy dust loads are present, and a non-sparking design should be used when explosive or flammable materials are present)
- Type of drive setting, such as direct drive or belt drive
- Space requirements
- Noise levels

Operating temperature (for example, oil bearings are suitable at 121.1°C/250°F and ball bearings at 287.8°C/550°F.

- Sufficient size to provide required volume and pressure with minimum horsepower
- Need for special cover or structure when working in corrosive environments
- The fan's ability to adapt to small changes in total pressure while maintaining the required air volume
- Requirement of an outlet damper to control airflow during cold start (if necessary, fan may be locked for gradual start until steady state conditions are reached)

11.8.1 Fan Rating Tables

After collecting the information mentioned in the above section, the actual selection of fan size and speed is obtained from the rating table provided by the fan manufacturer. This table is also known as a multi-rating table and provides a complete range of capacities for a specific size of fan.

The points that should be considered:

- The multi-rating table provides the range of possible pressure and speed in the range of the fan
- A particular fan may be offered in different structural classes (from class I to IV) related to its capabilities and limitations
- For a given pressure, the highest mechanical efficiency is usually observed in the middle third of the column volume
- A fan operating at a given speed can have an unlimited number of gauges (pressure and volume) along its characteristic curve; however, when the fan is installed in the dust collection system, the relevant point of the rating can only be at the point where the system resistance curve intersects the fan specification curve
- In a given system, the fan at a constant speed or at a fixed blade setting can have only one rating; this rating can be changed just by changing the fan speed, blade settings, or system resistance
- For a given system, increasing the output volume will lead to an increase in static and overall pressure; for example, for a 20% increase in output volume in a system with a 5% pressure drop, the new pressure drop will be $5 \times (1.20)^2 = 7.2$ inches
- For quick estimates, the possible output volumes available for a given engine size are used; actual power—an imperial unit equivalent to one horsepower but used solely as a scale of power for a motor shaft—can also be appropriate

In fan installation, fan ratings are used for fan discharge conditions according to volume and static pressure, based on tests performed under ideal conditions. Typically, field installation causes airflow problems that reduce fan air delivery. The following points should be considered when installing the fan:

- Avoid installing elbows in the discharge section of the fan, which reduces the performance of the fan by increasing the resistance of the system
- Connections such as elbows and mitered elbows (created by chamfering the ends of two pipes to be joined; the chamfer angle is typically 45 degrees, resulting in a 90° corner) or the use of square-shaped ducts, which may cause non-uniform flow, should be avoided
- Check that the fan propeller rotates clockwise or counter-clockwise
- For belt drive fans:
    - Ensure the motor sheave and the fan groove are properly aligned
    - The appropriate tensile strength of the belt should be verified
    - The paths between the inlets, propeller blades, and inside the chamber should be inspected for the accumulation of debris, obstacles, or trapped foreign objects

11.8.2 Fan Selection Procedure

When a desired operating point of the fan is selected, it is necessary to choose the right fan, so that when used in the ventilation system, it provides the optimal combination of fan static pressure and airflow. The process to be applied is to use a set of fan curves described in the previous section—a simple procedure is to select a fan curve that intersects the system resistance curve at the desired operating point. What is usually available to the designer is not a fan curve but a set of fan rating tables similar to those presented in Table 11.1. In this section, the procedure for selecting a fan from such a set of rating tables is described.

The first step is to choose the type of fan to be used. In general, axial flow fans are used in dilution ventilation systems. In LEV systems the particle-free air is passed through a backward-curved-blade centrifugal fan, and also in LEV systems, a radial blade centrifugal fan is used to move air full of particles. After choosing the type of fan, it is necessary to consult with an expert about the lists of fans that have rating tables for that type of fan.

In this section, the electroplating ventilation system is again used as an example. This system moves particulate-free air through the system ducts, so a backward-curved-blade centrifugal fan should be used. Fan information is available in Table 11.1, which lists these tables for a selection of geometrically similar fans of a wide range of sizes. Information is provided in the above table about the dimensions of the specific fan described; fans are selected according to the diameter of the inlet, which in this example is 29 inches.

These special sets of fans come in a range of sizes from 11 inches inlet diameter to 37 inches inlet diameter. The first step in selecting a fan from these sets is to try to match the diameter of the fan inlet to the diameter of the inlet duct to the fan, as determined by the system design procedure. The diameter of the final path of the duct for electroplating design is 28 inches. In this set, there are no fans with an inlet diameter of 28 inches. The closest values are 26 and 29 inches. Since 29 inches is the closest value to the desired value, it will be selected first.

The change of the final diameter of the duct has no effect on the airflow of the system and its effect on the static pressure of the system is negligible, but the change in the speed affects the static pressure of the fan due to the change of the pressure of the inlet speed. The surface of the duct is 29 inches, 4.587 ft^2; so the input speed will be:

$$V_i = \frac{Q}{A} = \frac{11230\,\text{ft}^3/\text{min}}{4\cdot 587\,\text{ft}^2} = 2450\,\text{ft}/\text{min}$$

The pressure of the corresponding input speed will be:

$$p_{v,t} = \left(\frac{V_i}{4000}\right)^2 = 0.38\,\text{in.}\,H_2O$$

The following equation is used to calculate the static pressure of the fan with an inlet diameter of 29 inches:

$$P_s = p_s - p_{v,i}$$

$$= 1.71\,\text{in.}H_2O - 0.38\,\text{in.}\,H_2O$$

$$= 1.33\,\text{in.}\,H_2O$$

Therefore, the static pressure of the ventilator (1.33 in. H_2O) has slightly changed compared to the value calculated during the system design procedure (1.28 in. H_2O).

Once the possible fan size is selected based on the inlet diameter, the fan table should be checked to see if the desired combination of airflow and fan static pressure can be achieved with this specific fan. Recall that airflow at a flow rate of 11,230 cfm in a fan with a static pressure of 1.33 in. H_2O was required. In examining the table for fans, as presented in Table 11.1, these values are placed between the four entries inside the square in the table.

A quick check of Table 11.2 shows that the required airflow is between the values in the table, i.e., 11,000 and 12,000 cfm, and the required fan static pressure between the values in the table are 1 and 2 in. H_2O. Therefore, the required rotation speed of the fan should be around 353 rpm to achieve an airflow of 11,000 cfm at a pressure drop of 1 in. H_2O. At a rotation speed of 454 rpm, the airflow will be approximately 12,000 cfm at a pressure drop of 2 in. H_2O. The proper working speed of the fan is determined by linear interpolation between the table values.

Interpolation should be done between airflow values and static pressure values. The resulting equation for the working speed of the fan will be:

$$N = N_{11} + \frac{P_{s.r} - P_{s.1}}{P_{s.2} - P_{s.1}}\left(N_{12} - N_{11}\right) + \frac{Q_r - Q_1}{Q_2 - Q_1}\left(N_{21} - N_{11}\right) \tag{11.25}$$

Similarly, the equation for the real power would be:

$$H_b = H_{11} + \frac{P_{s.r} - P_{s.1}}{P_{s.2} - P_{s.1}}\left(H_{12} - H_{11}\right) + \frac{Q_r - Q_1}{Q_2 - Q_1}\left(H_{21} - H_{11}\right) \tag{11.26}$$

In this instance,

$$\mathrm{N} = 353 + \frac{1\cdot 33 - 1}{2 - 1}\left(441 - 353\right) + \frac{11230 - 11000}{12000 - 11000}\left(368 - 353\right)$$

$$= 353 + 0.33\left(88\right) + 0.23\left(15\right) = 353 + 29 + 4 = 386\,\text{rpm}$$

And

$$\mathrm{H_b} = 3.10 + 0.33\left(5.21 - 3.10\right) + 0.24\left(3.60 - 3.10\right)$$

$$= 3.10 + 0.7 + 0.12 = 3.92\,\text{hp}$$

Finally, the static efficiency at the operating point can be calculated as follows:

$$\eta_s = \frac{p_s Q}{6343\,H_b} = \frac{\left(1\cdot 71\right)\left(11240\right)}{\left(6343\right)\left(3\cdot 92\right)} = 0.77 \tag{11.27}$$

Thus, the fan is operating at or near the point of maximum efficiency for this fan (Table 11.2 and Figure 11.24a), so this size is a good choice for this application. If the efficiency is lower, the next fan in the set can be checked and an operating point (rotational speed and real power) and new efficiency calculated. Finally, the size of fan that has maximum efficiency should be selected for use.

References

ACGIH. *Industrial ventilation: A manual of recommended practice for design*. American Conference of Governmental Industrial Hygienists. 30th ed. 2020; p. 20–500.

ASHRAE handbook 2020: Heating, ventilating, and air-conditioning systems and equipment: Inch-Pound Edition. American Society of Heating. 2020; p. 499–580.

Bearg D. W. *Indoor air quality and HVAC systems*. CRC Press, the Taylor & Francis Group, Boca Raton, London, New York. 2019; p. 20–170.

Burgess W. A., Ellenbecker M. J., Treitman R. D. *Ventilation for control of the work environment*. John Wiley & Sons. 2004; p. 15–372.

Goodfellow H. D., Tahti E. *Industrial ventilation design guidebook*. 2nd ed. Academic Press, Massachusetts. 2020; p. 76–380.

Halliday D. *Fundamentals of physics*. A John Wiley & Sons, Ltd., Publication. 2020; p. 1–200.

Khoshakhlagh A. H., Beygzadeh M., Golbabaei F., Carrasco-Marín F., Shahtaheri S. J. Isotherm, kinetic, and thermodynamic studies for dynamic adsorption of toluene in gas phase onto porous Fe-MIL-101/OAC composite. *Environ Sci Pollut Res*. 2020a; 27(35):44022–44035.

Khoshakhlagh A. H., Beygzadeh M., Golbabaei F., Carrasco-Marín F., Shahtaheri S. J. Toluene adsorption on porous Cu–BDC@OAC composite at various operating conditions: Optimization by response surface methodology. *RSC Adv*. 2020b; 10(58):35582–35596.

Khoshakhlagh A. H., Golbabaei F., Beygzadeh M., Carrasco-Marín F., Shahtaheri S. J. Evaluation of direct reading photoionization detector performance under various operational parameters. *Environ Eng Manag J*. 2021a; 8(2):123–128.

Khoshakhlagh A. H., Golbabaei F., Beygzadeh M., Shahtaheri S. J. Study of the dynamic adsorption process of toluene by a microporous copper metal-organic framework. *Saf Health Work J*. 2021b; 11(4):556–576.

Logachev I. N., Logachev K. I. *Industrial air quality and ventilation: Controlling dust emissions*. CRC Press, the Taylor & Francis Group, Boca Raton, London, New York. 2014; p. 280–360.

Mazzeo N. A. *Chemistry, emission control, radioactive pollution and indoor air quality*. IntechOpen, London. 2011; p. 15–447.

Nevers N. D. *Air pollution control engineering*. 3rd ed. Waveland Press, Inc, Long Grove, IL. 2016; p. 250–382.

Schifftner K. C. *Air pollution control equipment selection guide*. 3rd ed. CRC Press, the Taylor & Francis Group, Boca Raton, London, New York. 2021; p. 37–214.

Springfield M.O. Engineering cookbook: A handbook for the mechanical designer. 3rd ed., Loren Cook Company, Missouri. 2015; p. 48–140.

Theodore L. *Air pollution control equipment calculations*. A John Wiley & Sons, Ltd; Laserwords Private Limited, Chennai, India. 2008; p. 15–514.

Vallero D. A. *Air pollution calculations: Quantifying pollutant formation, transport, transformation, fate and risks*. Elsevier; SPi Global, India. 2019; p. 377–428.

Vedavarz A., Kumar S., Hussain M. I. *HVAC: Handbook of heating, ventilation and air conditioning for design and implementation*. Industrial Press Inc., New York. 2007; p. 25–520.

Wabeke R. L. *Air contaminants and industrial hygiene ventilation: A handbook of practical calculations, problems, and solutions*. CRC Press, the Taylor & Francis Group, Boca Raton, London, New York. 2018; p. 65–210.

Index

Pages in *italics* refer to figures and pages in **bold** refer to tables.

S

T

U

V

W

For Product Safety Concerns and Information please contact our EU representative GPSR@taylorandfrancis.com Taylor & Francis Verlag GmbH, Kaufingerstraße 24, 80331 München, Germany